Under A Libra God

A Scientific Guide to Spirituality.

NATHAN GUY HATCH

1st Edition, August 2025

for my Love

ABOUT THIS BOOK

This book is a companion to my previous previous work, *The Perfect Child*, and the concepts herein are best understood when the work of inventory therapy such as that book instructs is completed, the result of which is to resolve burdens of trauma and bias which impair powers of perception and experiences of a spiritual nature. Reading this work without having accomplished inventory therapy is putting the proverbial cart before the horse, and without resolving past experiences of trauma this work is unlikely to be very useful for the reader, whom should first precede this reading with ample inventory therapy practice.

Content

"One often meets their destiny on the road they take to avoid it—"
Jean de La Fontaine

The Butterfly's Riddle

A woman sitting in her garden was visited suddenly by a striking, sulfur butterfly that flitted down and landed on a nearby yarrow to sip its delicious nectar. "Hello butterfly," said the woman.

"Hello," replied the surprised butterfly, for humans did not usually speak to butterflies. "A lovely day, isn't it?"

"All the more, having been visited by you," she replied. The butterfly smiled with appreciation, "I have a riddle for you," it said. "If you can tell me the answer I will bless you with long life and lasting youth." The woman was not afraid of death, nor of growing old, but was intrigued nonetheless and nodded, "Alright," she agreed.

"Forbidden lovers, their love never sundered,
Only meet when one sees not the other."

The woman thought for a while. Finally she replied, "I cannot tell you the answer, but come back tomorrow and perhaps I will have it for you then." The butterfly then thanked her for the nectar from her well-attended garden, and flitted away.

The next day the woman waited anxiously in her garden for the butterfly's return, and soon it appeared as promised. "Have you the answer to my riddle?" it asked, again landing to sip nectar. "I have thought about it," said the woman, "but still cannot give you one."

"I shall return tomorrow then," said the butterfly between sips of yarrow nectar, "that we may continue our conversation."

As before, the woman again waited for the butterfly in her garden at the appointed time, and again the butterfly returned, but the woman still did not have an answer. For many days this continued, each day the butterfly asking for an answer, but every day the woman not providing one. After a time the butterfly and the woman began to speak of other things besides the riddle, about their lives, their travels, their loves and losses, but never did the woman give an answer to the riddle.

After many, many days the butterfly began to grow old, and the woman, seeing the butterfly in distress, picked yarrow blossoms to offer it nectar, and comfort when it had trouble flying. Finally, the day came when the butterfly could no longer fly and was drawing near its end. "All this time we have been friends," said the butterfly, "and you have yet to answer my riddle."

The woman gave the butterfly a gentle kiss and a grateful smile. "I knew the answer the day you told it me," she said. "It is the Sun and the Moon, the light they share is the love never sundered, yet forbidden because they rarely meet in the sky, except in eclipse when one hides behind the other."

"If you knew the answer," asked the butterfly, "why did you not tell it me when I was young and strong and could bless you with long life and lasting beauty?"

"You *have* blessed me," said the woman. "You brought me beauty when the world was ugly, life when I was sad, and friendship when I was lonely. The riddle was but a thought for us to become entwined with each other and I did not tell the answer as cause for your return."

The butterfly smiled, for the butterfly too had long ago come to love the woman, and likewise much joy in their friendship. The pair spent the remainder of the day together, quietly talking when the butterfly could find the strength, and as the night fell the butterfly took its last breath. Then the woman buried the butterfly beneath the yarrow it so loved, and every day of her remaining years visited the yarrow to remember her kind, bright friend whom had brought so much meaning to her life simply for being part of it.

1. The Beginning

I begin this work seated in a bare living room of a small apartment nestled high in the Mountains of the Basin and Range of the Western United States, not even a sofa upon which to sit, a television to pass the time, or a rug to rest my feet. Now is the midst of the coronavirus outbreak which has claimed the lives of more than half a million of our friends, family, and loved ones which would eventually call to Heaven more than seven million people throughout the world while political and religious leaders bickered for power and the pharmaceutical industry and alternative wellness schemed for personal enrichment, abandoning the unfortunate to their sorry fate in offense to all that is just and moral. I have just moved away from California, instead of pursuing material success and possessed of talents which have brought me some professional success I quit an unrewarding, meaningless career to instead live on a pittance, to help others and continue my writing while I resolve the health problems afflicting myself and mankind, to heal from a lifetime of heartbreak and guide others like myself back to a state of wellness and contentment, and the pandemic all but wiped out what income I was making to sustain myself, and it was only due to the support of long-time friends and patrons that I was able to even continue doing the work of the last half-decade, including such accomplishments as solving the Warburg Effect which is the primary debility which causes cancer, and writing my second book on human psychology, trauma, and parenting.

As was my own experience, I saw during this time how others' lives were sorely affected by those who preceded us, the mistakes and self-centeredness of past generations paid by the lives of the innocent and children, and me and my generation and those now growing to adulthood required to face an uncertain future of early metabolic illness, portended global calamity, and challenges never before seen by man, the first in all of history to contend not only with competition for survival but also the separation of humanity by mean and hateful infrastructure, early loss of bodily functions for the toxicity of industry, and the threat of extinction of the natural world and the many amazing other animals which also share our tiny, beautiful planet caused by creatures whom, unlike past factors of mass extinctions, are entirely capable of not doing it.

Tormented and deranged, the generation before mine seeks not to correct the past and save ourselves from this fate but delusionally fantasize about leaving this planet altogether, preferring to live on the dusty barren wastes of Mars without an atmosphere and the freedom to ever again go outside, only one-third the gravity which will cause human bones to deteriorate and deform. Though the Earth is not even close to its end such is their trauma they prefer never again to breathe fresh air or play under a Summer's Sun, their waking horror wrought also in turn by abuse at the hands of their own parents and their parents' parents and their parents' parents' parents, forever corrupting their conception of life and reality and the value inherent in the self, failing to recognize these plagues of humanity as something borne not from inevitability, but *greed*, in self-preservation and stoking of geopolitical disharmony for nothing more than service of their fear and desire for money.

One of the greatest mysteries of the current era is whatever happened to the anti-war hippies of the 60s and 70s. Children whom in their young adulthood gathered for free love and protests of peace and disarmament later grew into small-business owners supporting genocidal and violent military-industrial hegemony because the safety of majority membership is easier than fighting it, the only remaining part of their once idealistic youth the long hair and drug addiction, because as we grow from childhood naivety and ignorance into adulthood the stresses and horror of life easily rob any unsteady conviction of its shaky foundation. New generations too begin their life with wonderful dreams of the power to change the world but rudely find their ostensibly altruistic motives nothing of the sort, all along a guise for the desire only of power, then discard their lofty ambitions at the slightest of inconveniences. If morals are so easily discarded they were never great to begin with, and the delusions by which we navigate life begin to accumulate, failure after failure and mistake after mistake until we no longer recognize the old stranger staring back at us in the mirror, which instead of realized childhood dreams is become a childhood nightmare.

Yet while religion positions itself as the antidote to pain and suffering it subversively divides friends, families, and communities, turning neighbor against neighbor and father against son in the struggle for control and power not only of each other but that also of our very reality. As mortal humans we are not capable of such control, and the incongruity of such fantastical, delusional worldviews and their real effects on our lives, health, finances, and personal relationships betrays any possibility of real spirituality as adherents demonize children and persecute minorities simply for not believing as they do, servants of fear who can never know a day's peace for the love of chaos and addiction to anger which arises from the subversion of humanity.

I realize the absurdity of attempting to explain the particle-wave duality of light, the science of time, and the origins of consciousness while possessed only of a High School education. While well-read and accomplished I have also hardly touched the intellectual reserves of history, being so preoccupied my entire life with simply surviving. But my experience has also shown that many things we have heretofore believed frustratingly complex are not nearly so, having pulled the eternal thread of human mortality found seeming complex diseases as cancer and addiction all connected by simple, fundamental, and consistent rules of biology, biochemistry, and cause and effect, and I see now the irony required of a person with limited academic accomplishment to divine the nature of biological and exis-tential realities because those whom are more accomplished instead overthought and overwrought matters which only *appear* complex, misdirected also by shared collective biases, prejudices, and trauma and the desire for profit and fame which obscure humble, simple truths.

As great a liability as the ignorance of biology which promoted epidemics of diabetes, cancer, and depression in our era in spite of other great scientific and technological advancements like particle accelerators and the doorstep of nuclear fusion is man's ignorance and insecurity of his meek position in this grand Universe. Being limited in my academia and life experience has not also limited my access to discovery, coming of age alongside the internet when more infor-mation is available instantly at my fingertips than for the previous entirety of recorded human history, enduring (not by determination, mind you) indescribable and relentless suffering my path in life has given in return an ability to peer not only into the hearts and minds of humanity but reality itself, so I find myself in the conflicted state of being unable to describe mathematically that which I perceive to underlie reality every bit as what led to my elucidating cancer, addiction, depression, etc., and as such burden others to the responsibility of the maths and mathema, where instead here I lay a foundation on which these concepts may be expounded upon in the future by those more capable than I and thus together chart a more successful path not only for all humanity but everything which exists upon this wonderful Earth.

We are in fact as much a part of the great story of reality as reality itself, our perception of our place in time and space limited only by our biases and preju-dices and the trauma of abuse and heartache both collectively and individually. Though every one of us every day from moment to moment commands our life and constantly proves ourselves quite powerful in the constructing of our identity and world will still believe ourselves incapable and incompetent and thus unable to do the work that is required to get the kind of life we want, believing ourselves capable of only specific herculean tasks as surviving every day with burdensome trauma, showing up for work or our families while also struggling with alcoholism or addiction, or staying on this Earth in spite of debilitating depression and other metabolic disease. But we are all of us capable of the greatest feats, and under-standing the deepest mysteries of the Universe and communing with the divine is prevented only by those many distractions and poor conceptions of self worth which burden our mortal existence and distract our considerable human faculties with unreasonable and unkind conceptions of fear and fate.

Seemingly ironic (though it is nothing of the sort), spirituality is best understood through rational thought, and is more rational and part of our physical existence than any ideological conception of life heretofore conceptualized. God is found in science, and science can and does describe the nature of God and our innate spirituality as a living creature, and the choice between spirituality and rationality is an illusion borne of fear and the desire to put within our hands the control of destiny. Yet anyone can see that we cannot command the circumstances of our birth, nor save other humans and creatures from their mortality, and as such becomes plain the reality of our station which is that we do not command destiny any more than we do the wind and the rain. The truth is that neither spirituality nor rationality need triumph one over the other, for they are both entwined in the workings of the Universe, as you will learn throughout this book, for we are nothing if not beings of spirit borne from the indefatigable foundations of reality itself, which is in its nature incomprehensible and wondrous as required by such ethereal concepts as spirituality yet also organized by laws of physics and science which construct it, and all are plainly observable in the reality which surrounds us if one simply takes the time to do so.

Our parents and those others who came before us were limited by their understanding, and because of their personal traumas saw division where there was none, conflict where there was only misunderstanding, and hatred as ignorance. Though much challenge awaits us our future will be the brightest ever seen on this Earth because of the collective enlightenment so many of us are now seeking, empowered by the connectedness of the internet and technology which facilitates the needs, desires, and curiosities of humankind, and with a little care we may all yet find the kind of lives we have always told in story and mythology, for salvation is always found in knowledge.

This book is written with the assumption (and requirement) that the reader has performed an abundance of inventory therapy as discussed in my previous works (especially as instructed in *The Perfect Child*), and may otherwise be difficult to comprehend if this is not the case. If the reader finds the scientific concepts discussed in this book frustrating, it's okay to skip to the discussions on spirituality in latter chapters. Unlike my other works this is almost entirely philosophical, but do not mistake its content as proselytizing—the purpose is not to extol truth, but to seek it.

2. Metabolic Health and Powers of Perception

As my health declined in my early thirties I was often aware of an inability to effectively sense the world around me. While my mind could produce sight and sound they often seemed a pale facsimile of what was really there, fuzzy in their details and abstract in their certainty, and required much more thought and attention than what was previously required in my youth.

The reason for this dulling of my perception of reality was firstly because we do not truly and directly experience the world around us—the *brain*, which is the organ that creates the mind and the very essence of our being, lives inside a dark shell of skull separated and protected from the outside world, and rather than interacting directly with the world around us instead compiles information received by the senses of sight, sound, smell, taste, and touch to then give an *interpretation* of what is there. This means that when senses like sight, hearing, or touch are impaired by aging, disease, injury, or chemicals like drugs and alcohol the brain then has less information from which to construct our reality (that is literally the point of drinking—to help relieve us of stress by blocking perception). But because we are still alive when the brain ages or our senses dull the brain continues to produce interpretive compositions, even when there is less data, and fills in the missing data with assumptions based on past experiences, and when there are gaps empty of information the brain also can and does entirely conjure things which are not actually there, a function of the mind which is even more perilous when the infill is informed by fear due to past experiences of trauma and abuse.

This function of how the brain process information and fills in gaps where information is missing is easily demonstrated by visual illusions such as the *Zollner illusion* in which cross-hatching makes it impossible for the brain to see lines which are actually parallel, or the *Hermann Grid illusion* in which the brain creates dots where there are in fact none at all. Especially in terms of human eyesight, the region of the retina where the optic nerve connects has no visual receptors at all, so all light sources that would otherwise impact that area are entirely contrived by the brain as it intuitively invents missing data, and while the brain is very good at constructing non-existent data it is still entirely fake, which then also exposes us to errors in perception that are directly proportional to our metabolic states and health.

Likewise, film and animation are also entirely illusory, made from a simple construction of still images that are displayed in rapid succession which tricks the mind into seeing movement, a function which is also entirely dependent on the rate of input of the brain and eyes. Even video games are a rapid succession of still images, rendered by computers at rates that so fully trick the mind into seeing fluid and seemingly real-time motion which in reality is nothing more than sequential still images. An organism which processes visual information at many times the rate of a human's would oppositely not see motion in film, animation, or video games but instead a series of sequential still images, and if somehow the rate of visual input for humans could be increased we too would not actually view film and video games as anything more than a slide-show. This is also the reason why my writing, which entirely lacks editorial support due to a disinterested publishing industry, so often has spelling errors and grammatical mistakes, because in spite of continuous proofreading my mind (like many people's) will simply imagine a word that does not actually appear on the page, or correctly spell in my mind even when it is not correct on paper, and the mistake is difficult to even spot because the unconscious mind continuously corrects the error no matter how many times I go over it.

One phenomena I experienced during the decline of my health was seeing visual illusions with increased regularity, plays of shadow and light or objects which resembled things they weren't, most especially when waking first thing in the morning or during the night when I could not sleep, when my brain was not yet up and running would see people or objects that weren't actually there but were patterns or shapes that resembled those things. Most people intelligently recognize such illusions as tricks of the mind and will jokingly talk about sleep paralysis demons, but others with less intelligence and greater superstition and trauma believe these apparitions to be real manifestations rather than artificial constructs of the brain in response to fear and poor cognitive ability, and when beset by intense fear and anxiety these delusions and illusions can result in yet greater stress to an individual who does not understand how the mind and body actually works.

Illusions and delusions of the mind which occur during metabolic decline are more frightening than normal experiences also because the stress of metabolic disease and poor health considerably raises stress hormones like *adrenaline* and *cortisol* which mediate the flight or fight response. Phobias and neuroses exist in fact because those with such extreme reactions to things like spiders, flying, the dark, loud noises, etc., simply have a much more pronounced release of and response to adrenaline and cortisol than those without phobias (as discussed in *Fuck Portion Control*). Though I was able to recognize fleeting illusions or pronounced experiences of fear such as in social situations as a product of my deteriorating mind and body the experience was still disturbing because it felt there was nothing which could be done to get control of my health and stop my rapid descent toward death. Luckily my time had not yet come and I discovered the work of Dr. Ray Peat which set the course for my recovery and later research and discovery, and though it took much effort, knowledge, and consistent behavior to rectify my condition it was far less than what I expected and never once included exotic cures or expensive supplements or interventions, but at all times rational and sensible therapies primarily rooted in food, dietary behaviors, and the environment, with some understanding of complex biology, because our biology is as much a function of our environment (including nutrition) as anything.

Similarly, the condition of cystic fibrosis at the root of my poor health which caused these health problems also made it difficult to breathe, as it directly affects mucosal tissue function and thus also that of the lungs, and the rare times I could actually sleep would wake suddenly and unexpectedly, gasping for air, on rare occasions even choking on aspirated stomach content due to catastrophic failure of the esophageal sphincter muscles. Among other reasons, cystic fibrosis causes an inability to properly metabolize and maintain electrolytes like *sodium* and *chloride*, and sodium is one of the primary inhibitors of adrenaline since one of adrenaline's functions is to increase sodium flux through cells (and is thus inhibited when sodium flux is sufficiently achieved, to avoid excess), so the condition specifically exaggerated my adrenaline release due to poor sodium status and thus also caused an increase in discomfort, fear, and disquiet that was made all the worse by an unsupportive and judgmental family and partner whom blamed my problems on not going to the gym enough even though I was the only one of them who trained competitively after high school and exercised more often than any of them and dieted daily.

This relationship between sodium and adrenaline exists because sodium is the primary driver of cellular metabolism which in turn primarily runs on *carbohydrate*, so increased sodium flux whether mediated by adrenaline release or good health in which there is plenty of circulating sodium also increases the rate of metabolism of the brain and thus also perception, since our brain cells which handle perception and information processing are also cells which function on these principles. It is also why those who chronically diet, starve, fast, and overwork and abuse their bodies lose not only fat and muscle mass but also cognitive function and awareness (sometimes quite rapidly), and oppositely become insane and dissociated because we literally lose our perceptive functions because of the stress of poor energy production which is a consequence of losing electrolytes and carbohydrates, then the body attempts to compensate by increasing adrenaline to increase sodium flux from a smaller and smaller pool of sodium to the point stress hormones rise so high as to entirely impair normal human behavior.

   This is one of the reasons why discovering the work of Dr. Ray Peat, who advocated the use of supplemental salt to raise the metabolic rate, had the effect of improving my wellbeing and why most of us crave salt and carbohydrate because extra salt along with behaviors that improved salt retention (like not going hungry) reverses this metabolic decline to then lower adrenaline and thus also the constant sense of overwhelming fear and give tissues a much needed break to repair and heal. Adrenaline is the very hormone of fear (in very small doses, excitement) and many people, including some I know, present with symptoms of undiagnosed cystic fibrosis and have life long histories of panic attacks, depression, or outsized obsessions with physical health since we intuitively recognize when our health is impaired (there is no such thing as a hypochondriac—they realize something is wrong but lack the ability to identify exactly what), since deficiencies of sodium and carbohydrate raise adrenaline and cause rapid heart rate, even to the point of alarm or causing shortness of breath, but biased and prejudiced toward me and my work because of hateful, homophobic upbringing refuse to listen to my advice no matter how many times I have been helpful to them, and instead choose to continue dieting and harming their body and thus sustain unending suffering and cognitive, sensory decline to overburden the mind with yet more fear and anxiety and distract us from accomplishing our goals and dreams and living truly the kind of life we desire.

   Many of us are familiar with the concept of insanity, and often see it in others as they experience decline in their health and wellbeing, which occurs because the brain and nervous system which transport and receive the information that allows us to interpret the reality around us are physical organs which must themselves be healthy in order for the mind, which is a product of the brain, to function correctly. So metabolic disease whether caused by dieting behaviors, environmental intoxicants, or opportunistic pathogens greatly affects our ability to literally perceive the world around us, and decline of these systems can and does then lead the brain to form incorrect constructs due to a deficit of information, and because the mind is what the brain does this dysfunction leads us into insanity and other mental illness. Those of us experiencing this limitation of human mortality are often those whom adopt odd and unhelpful obsessions such as obscure or fanatical beliefs, driven by fear and malfunctioning faculties seek psychological refuge in any coping mechanism which might bring relief, even if it is false. Even though I did not have any such mental coping mechanism I developed quite pronounced tics and twitches of my speech and mannerisms which, when I first witnessed them on my first video recording for social media was not only disturbing because I had no idea this was happening but that nobody, not even my old partner, had sought to help me. In fact, it was likely part of his resentment toward me and justification for leaving, since many people feel contempt for those whom exhibit evidence of our shared mortality, so great is our fear of it.

   For similar reasons, those belonging to religious faiths who demonstrate the most flamboyant, outward convictions, advocacy, and evangelism are in fact the least faithful because such behavior is similarly driven by insecurity, doubt, and fear, to motivate fanatical and actionable behavior in opposition to that trepidation, while those who are most content in their faith are those whom likewise demonstrate contentedness. I know this also from personal experience because when I was an insecure teenager logically doubting the religion of my family, believed as told that my skepticism and doubt (which I daren't express openly) was a product of not being faithful enough, and so then engaged in greater proselytizing and participation in the misleading of both participants and non-participants alike. It was no wonder that someone so involved in the religion like myself should also then so quickly abandon it once its apparent deceit and malevolence toward those like me became so great as to finally break the indoctrination. But such uncommon beliefs and commitment to religion are not simply a product of desperation or faith, but also a strong component of mental illness such as what I suffered throughout my youth because this state *increases* fear and anxiety, and with limited tools by which to handle overwhelming fear and anxiety and likely having been taught that fear and anxiety are caused by an absence of faith then become incredibly irrational as we desperately try to arrest those overwhelming and uncomfortable emotions any way we can. Yet because those feelings are primarily functions of hormones, which are physical chemicals made by pathways also affected by factors such as blood sugar, interpersonal conflict, and pathogenic colonization those wrestling with religious doubt and uncertainty never actually find relief from their stress and thus get worse, not better, and plunge further into insanity, broken relationships, obsessive control behaviors, and dependency on substances in desperation for relief.

This function of the brain and nervous system is in fact the primary reason why mental illness is associated with substance abuse, because neurochemicals like *acetylcholine* which mediate consciousness, fear, anxiety, and hormone release are also in turn influenced by exogenous chemicals and nutrients. As I discovered near on a decade ago the condition of alcoholism and addiction which I also suffered is a neurological disorder mediated by acetylcholine dysregulation which just so happens to be treated by alcohol and regulated by drugs. Yet even those in the recovery community were not interested in listening to me, such is human indoctrination nature that they do not expect relief to come because of the wording of their recovery program and validation of authoritarianism. Many religious people also secretly (or not so secretly) harbor substance abuse problems because of the intense stigma concerning dependency and lack of compassion for the mortal condition which is the root of their religious dogma and institutions.

Alcohol and drugs also have similar effects on microorganisms as they have on ourselves, and also offer therapy by inhibiting parasites, bacteria, and fungi which then has considerable influence on our health and wellness—For instance while cocaine alters expression of hormones like serotonin and dopamine it is also *antiparasitic*, so widespread parasitism such as exists in industrial food systems also underlies cocaine abuse, but by inhibiting serotonin (which is the hormone of shame, guilt, and remorse) cocaine also inhibits the moral function of serotonin stimulus and thus helps facilitate antisocial, heinous, or even evil behavior (studies show a rise in circulating serotonin during cocaine use but this because it causes cells to dump their serotonin which then temporarily relieves them from its influence). Because our mind is logically aware of the context of our behavior we do not require serotonin to be good people, but it helps to have hormonal feedback, especially for those whose minds are so addled by trauma and stress they do not think much about their behavior, and we are never actually relieved of the consequences of our actions simply because we don't feel guilt, and through drug use ironically become more and more dependent upon them to exist while living with so much unresolved trauma.

Because of this misunderstanding, many people pursue drug use (recreationally but also scientifically and medically) as a means to escape the confines and limitations of the mortal condition and heal disease, but the problem is that while drugs can be used to treat symptoms they are not actually a natural part of our human biology, and only ever treat *symptoms*, not the underlying cause, and because they usually function by antagonizing normal human systems are also never long-term solutions to problems like mental health, depression, and trauma since they do not actually address the root causes. Over time the metabolic deterioration caused also by the very factors which also motivate drug use, and the abuse of substances too (cocaine for instance also disrupts normal fat deposition patterns and causes the characteristic round, puffy, fatty face), eventually cause undeniable and detrimental physical deterioration as well as that which is psychological.

Because the combination of mental trauma, physical stress from dieting and excessive exercise, pathogenic colonization, and substance abuse is so prevalent an experience it readily occurs all around us, and many young Adonises often descend from heights of beauty and fitness to extreme mental illness and physical wasting in demonstration of this connection of mental health to abuse of the physical body, because dieting destroys not only our body and immune system but also our microbiome upon which we are entirely dependent for nutrients like short chain fatty acids and B vitamins and which also protects us against opportunistic microbes, and the behavior of many athletes and fitness advocates or those otherwise traumatized about their body image is the fastest way to induce mental illness due to derangement of neurochemical and endocrine function. For instance an entire class of microbes produce biologically active *amines* which are more bioactive derivatives of amino acids which have strong and extreme effects on cellular, metabolic, and neurological function. *Histamine* for instance is an amine produced from the amino acid *histidine*, and is required in healing, immunity, and inflammatory processes, but constant exposure to histamine producing microbes in the gut due to poor diets and poor immune function also cause a state of constant inflammation which affects the nervous system and the immune system.

Tyramine and *phenylethylamine* are also produced from tyrosine and phenylalanine and are such powerful adrenaline analogs that the tiny amount which occurs in cheese and bread excite the mind by causing a small elevation in sodium flux and energy production, but in large quantities produced by microbes as a result of sodium malabsorption (meaning sodium remains in the gut rather than absorbed into the body) cause a person to experience what feels like constant and unceasing adrenaline excess as if hooked up to an intravenous supply of steroids. Most amine producers are microbes of *putrefaction*, which is the process of decay by which microbes break down dead and dying organisms, and thrive in environments high in both protein and salts which does not occur naturally in any other environment except states of decay such as when our ability to absorb salts becomes impaired by pathogens and illness. Ammonia itself which catalyzes this environment is a common product of opportunistic microbes like *H. pylori* and parasites and is itself a strong neurotoxin which causes brain fog and cognitive impairment to make the world feel like a vague and rapidly passing nebulae of apparition and delusion.

Because the mind is the function of the brain which is directly affected by these factors it is often very difficult to help and rescue those who are mentally ill because their very ability to comprehend reality and act effectively is impaired by the very illness they need to resolve to regain that mental acuity and awareness. Indeed most of the people I met in the recovery community while participating in Alcoholics Anonymous continued suffering from paranoia, delusion, rage, insomnia, irritability, impatience, contempt, fear, and hatred amplified by these underlying causes of addiction and alcoholism and though I would present to them the solutions which made my recovery not only rapid but permanent and demonstrated complete and total recovery refused to even read my work or try the information I presented, several of whom later overdosed and died or otherwise persisted for many years struggling with relapse and sobriety in a never ending dance of delusional psychosis enabled by the very community supposed to assist them by insisting there is not and never will be a cure.

When I was a teenager there was a period of several years in which my parents heard that not peeling potatoes was healthier than peeling them, so our weekly consumption of mashed potatoes with Sunday dinner became peppered with chewy potato skin (which are fucking gross in mashed potatoes you weirdos). Because peeling potatoes also takes time this idea was especially appealing (yes pun intended) to my parents who were as constantly irritated and impatient as they were obsessed with physical health and beauty. But this was also the most emotionally tumultuous time in my childhood, with my parents growing increasingly erratic and fanatic about their children and religious beliefs and impassioned in their hatred and resentment of each other and supposed political adversaries—A few years after being kicked out of the house when my family was away on vacation I once snuck my first boyfriend there to see where I was from, and he remarked how strange and creepy were the number of religious paintings hung throughout the house. Having grown up in that environment at the time it seemed normal, but it turns out that potatoes are part of the nightshade family, and the skins are extremely high in a glycoalkaloid poison called *solanine* which is meant to kill nematodes (tiny soil parasites that feed on plants) by destroying *acetylcholinesterase* which regulates levels of acetylcholine and kills organisms by overstimulating their own nervous system. Because we are so large it does not obviously affect us, but solanine is so poisonous that eating green potatoes can and has killed people unaware of its toxicity, and chronic consumption of potatoes and other nightshade plants is, as I discovered, a primary cause of addiction and alcoholism due the effect of glycoalkaloid poisons which overstimulate the nervous system and promote relentless discontent such as is commonly experienced by obsessive religionists like my parents or ancient, violent Mesoamerican Mayan civilizations from where nightshades like potatoes, peppers, and tomatoes originate, whom in turn treated neurological poisoning from a diet high in glycoalkaloids with complimentary therapeutic use of cocaine (coca) which, like antihistamine, blocks the stressful effects of elevated acetylcholine without blocking acetylcholine itself. Nightshades from Europe like eggplant and belladonna were not as commonly used as food (eggplant is particularly potent if not pre-drained of its juice), and the arrival of potatoes (which are so fun to eat not only for their carbohydrate but also their solanine neurostimulant) in turn increased rates of alcoholism in any populations which ate potato very frequently, and to this day I cannot eat potato more than three times a week without eventually having cravings again for alcohol, which destroys acetylcholine.

Years ago a young person I was trying to help recover from alcoholism continued to drink heavily in contempt of my advice and tools like sodium acetate (they would purposefully not take sodium acetate when they wanted to drink because they knew it would satisfy their cravings for it) and because acetylcholine is also required to form memories and learn they then hardly remembered one session to the next nor the directions required to recover, often misunderstanding or mistaking directions for recovery and choosing and picking which were followed and which not, as if that was somehow an effective way to approach recovery. Though briefly trying inventory therapy it also requires abstinence from substance use in order to form new learning experiences, which they were unwilling to do, and so did not experience the positive effects as what is usually the case for most people whom earnestly put in effort, and misunderstanding my instructions due to the mental impairment caused by both their trauma, condition, and substance abuse gave up entirely in short order. In later interactions I was forced to simply admonish them to join a recovery group, since sobriety is a primary step required for recovery in order to facilitate neurological regeneration and new learning experiences, but which as far as I know they never did.

This neurological biology of the mind which is entirely based on physical health is thus why people often cultivate odd, extreme, or unhelpful beliefs or delusional ideology, because the very ability to perceive life around us is entirely a construct of the human mind and not reflective of the actual world around us. In that sense, those who believe reality to be a simulation or illusion are technically correct, in fact, that each of our realities is nothing more than an interpretation by our mind of the one which does actually exist around us, which is filtered through human biological evolutionary psychology and metabolic function, which can very easily lead us to err if we are not aware of this fact in the first place, and it is ironically through metabolic deterioration that we begin to sense that facade as our faculties in turn lose the ability to construct it. As my own metabolic condition continued to deteriorate I was alarmed when returning home from visiting family one winter holiday that I had almost no memory of the entire year. I knew it happened, because we were now at the end of it, but it was as if I had been asleep for the entirety of it. This occurred also at the height of my drinking and since alcohol neutralizes acetylcholine it was literally as if I were asleep even while awake, and not yet aware of this nature of biology or function of addiction I resolved, as was so often admonished by stupid social-media influencers and celebrities to be 'more present' the next year ahead and to purposefully make more of my time and ambitions. While I did in fact do quite a lot of things during that time such as develop my first video game, a process that took six months of daily work, when the next holiday came around I could again only remember about a month or two worth of time. Alarmed at having essentially lost the previous two years of my life, in addition to having cancer and the imminent dissolution of my relationship the absence of memory was a major contributor to the urgency which compelled me into Alcoholics Anonymous shortly after, and finally making a recovery and then after solving the biology of addiction and alcoholism to then elucidate these principles of human biological psychology and the literal perception of reality.

At the nadir of my mental and physical deterioration I sensed reality with literally less dimension than when I was healthy, or would again later as I recovered, specifically that the world around me seemed literally flatter, *not* metaphorically flatter, with less depth and perspective and complexity, and later during my research realized that because reality is literally constructed of dimensional divisions as described in quantum mechanics that a person's ability to actually perceive the dimensions out of which we are created does decline during mental and physical deterioration to contribute to delusional conceptions of reality such as believing existence to be flat, as if the human mind loses the ability to exist in and perceive the third-dimension and therefore confined or restricted more or less to the second before it. Meaning that these people are not necessarily delusional, in fact, but simply unable to understand what it is they are experiencing, the reduced ability to perceive the higher dimensions of reality in which we exist.

While this sounds like a metaphysical argument, our mind and biology exist in the construction that is the natural world which is, in fact, built of these dimensions, so its function is also part of and composed of those dimensions as well and not separate from it, and the brain thus literally can and does perceive the underlying dimensions of reality without knowing consciously that is what is occurring since our psychological orientation is a result of evolutionary biology designed for us to succeed as an organism in all dimensions of spacetime, not a scientific instrument. But this is why those who decline physically and mentally often experience psychosis, delusion, and illusion which they believe to be reality, because it is reality, but only a very limited part of it, since the ability of the mind to perceive and reconstruct all of reality is what accomplishes mental health, stability, and the ability to comprehend higher, not lower, planes of existence. It is also why we argue, as I have heard previously stated, to more acutely feel the presence of others in our lives, when we are less emotional or mentally able to experience them and the present in which we all exist, because mental and psychological deterioration impairs the mind's ability to construct reality the feeling of intense feelings such as rage and hatred can be a temporary antidote to treat that illness as potent as any drug and help us feel more alive than is typical in that deteriorated state.

Restoring at least some of the health of the body as discussed in my previous works is thusly required to restore the health of the mind and the ability to perceive and exist in reality, because the body is what constructs the mind and facilitates neurological and endocrine function, without which it is impossible for the mind to be whole, as a product of the brain. After that, practice of trauma therapy such as is accomplished by inventory can help clear the mind of fear, trauma, and insecurity which motivates self-destructive behaviors detrimental both to our physical wellness and mental and emotional health, to then make space for greater psychological capacity and awareness which then includes the ability to comprehend and experience greater spiritual and psychological understanding as discussed in the remainder of this book.

3. Knowing The Unknowable

During the Summer of 1996, returning home from a vacation at Lake Powell in Southern Utah, my parents put a tape into the cassette player and out of the speakers came the voice of a supposed intellectual from their religious movement whom began to speak on their religious mythology and ideology, masterfully weaving tales and intrigue of prophecies and revelation and the last days. Their religion was one of the first in history to attempt direct incorporation of scientific concepts such as planets and non-geocentric theology, claiming that heaven where God lived was actually out in outer space near a great planet, which they even named, and the man married old, familiar passages from the Bible with their new-age religious scripture to in turn describe how the Earth itself would be hurled through space near to the planet near where God resides (he's not on the planet, but next to it) when the end of times comes which would cause the stars to appear as if they were falling such as what is described in the Bible.

My parents' religion was born during the Industrial Revolution when the world began a romance with scientific discovery and progress. The first hot air balloon had been flown at the turn of the eighteenth century and the first steam engines rolled onto the world stage, a new and prophetic harbinger of industry and progress to herald the integration of peoples and spreading of prosperity that would birth the first World's Fairs and authors like Jules Verne and his *Twenty-Thousand Leagues Under the Sea*. Scientific revolution was proceeding at such a

breathless pace it alarmed religionists with portended antiquation of their super-
stitions and traditions. Religion had no more answers to satisfy man's fear and
curiosity, overshadowed by knowledge and stuck in the archaic mythology soon
to be forgotten those whom found comfort in the cradle of theocratic institution
thence summoned a so-called Second Great Awakening from which sprung many
new-age religious orders which sought and fought to bridge the widening gap
between scientific progress and religion. Yet while this new religious ideology to
which my parents in the twentieth century (and thus myself) belonged was quite
fantastical it also put the concept of God in closer reach to the average person,
no longer presiding from some unattainable, heavenly dimension he in fact sat
upon a throne in outer space. Their religion also claimed members could themselves become Gods one day if they remained faithful and obeyed, a truly alluring
promise for simple mortals who desire above all to have power over mortality.

This proposition was so preposterous, however, that even as an indoctrinated
teenager who knew almost nothing of the outside world or other religions I could
not help but think how absurdly contrived it all was, how self-serving and desperate, and that every bit of dialogue and discourse of their religion mimicked works
of science fiction and fantasy like *The Lord of the Rings* and the tale of Frodo and
Bilbo's battle against the evil Sauron, or the interstellar travels of Paul Atreides
in *Dune* whom possessed the power to compel men against their will and ride
atop giant sand worms. The obvious coping mechanism that was their religious
beliefs was more than enough to dissuade me from its legitimacy. The authors of
their religion perhaps never supposed it possible to take a photograph of God and
Heaven with the James Webb telescope, which can see to the very edges of the
Universe, if we but knew the exact coordinates.

Leaving their religion was all the easier a decision, however, once I extracted
myself from the immediate indoctrination, because of their also unbridled hatred
and bigotry toward people like me whom were gay, not white, or transgender or
female, because if God really was the father of mankind as they and other religions
claimed their simultaneous abuse and hatred of spiritual siblings only served to
undermine and invalidate such a hateful worldview and exclusionary religious
ideology. Ironically then one of the reasons I left their church was because of the
very lessons in honesty, love, charity, and service they taught me as a child to
which as an adult I was then expected to discard, to uphold their hatred, suspicion,
and love of money and thus rudely alarmed to find they did not actually practice
anything good or useful. I was so done with religion because of my experience in
it I did not even explore any of the secular criticism of the church itself, which it
turns out is more than enough on its own to condemn the entire establishment as
rooted not in spirituality and higher learning but banal sexual predation, exploitation, and opportunistic financial chicanery (as so many are).

Yet the desire of religious people and others to understand the mysteries and
purpose of existence is understandable. Terrible things can and do happen in life,
which seem to entirely invalidate our worth as individuals or any semblance that
the Universe even knows we are here, and suffering can rob us of the very will to
live and suck every ounce of joy and hope from even the most starry-eyed child.
Many of us harbor burdensome experiences of trauma which inflict ceaseless worry,
anxiety, and fear, and until I began coaching and helping people with trauma I

was, like most people, fully ignorant to the magnitude of just how many fathers rape their own daughters in complete defiance of all human virtue. Life can be truly horrifying and it is natural for us to seek relief from our burdens, even sometimes at any cost, which is the primary reason also for conditions like alcoholism and addiction though many people (especially those whom are religious) choose instead to condemn, rather than commiserate with those who suffer.

One of the reasons we seek enlightenment is because we (and likely other animals as well) innately understand that knowledge often brings tools which can bring us relief and liberation—not necessarily existential in nature, although it can, but mostly those which are practical and applicable to problems we face. For instance, all throughout my youth I had been indoctrinated to believe that sugar was harmful for our health and cravings for sugar were a temptation by a weak body which must be resisted and disciplined, which was an idea not actually based on any sound science but because we have been indoctrinated over many centuries in religious desecration of the human condition to believe pleasure as a weakNesses of the body which should be conquered. Sugar, however, is a primary metabolic fuel from which the body produces energy which keeps the body alive, and the body craves it not because the body is weak but because we will literally die without fuel, and because energy production lowers stress hormones sugar then makes us feel better due to release of reinforcing hormones and neurochemicals that enlighten us to our nutritional and dietary needs as an animal which, when fulfilled, promote life, or when deprived produce pain and suffering not because it is good for our character but because it literally causes illness and death.

Most religionists are also completely naive to the obvious homogeny they share with other religionists, regarding as deranged or even pagan those of the past whom worshiped Jupiter or Odin with the very same tenacity, devotion, sacrifice, and passion for their personal beliefs as any Christian of today, regardless of specificity, which reveals in fact the nature of man, not truth, as given to passionate ideological commitments, especially when indoctrinated as children irrespective of those actual beliefs. A child raised to believe the God of the Universe is a gay whale that lives deep in the ocean which appreciates being sung songs by Britney Spears will defend that ideology with the same fervor as any other religion, because this is how the human mind works, and is a function of biological evolution and survival mechanisms which intends for offspring to passionately retain lessons learned from parents which is normally meant to guarantee survival skills such as what even is food and how to find and use it, but since we now as humans have language and culture this biological survival function is corrupted to also instill hatred and division among humans and burden children with wild and frightening ideas about the nature of fate and the Universe which then burden us with fear, suspicion, even paranoia.

When I was young and trying to find answers to my homosexuality the members of my parent's religion and my parents themselves insisted I wasn't really gay and was just confused. But I *was* gay, so my personal experience conflicted existentially and invariably with the prejudices of my rearing and the way I was told reality operated, so when I went to these religious authorities who supposedly communed with God and was told simply to "not be gay," as if that was actually something within my power to be, or even an answer, it expressly contradicted what I knew to be real and thus invalidated itself in the process. If the God they claimed to speak directly was all-knowing surely he knew why I was gay, and if these people supposedly had direct access to him why couldn't they provide clear insight? What exactly were the steps to become straight, as God himself should know that? I was more than willing to do them in order to fit in, and had already proven it by following what they so far had admonished such as avoiding pornography and not jerking off for thirty days. But no matter how much I obeyed remained every bit, if not more, aroused by other guys just as any straight young woman would be. How exactly was avoiding pornography meant to make me straight anyway? I hadn't used pornography all my life and still turned out gay, so how could that be a possible cause, or even a factor? Those pursuing a career in medicine attend medical school and go through the steps to become a doctor or a nurse, and building a house starts with blueprints which, if followed faithfully when assembling materials, results in a home. Getting married starts with meeting and courting a suitable partner, asking them to be married, then going through the ceremony to be wed, and thereafter caring for each other's wellbeing. But there was no such prescription for becoming straight, even though the admonition to do so supposedly came from the supposed creator of the Universe. Maybe for instance I had to memorize certain passages from scripture, or spend time with several women at once, or play sports or develop a deeper relationship with my father, I could then win the heterosexuality that was so apparently important (all of these things are various strategies gay boys have tried, in fact). In complete and utter opposition to his religious beliefs my own father in fact urged me to have group sex with women, betraying his own lack of faith in his religion and instead the underlying hatred and bigotry which actually motivated his abuse of me. A proven willingness to transgress his religious beliefs, just not for the benefit of his own son.

But the church lacked guidance more actionable than "pray about it," which I had done since the age of thirteen when I first realized attraction to my own gender endangered my safety among those hateful people, and in spite of literal years pleading with God still turned out gay in spite of also having never seen pornography or even known a single other gay person. Not only was I never magically turned straight, the only consequence from all this struggle was nothing but an endless abyss of despair made all the worse by the rejection of the very people who were supposed to love and support me whom instead did not hesitate for even a moment to kick me out of their family and community, and yet I was the one expected to accommodate and respect them and their hateful and absurd beliefs? Go fuck a donkey.

Observing religious people there is often an increase of insanity as they age and approach death, not simply a process of aging but that which is driven by intense fear and the insanity of a life spent in constant dissociation from reality and the people in it. One would think religious people would not have such pathological obsession with the mortal body, believing as they supposedly do in the divine and the afterlife, but membership in religion is a reaction to an ironic and devastating fear of death and the unknown, not belief in higher purpose, a fear experienced through the body and so absolute as to undermine everything of real value in their lives and to live from day to day despairing about one's soul in a cycle of endless existential self-recrimination and hatred of the mortal form.

Yet what is most striking about all denominations and indeed religion in general is how they all utterly and without exception fail to unite humanity in our similarities. Most of them actively engender division and hatred for the Other, and growing up I was expressly taught that ours were the only correct beliefs, and that other people were not only wrong, but wicked and evil and deserving of God's wrath and punishment. Did this even need to be a thing? Why must we actively teach and engender such glaringly hateful conceptions of God's other children, if the need does not in fact arise from intense feelings of fear and insecurity rather than righteousness? If any religion truly was possessed of absolute enlightenment shouldn't it *improve* the lives of everyone it touches, even those whom are not its members, rather than catalyze and enflame interfamilial, interpersonal, and geopolitical tensions?

In many ways the macrocosm and microcosm are the same, and characteristics of an individual can be mirrored in vast, bloated institutions since they are constructed by humans possessed of those very same characteristics and fallibilities, and the reason religion has not only failed to alleviate human suffering, but actively engenders it, is because *all* religion is born from fear and the desire of human beings to subjugate reality to our will, to protect ourselves from the abyss of darkness and the pain of loss which can and does occur at so many intersections of life. The conflicts engendered by religions are thus also those within man as a human, recognizing the potential for adversity with those who are superficially different than ourselves, this fear both explained, justified, and perpetuated by the theology which is its progeny.

Life also used to be all the more despairing—filled with disease, death, and rampant geopolitical conflict. Further back in our shared history as human beings, before we possessed much technology at all, religion originally served the purpose of passing on useful information for survival and wellbeing to children, such as sex education and that disease can be acquired from others, or that being disloyal to marriage partners can cause immense pain and suffering for those we ostensibly love, or that abandonment of children was a burden to society and parents must maintain their responsibility and procreate responsibly. But what better way to motivate precious offspring to avoid potential loss and heartache than with fearful mythology, to frighten them of the very fate of their souls? Fear is a powerful motivator, and much ends can be achieved by engendering it. But then of course those children grow into adults, continuing to believe that God really hates them for the very qualities he has made in them, which further deepens fears of death, the unknown, and loss and thereby impairing our very ability to be effective parents and members of society by perpetuating the very trauma which religious ideology ostensibly seeks to relieve.

Too often we get lost in the minutiae of life, and perceive things which are exactly the same to be different when they are not in the least—such as those with nearly identical religious beliefs hating each other and not realizing their only difference is the title of their religion. This perpetual conflict is exactly why the founders of the United States enshrined in the Constitution a separation between Church and State, to prevent the use of religion as a means to power that is so often used as an excuse to destroy and persecute, yet even with this separation there still has been immense conflict between religious groups which would have otherwise led to similar intergenerational conflict between people who share only the most trivial of differences, tensions stoked in truth by nefarious actors which profit from division and hatred and seek control, domination, and riches, each camp in fear of the other emboldens their position with yet more religiosity which serves only to deepen fear of the other and further separates themselves from the family of man, even and often to their own demise. Great was the irony when I moved to Los Angeles and found the people to be more warm and welcoming than the people of Utah whom claimed to instead be the ones that were, and the irony of those who reject my work because of their deeply held bigotry whom characterize me as an adversary simply for being different even as I actively advocate for their wellbeing persist in thyroid disease, obesity, depression, anxiety, insomnia, erectile dysfunction, and cancer as if the person they are spiting is the one whom suffers, persisting in ill health and traumatic conceptions of self worth simply for the petty vindication of feeling right.

The desire for religion does, however, originate from one constant truth of mortality, which is *fear,* the child of ignorance, and in fact the quest to find answers to assuage fear and defeat ignorance is a natural trait of all human beings since things we fear are often real liabilities to our wellbeing or survival, and the function of fear is in truth a biological motivator to get us to take care of our wellbeing, anticipate potential threats to our survival, and to gather resources both material and immaterial so that we may better survive. Without fear we would not be a species on this planet. But fear can also be overpowering and irrational, and can take control of our reason and our humanity to then rob us of the very things which make us human and life worth living, and answers which are superficially analgesic to fear such as what religion manufactures continue to fail to provide comfort because our conscience knows, deep down, that we are simply accepting an answer because it is *convenient,* not true, which is why those obsessed with religion never find real peace and continue seeking dominance over life, fomenting hatred, anger, and hiding addictions and secret schemes for money because the subconscious cannot be deceived and continues to fear for the inadequacy of the answers we choose.

A few times in my youth I had what I thought were spiritual experiences, or more accurately they were but I viewed them as evidence that all the fruitless effort and pain required of my indoctrination was evidence of my unworthiness. Once when attending a religious sleep-away camp for teenagers hosted by an organization affiliated with our religion (which, no surprise, was also a hunting ground for sexual predators) I had an intense but brief moment of feeling at peace during a speech given in a large auditorium while seated high in the top seats. It was so profound that it stood in obvious contrast to the chronic tension and anxiety normally experienced at every other waking moment of my life. But the feeling only lasted for a few minutes, though I prayed desperately for it to linger, and in spite of herculean efforts thereafter to experience it again, never did.

My despair became so overwhelming that one night at the age of twenty-one I would take a large knife to my wrist in a sloppy attempt at taking my own life. Luckily I survived, but it was not until many years later after struggling with alcoholism, poor health, and terrible at life and joined a recovery program that I would realize my unanswered prayers to be relieved of being gay so long ago as a child had, in fact, been answered, and instead I been deceived by those around me to believe they had not. The answer God had given to my ceaseless pleas to be made straight was a resounding and undeniable, 'No.' Yet because humans are often selfish, myopic, and prejudiced my family and religion chose to believe that God had not in fact answered my prayer to be straight, when clearly it was just an answer they didn't like, because if God is the Master of the Universe everything that occurs within it is his will, which includes whether I and others are gay or not, and it was my family and our religious community, not God, which had failed me.

This is the great irony in the quest for knowledge and understanding about existential questions, is that the answers are all around us and very often plain to see, but human bias, trauma, culture, and fear often prevent us from recognizing them simply because they are not the answers we want. So we instead conjure complex mythology and fantasies of sacrifice, divine intervention, and superstitious intrigue so that we may feel a sense of drama and purpose to distract us from our plain and simple existence, to feel in control in a world in which we are not the ones in charge. It is ironic that through my quest to recover from metabolic illness I found also answers to those existential questions which burdened me for the entirety of my younger life, because I was not by then looking for answers to those problems, rather instead simply trying to get well, but they were there nonetheless because they are in fact inextricably connected to life, biology, and reality, and one of the reasons I found them when others have failed is because those who came before refused searching in science, nature, and biology, ignoring the very evidence which exists all around us because they refused to believe the answers were either so simple as to defy expectations, or different than the answer they desired.

When in my twenties and still profoundly conflicted about my worth as a person and purpose and place in life and having been cast out from my family, friends, and society of my youth was thus also robbed of the excuses and beliefs which accompanied them. Unmoored and drifting in the vast world alone and without intimate relations I experimented with new alternative explorations into spirituality and explanations for the Universe. Yet religion had offended me to the core, now an adult capable of witnessing the morally repugnant abuse, neglect, and harm caused by these institutions and their adherents wanted nothing more to do with it. So one of the first things I tried was simply learning yoga and adopting a regular practice. Yoga was calming, for the most part (or at least at first). I could spend time alone with my thoughts, focusing on my breathing and poses and get lost for a moment from my overwhelming despair and stress, and practicing with other people gave me a wonderful albeit temporary sense of belonging.

But during parts of my practice I would have existential terrors of being alone in space, falling through the void left behind by an Earth which suddenly vanishes from its place in the Universe, wondering what it would be like to fall through an emptiness with no point on which to orient myself. This sensation was all the more real to me since I had several years prior gone on a night dive offshore from Honolulu in Hawaii, and diving in water at night without light removes any point of orientation, and because being in water makes a person feel weightless, direction suddenly becomes alarmingly arbitrary and instructors teach divers to breathe out from their respirator and watch the direction of the bubbles so as to achieve an orientation on which the mind can grasp. At one point though, descending on the anchor line was too far from the boat to see it and not yet near the sea floor, and suddenly experienced this debilitating sense of vertigo which was unlike any experience I have ever had before or since, and nearly panicked. Remembering the instruction I breathed out air and saw the bubbles float above me, and I had not changed my physical orientation at all and was still right-side up, suddenly felt my reality snap back into place, then eventually made it to the coral reef below and the secret night sea creatures which inhabit it (though pollution from Honolulu had long ago ravaged the reef of its vitality). I imagined falling through the void of Earth as much like that, and was often visited by that existential nightmare in dreams and the space between sleep and waking.

But yoga still gave me a better sense of peace than all my years in the religion of my youth, and I would feel energized and empowered after nearly every class, feeling like I was always on the verge of finding my community and thus hopeful for the future. Passing the median of my twenties an acquaintance introduced me to the practice of meditation. I was skeptical at first but learning more about it appeared to be something which could further promote relaxation and healing from the traumas of my past and lead me to the acceptance and peace and the satisfaction of questions which haunted my nightmares I so desperately needed. The white, middle-aged man who traveled around the west side of Los Angeles teaching this practice behaved as if he were an aged Indian guru, and entered the introductory class with all the pomp and circumstance of an aspirational cult leader, his helpers "asking" quietly for donations and treating him as an enlightened being rather than an ex-doctor superficially disenchanted with Western Civilization whom continued wearing expensive, couture scarves and knee-length shirts and probably lived in an expensive house while exploiting the fears of people like me searching for answers to life's big questions. During his talk he tried to draw clunky parallels between meditation and quantum string theory, which made it painfully aware he had no idea what it was but which most people in the room seemed to eat up since, like him, they had no idea.

Yet, when he described the meditation practice it was one of the most simple I had ever heard of, and having wanted to learn a meditation practice signed up. Our first class was at the home of one of his followers, and before he even arrived we were each taken aside and "urged" to donate an entire *week* of our salary to his organization. I tried to resist their insistence that I "donate," but still being psychologically vulnerable quickly gave in to the pressure and wrote out four checks for $200 each since I could not afford the full amount without severely over drafting my bank account. I was then given a special word to memorize which I would use in my practice, but told not to write it down and only commit it to memory. Finally, the promised guru arrived and he began to teach us how the practice was done, why it worked, and what we could expect, then led us through an actual meditation. In less than ten minutes I suddenly lost my mind into a state of peace like I had never, ever experienced nor thought was possible, even greater than the brief moment as a teenager, like my spirit disconnected from my body, peacefully, in a quiet darkness which was neither frightening nor disorienting. My mind found it difficult to stay there, however, and would repeatedly drift to topics of concern which normally dominated my thoughts both waking and not. But the technique of the practice, repeating this single *mantra* they had given me would bring me right back into that peaceful, homogenous state of being.

Not for one minute did I buy into his mystical, pseudo scientific explanation of why this meditation practice worked, but work it did and after several classes being guided through it, requiring only twenty minutes a day to achieve profound and lasting relief from the tension of my everyday life I was able for the first time in my seemingly long life to achieve some sense of relief from my existential crises and constant stress. I did afterward realize the amount I had given them was exploitative and so canceled the last post-dated check, though very grateful for this new tool in my life which had proved far more useful than all the many years of studying scripture and worshiping in the houses of religion, even if their organization and intention was every bit as opportunistic and cynical.

There are many forms of meditation or trance-like traditions in both religious practices and non. Some groups practice trance-inducing traditions which both excite and relax their practitioners, regarded without a sense of irony by other religions like my parents' as crazy or associated with the devil though they are also in turn regarded exactly the same by those they judge for their own quixotic behaviors. Hypnotherapy is a similar and somewhat popular tool used to treat people who struggle with a wide range of issues, and meditation is, in reality, simply the act of self-hypnosis which through repetitive motions or practices like repeating a mantra it shuts down other areas of the mind to tune out all other sensations and informations. In fact, self-hypnosis is exactly what other spiritual practices are— the repetition of words, sounds, songs, ritual, and tradition are all mechanisms which can distract the conscious mind and allow the subconscious to take control, and while this meditation practice brought me even more peace than I experienced for only a fleeting moment at that sleep-away camp I could reproduce it on command and will it to happen any time I sat down, folded my arms in my lap, closed my eyes, and repeated that mantra in spite of also being an unapologetic homosexual often having homosexual sex who drank alcohol and never prayed nor read scripture nor attended church. Being able to achieve a state of transcendence, albeit temporary, through methods diametrically opposed to those with which I was indoctrinated in my youth which utterly failed to do the same convinced me further of the futility of my family's unquestioning zealotry, but also more curious to further explore the bounds of reality and discover for myself the truth of our existence.

However, the progress I made during this time deluded me into believing that my long suffering would shortly end, but it would actually be another ten years and much more pain and heartache than I ever imagined before I would finally find answers which truly resolved those conflicts, though all the while I would be developing conceptions of matters both social, historical, psychological, and natural through seeing truth's shadow in the causality of life which would later inform my enlightenment and lead me to discover solutions to problems like depression, cancer, or the endocrinological origins of institution and geopolitical conflict as described in my first book, as well as the conflicting duality of coping and defense mechanisms which frustrate our efforts to find success in relationships and fulfill our goals in life, and eventually on to fundamental existential dilemmas such as now contained in this book.

Once while meditating at the nadir of my young life, having been through recovery and learned new skills of healing like inventory therapy, the realization suddenly came to me that my fear God could not love me was a lie rooted in the indoctrination of my youth which had even placed that doubt there in the first place, borne from such chaos, anger, hatred, and strife at the hands of those who have never known God and did not in the least demonstrate any familiarity with the peace which comes from true spirituality. I saw in their preoccupation with doctrine, lore, and mythology the distraction it truly is, a distraction from their own pain and heartache and experiences of loss and horror. It was an epiphany which freed me not only from the consequences of their abuse and neglect but also helped me begin understanding the reality of life and our connection to that which was greater than our individual lives. While I had not yet come to a consensus on my purpose on Earth nor my value as an individual, it was my first glimpse into eternity, and the result was a partial answer to one of the big questions I had been asking all my life, through newfound tools that could help me dive further into the beautiful darkness of mystery, and while a crazy old white man who wanted my money was the person responsible for facilitating this experience he was not the arbiter of truth nor the profundity of my experience, which instead came from my own biology and the function of the human mind and spirit which such a person, like every religious leader I have never known, simply meant to exploit for their own profit. I had made a connection finally with my true self and thus those things we so often think greater than ourselves or beyond our reach.

This is why inventory therapy is requisite for all healing, including of physical health, because we often do not even know how our unconscious coping mechanisms direct our behavior, taking us in directions which are in fact antithetical to recovery. Some frustrated with my work when not seeing the kind of success they want act on that frustration and return to behaviors they did before which clearly did not work, having arrived at a final judgment simply because they got frustrated, not realizing there was probably something they were continuing to overlook, misunderstand, or had not yet learned (or which I did not communicate effectively), but having made their judgement never again return to the responsibility of educating themselves on biology and how exactly to take care of the body, instead seeking control of it, which is not something any of us can do.

To graduate to exploration of higher purpose and enlightenment are far beyond the concerns of mortality and control of trite biological functions as hair loss and weight. As I have often mentioned, the entire reason I have done any of this work and that anyone knows my name, my life story, and my academic research and science is because I did inventory practice and through it saved myself from self-destructive control and coping behaviors borne of trauma, which then not only empowered me to be more effective in my own life but swung wide the doors of enigma, to walk with the Muses and converse with Eternity, as may all who in turn take the time to do the same, and thus come to know that which is otherwise unknowable, such is the power of an unburdened mind.

4. Explaining the Particle-Wave Duality of Light

In 2015 when I was beginning work on *Fuck Portion Control* and making great strides in reconstructing my health I was one day hard at work at my job as a 3D motion graphics artist designing and building an animation in which I was causing brightly colored spheres to oscillate while traveling through the animation. While setting up the system I started with a single sphere in the center of the viewport and began to animate it up and down along the y axis. Suddenly as I began to animate the sphere forward, and without intending to concern myself with such complexities of reality, I realized I had stumbled upon the true nature of light and solved the particle-wave duality which has befuddled scientists for the last several centuries.

Sat in my tiny apartment staring at my computer screen as this particle animation played on a loop, with not a single person to share such a profound discovery, I further researched existing science on the nature of light and would find that, no, in fact I was not the first person to actually propose this solution, though no prominent or leading scientists have done so nor even attempted to explore this idea empirically, whom rather uncritically accept the particle-wave contradiction as a paradox of nature rather than a simple, human observational fallacy.

As I have repeatedly mentioned in my works there are no paradoxes in nature. Paradoxes cannot actually exist else, if real, would invalidate reality and we would not be here, because the fundamental laws which govern reality would be in conflict and either cancel each other out or create a reality so chaotic and incoherent that existence of organized matter such as ourselves or the planet on which we live would be impossible. Paradoxes only appear to humans because of our limited powers of comprehension as a mortal animal species, and are borne of ignorance of the laws of the natural world, since we are only just now learning about them, which is the entire point of scientific discovery, to divine those laws and come to a rational understanding of nature, not imposing our ideas upon it, which of course is impossible anyway. Light is no different, and the particle-wave duality of light has always been a simple observational trick of natural laws confusing those who did not as yet fully understand them (and, frankly, overthink them).

Newton, again, was the more correct in his description of light, as light is in truth a particle or packet which, traveling through spacetime, also oscillates for the very energy it received from its place of origin that firstly compelled its exaltation, and an oscillating particle traveling at a constant vector is nearly indistinguishable from a wave (*figure 1a* and *1b*). This is more true of light particles because, unlike matter, light generally maintains a constant velocity (depending on the medium in

figure 1a.

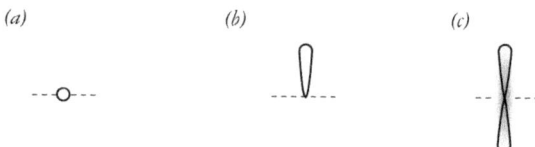

(a)   (b)   (c)

A particle at the center of oscillation (a), moving toward a limit of oscillation (b), and the entire field of oscillation (c).
* The tail is for illustrative purposes only, meant to convey motion, not describe the shape of a light particle.

which it travels), which thus results in equidistant minimum and maximum limits of the oscillation appearing then as crests and troughs of a wave, separated or stretched along the direction of travel precisely *because* of its travel.

This easily explains the otherwise confounding peculiarities in double-slit experiments and behavior of light where light particles, while traveling, exhibit wave-like behavior but when colliding with matter such as photosensitive film or intermediate barrier like a screen, thread, or hair behave instead like a particle, and the collision with matter stops both its oscillation and travel to then reveal the particle but no wave, because there was in fact never any wave, just an oscillating particle traveling along a vector, and where particles also appear in a field of probability, which is created due to the distance from its center of oscillation, may be anywhere between those limits at the time of collision but also does *not* appear in other locations in the same field of probability because it is a particle which occupies only the space it does and not the entirety of the field as does a wave.

It has in fact already been demonstrated that all subatomic particles oscillate or vibrate, so this conception is not even remotely outlandish. Incidentally, I suspect that at least some of the supposed disparate parts of the atomic structure are not actually unique and disparate particles but merely the same particles in different states of oscillation which are dictated, influenced, and changed by their environment and energy, including by that of other particles with which they interact.

This problem of incorrect characterization of light arises because science often makes science and the natural world unnecessarily complicated and often gets in its own way by naming and categorizing things as if they are fully independent and unique from everything else when in fact they are simply fractionally different versions of something that is in fact unified and singular, which then especially confuses lay persons and the uneducated and makes our brains hurt but also prevents new researchers entering the field from seeing the forest for the trees. For instance while many fatty acids have unique names that make them sound especially disparate there is actually only one single universal fatty acid which is then

figure 1b.

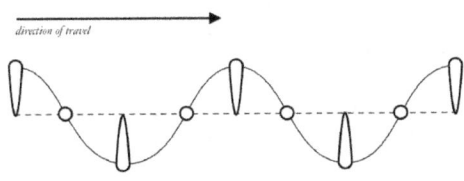

The apparent wave behavior created by an oscillating particle traveling on a vector

lengthened, shortened, saturated, desaturated, branched, etc. This problem is an unintentional consequence of human pride, of those doing this work and making such discoveries rightly and understandably desiring to make their mark upon the course of history, but in the process also causing overwrought and inconsistent jargon to make things unnecessarily complex which then incidentally later imperils research and progress through overcomplication.

Nowhere is this more consequential than biology and thus practical problems and solutions to human health, where even very intelligent and educated experts (let alone lay persons) do not recognize the presence of short chain fatty acids within cholesterol or steroids, nor that all starch is actually sugar, nor that fat is not burned by the human body as energy but instead converted to gas and water during metabolism which is then breathed and peed out of the body. Because of this limited understanding of the simplicity by which reality is constructed, even the most accomplished scientists will not see within the longer atomic chain of certain molecules those upon which it is based and recognize the intrinsically interconnected nature of the physical world such as the short chain fatty acids nested within longer chain fatty acids or even steroids. In fact, it would be more concise, but admittedly boring, to name elements of nature, physics, and biology as what accurately reflects this law of simplicity, such as naming the short chain fatty acids 1, 2, 3, etc., instead of formic, acetic, propionic, butyric, etc., which makes them sound terribly different from one another when in fact they differ simply by the number and type of hydrocarbon bonds.

The very elements themselves demonstrate this fundamental principle of simplicity as well, where each element essentially differs only by the number of protons it contains, so instead of hydrogen, helium, lithium, beryllium, molybdenum, iron, copper, and fucking Tennessine (really???), there is in reality just element 1, 2, 3, 4, 5, 6, 7, 8, etc., which are their atomic numbers, by the way, which more adeptly communicates the often supreme simplicity and basic math at the very foundations of the natural world.

This is even more relevant when considering that elements can and do change from one into the other and are not as undying as is often implied. For instance, add one proton to a carbon atom and it changes to nitrogen. Add one more and it becomes oxygen. Forms of manganese regularly decay into iron so much that some of our planet's supply of iron actually originated as manganese. This relativity of every element is a reason why supposed paradoxes of having not been visited by advanced extraterrestrial beings, if they indeed exist, whom would never come to our Earth in search of resources because technology sufficiently advanced to achieve interstellar travel requires complete understanding of physics and building machinery capable of bending spacetime and thus also an ability to simply synthesize atoms and molecules as desired (never mind the fact that an infinite amount of water and minerals already exist just floating out in space anyway). In reality the number of each element more correctly describes what it is, and as such also everything else which makes up the universe, such as light, which is nothing more than oscillating packets of energy which simply vary in the intensity and frequency of their oscillation.

A supposed paradox revealed by a triple slit experiment called 'Hardy's Paradox' normally explained by the quantum mechanics concept of superposition, which is extremely over-complex, is instead easily explained by oscillating particles—The experiment adds a third slit and two sources of electrons and supposedly demonstrates apparent interference patterns which seem to defy established principles of wave dynamics wherein a combination of two or more waves will either constructively add to each other's amplitude or oppositely cancel out their amplitudes depending on their relative position to each other. Hardy's paradox, like all paradoxes, does not actually exist and instead demonstrates the complete absence of destructive interference for no other reason than *light is not a wave*. Because light is not a wave, but an oscillating particle, there is also no such thing as wave function collapse, which is instead oscillation collapse, as particles strike matter and are captured by the atoms in that matter (to which their energy is then transferred). Because the patterns seen in the triple slit is not really a paradox, because none exist, scientists incorrectly believing light to be a wave must then explain its behavior as a phenomenon of quantum superposition and creating yet another problem to solve a problem that does not actually exist which is in reality no more than another human observational fallacy. Hardy's Paradox gets even more convoluted because the concept of antiparticles is required to account for the phenomena in quantum mechanics responsible for perceived destructive interference of *pure energy* and the sheer convoluted nature of these kinds of theories should instead suggest potential errancy because nature is efficient and exacting and not so disordered as to create what in practice would be actual paradoxes.

Superposition is also not an adequate explanation for the results shown in Hardy's Paradox because if superposition were occurring the frequency (energy) of the lightwave would *increase* from the third slit, not stay the same as it does. All that is occurring in these experiments is that particles show up in areas not expected of wave behavior simply *because there is no wave*, and the whole thing is an attempt to solve for a problem that does not actually exist, and patterns of supposed destructive interference appear because it is an optical illusion caused by the fact that there simply is no particle in those positions in the first place, because light and particles are simply oscillating, and appears to have constructive interference (when it does) because that is simply where most of the particles are located at the moment of observation.

The consequence of not understanding these realities of nature is that much time and resources are wasted in sunk-cost fallacy, and it is understandable that people would want not to be wrong (I'm sure much in this book is wrong!). But what we observe in quantum mechanics which is explained by antiparticles, antimatter, and the uncertainty principle arise from misunderstanding electromagnetic radiation as waves rather than particles in oscillation. If electromagnetic energy were in fact a wave the entire function of Faraday cages would not work, as the cancellation wave of a Faraday cage which interferes with oncoming electromagnetic radiation would then entirely annihilate that energy, which instead flows over and around the age even after reacting with it, which a wave would not do if it was collapsed or annihilated. In fact, the entire suggestion that electromagnetic energy is a wave which collapses implies that energy can in fact be destroyed, which is entirely antithetical to all accepted laws of physics and the very construction of reality itself, and viewing light as a wave thus breaks this unbreakable law of physics that energy can never be destroyed. But the oscillation of a particle can be collapsed without actually annihilating the energy that makes it real, due simply to the rules of the Law of Motion, and thus satisfying other principles of reality in congruity as is required for the explanation of all things.

Though we also talk about the different wavelengths of light there is also no such thing as different types of light—There is only *light*. Or, more accurately, only one *electromagnetic force*, and the colors in light we observe (both visible and not) are simply different subjective interpretations of one single phenomena of electromagnetic energy radiation which simply varies in wavelength and intensity, not a collection of different types of that energy, which is instead a useful tool for us to describe and categorize nature for our own understanding. There is not red light and blue light nor gamma rays and ultraviolet rays, only *light*, which appears red and blue because of the way our brains interpret the different wavelengths, or that our machines and devices measure from what is in fact one single phenomena. Hardly anyone, even academics, also know that nobody has ever actually measured the speed of light. Einstein knew this, and it is reflected in his characterization of the speed of light using relativistic terms which are often overlooked. The speed of light as we know it is instead an accepted convention which is as close as we can currently use to describe its behavior, because it is in truth impossible with any current technology to properly account for the relativistic nature of spacetime and accurately confirm the speed of light.

The oscillating particle model of light also conveniently and, most importantly, *simply* explains how *color* works—It is well known that we see color because those are the wavelengths of light *not* absorbed by the matter it strikes, while all the other wavelengths are in fact absorbed, and one day while peeling an orange and admiring the pure intensity of its color I realized that, because all particles oscillate, the wavelengths of light which reflect from matter such as the peel of an orange or the petals of a flower are those which have a harmonic oscillation resonance with those molecules from which they reflect, since because they are an oscillating particle and not a wave their harmonic resonance means they thus never collide, the frequency of their oscillations being such that a collision never occurs in the time which the light passes through them and so the particles never meet, where a wave will meet other waves regardless of their frequency, then remitting those photons back to us while other wavelengths will instead collide with the particles whose phases of oscillations are different enough to catch up with each other and collide. If light were a wave color must be explained in terms of bouncing or reflecting off a surface, off of the atoms which make up matter, but how this could possibly explain color or reflectance would require an overwrought explanation of how waveforms maintain their waveform even after striking matter, which is itself a paradox when considering the entire atomic universe also has oscillation. An oscillating particle's reflection, however, is easily explained because, just like a space probe being shot around a celestial body using gravity and momentum, photons of light which fail to collide with matter because of their resonant phase oscillation are thusly slung around the atomic nucleus to jettison it away at the very angle of reflection, and thus the macro and micro are married through laws of physics unified without paradox, because we know the force of gravity does affect light, which it should at the micro and not only the macro.

This is also how mirrors, visual reflections, and diffusion also likely function, because the atoms in the material such as what construct a mirror have an oscillation that is sufficiently in phase with the visible wavelengths of light that far less collision of particles occurs and so almost the whole visible spectrum (which is in fact a very tiny sliver of the entire electromagnetic spectrum) are reflected back at us as they are flung around atomic nuclei at angles of reflection and refraction. Glass famously lets nearly all visible light though, but does block UVB, UVC, and some infrared radiation if thick enough, which means that the atoms which compose glass have resonant oscillations which collide with the electromagnetic spectrum at a regular interval which includes the UVB, UVC, some infrared, and even parts of the outer spectrums but which skips over the UVA, visible, and near-infrared, due to its "wavelength" derived from its oscillation period. This is also why collision with electromagnetic radiation heats up matter, because of constructive interference, when the oscillating particles do collide their combined energy always increases, such as from the conversion of a photon to an electron when colliding with an atom, where when waves collide they can experience either constructive or destructive interference, and if light truly were a wave it would also demonstrate destructive interference when colliding with matter and actually sometimes cool it down, but instead always heats it up in collisions (technically there is laser, or doppler, cooling of matter, but that is achieved by harnessing doppler principles and the effect of photon *emission* from atoms to lower their oscillation frequency, not the effect of the light itself which would otherwise heat and excite the atom if it did not release that photon as occurs in this specific effect).

One of the problems in conceptualizing such things as particles in nature is that scientists and "communicators" often attempt to construct analogies to explain physics concepts, but because they are too intellectual and not artistic tend instead simply to further obfuscate that which should be easily described without analogy, since reality operates on logic and reason. String theory analogies are some of the most absurd, but even simple, real world physical demonstrations such as a taught sheet with a heavy ball weighing down the center as a substitute for simply explaining gravity are at best ineffective because of the fact that they are an analogy and not the reality and mislead people to think that gravity is in fact like a flat sheet. Particles flitting about in the universe and forming matter is likewise quite difficult to comprehend but the concept of rain is not only easy to understand, since every human has seen and experienced rain, it is actually not an analogy at all because rain is literally the macro-scale manifestation of the same underlying particle physics which occur at the subatomic level, which is exactly why rain behaves as it does, because it is made of those very particles which create reality in the exact same way as how rain forms. In fact, most people do not realize that nearly all matter even as large as rocky planets like our Earth obey fluid physics just as does falling rain, except on much longer time scale, because that is how the energy of which reality is constructed actually behaves. Rain only occurs when clouds build up enough water vapor to condense, then form smaller droplets too light to fall to the ground, but these droplets merge and grow and merge and grow until they are finally large enough to become rain, which is exactly how particles in the universe form, in a "rain" of energy, the smallest being quanta which erupts in very, very tiny sizes from nothing which then coalesce into progressively larger and larger particles. The idea of nothing is also impossible for the human mind to conceive, however, but in fact there is no such thing as nothing, which is only how we perceive the apparent emptiness of space, and instead there is such a thing as *everything*, but it appears as nothing until it is excited by disturbance, and when disturbance excites the apparent nothing this then gives off the very particles of energy which coalesce to make up the most basic foundations of our Universe.

Further supporting the oscillating particle nature of light, energy is also always constructive and quantized, unlike waves which can also be destructive and continuous, which also accounts for the logical behavior of particles as oscillating and not waves, otherwise quantized atomic properties like electron valences would not exist. This in addition to the idea that energy such as light could ever be destructive is an absolutely preposterous notion that long ago should have invalidated any wave property of light as revealed by the slit experiments because *energy can never be destroyed*, let alone destroyed by other energy, yet researchers describing light as a wave do just that when trying to characterize the supposed destructive interference patterns supposedly demonstrated by light which in reality is simply the absence of light particles, not destructive interference, because energy cannot cancel itself from existence. This is a fundamental law of mathematics in that 1 x 0 = 0, not 1 as those models are claiming (meaning that there cannot be nothing where there is currently something), wherein if destructive interference of literal fucking energy were a real possibility there would thus be no reality.

Furthermore, the quantized nature of particles (including light) likely accounts for the phenomena otherwise currently described as antimatter, wherein every oscillating particle has an other-dimension counterpart which likely balances every particle in its oscillation as the energy reflects off its minimum and maximum boundary to create quantized patterns and a minimum of two opposing factors of every particle (otherwise the particle would shoot off into the perpendicular direction of oscillation) thus giving reality a structured, quantized state and eliminating the possibility of true chaos which would invalidate existence if energy really were able to destructively eradicate itself.

Many truths about life are in fact different than how they might readily appear, and while it may seem like such esoteric problems are not directly consequential for the lives of most people, it in fact promotes and sustains seemingly disparate problems which do. For instance the observation that high blood sodium is associated with disease led many uncritical researchers and medical professionals to conclude that dietary salt, not biology intending to retain salt to keep water in the bloodstream and help sustain blood volume, was the causal factor as discussed in *Fuck Portion Control* which is a purposeful function of the cardiovascular system to help maintain blood pressure so the heart can continue pumping. Not understanding the true nature of light prevents us from solving yet unsolved problems like the origin of the Universe, effective interstellar travel, antigravity, etc., and ultimately the purpose of existence.

5. The Nature of Time

The laws of relativity first described by Einstein dictate that laws of time and space are relative, meaning that its nature depends on where you are, what you're doing, and what you're looking at. But certain phrases and metaphors used in describing time and space such as *time-dilation* is not only commonly misunderstood, but is technically an incorrect way to describe the relativity of time, which also makes comprehending science and physics much more difficult because it actually presents incorrect information, and then people think they are dumb for not understanding something when in fact the construct they are trying to understand is in fact incorrect.

For instance it is often said that time dilation occurs for objects traveling at lightspeed. In fact, time-dilation (the relative slowing of time) does *not* actually occur, but is only an *observational* effect for the person doing the observing. If we for instance were traveling at lightspeed in a spaceship everything around us would feel the same as it does here on Earth, with no observable difference in how time passes for us or the ship and anything in it also traveling at lightspeed. If we did the opposite and slowed to as close a stop as were possible time too would still feel and pass the same for us as it does anywhere else and at any speed of acceleration. Time-dilation therefore is not something that actually occurs, ever, in any part of reality, and its use implies that it does occur, but which is only an effect of observation and relativity, where in fact time is *always* the same for every entity in the Universe no matter where it is or what it is doing.

What does occur, however, is that time is *relative* (the entire point of Relativity), so when accelerating to lightspeed (i.e. such as light) everything that is *not also* traveling at lightspeed (i.e. the rest of the Universe) will in turn appear to have its time *also accelerate* such that all events in all of existence would occur in the same moment, and time therefore does not stop when traveling at lightspeed because it slows down, it "stops" because the time scale of all existence also accelerates to lightspeed such that everything is experienced in a single moment so fast that time functionally stops existing, the apparent time of everything not also traveling at lightspeed accelerating in proportion and relative to our frame of acceleration.

This is, in fact, why light has the unchanging nature we observe as it moves through space, unless acted on by other forces, not because light is inherently stable or unchanging but because its time scale relative to ours as it travels at lightspeed stretches from the beginning of its origin to the end of all time, and would exist forever and ever without end if it did not collide with other matter, to experience infinite time, where we in comparison thus observe the light in an infinite fraction of a moment of time, which as such appears frozen to us so great is the timespan of the light that we cannot observe even any hint of time.

This is why we thus observe light to have decelerated time (frozen), because it travels at the maximum acceleration, and time dilation does not occur but *appears* to occur, so the use of time-dilation is not correct and why that term is so confusing and useless and, as commonly mistaken, the laws of relativity as such do not mean that time stops when traveling at lightspeed, and if we were traveling at lightspeed and surrounded by clocks in our light-speed spaceship the clocks would all tick at the same speed we always experience them, but not only would the galaxy zip by at lightspeed around us, every event in the entirety of time also passes during that acceleration, and the entire Universe around us would pass in an instant, yet our reality in the spaceship would not seem *any* different than if we were on Earth sitting in a chair, because time is *relative* (that's the whole point of Einstein's laws). So the term time-dilation is not very useful for helping people to understand time, because time is *relative* to the observer and time dilation *does not* actually occur as a principle of reality.

When we think of light travel too and reference other celestial bodies and how many light years it might take to travel somewhere at the speed of light this is also mistaken. Because time is relative, acceleration to lightspeed would mean we reach *any* destination in the Universe in the blink of an eye, as the speed of light is how we measure lightspeed from our much slower frame of reference, but to the light itself is *instantaneous* because the time of the rest of the Universe not moving at lightspeed is also experienced as accelerating to lightspeed. This can be demonstrated by the fact that for us it takes 323 lightyears for light to reach us from the star Polaris, but if we accelerated to half the speed of light we would experience that amount of time as 161.5 lightyears, and if we accelerated to 3/4 the speed of light it would be just 80 lightyears, at 90% it would only be 32.3 lightyears, and accelerating to full lightspeed would thus be 0 lightyears. As such, light is actually a time traveler, visiting us from the past rather than also evolving along with the Universe, which is why light doesn't change and evolve as it travels through space (unless acted on by other forces like gravity and matter), and why astronomers can look across the Universe to observe the ancient past and struc-tures like early galaxies which in fact are no longer in existence because that light they once emitted visits us in the future in so rapid a time that what appears to us as billions and billions of years is less than an instant to the light itself. If light was not like this it would itself evolve and change over time as it traveled, to reflect the evolution of the planets and galaxies long dead from which it came, just as planets form and decay and stars burn through their fuel and eventually evaporate, and it would be impossible to look back in time at all, but it is possible because the lightspeed of light freezes its time relative to our point of observation which to the light is less than a blink in time.

It is thus more helpful to consider time in terms of *percent of lightspeed accel-eration*—When accelerating to 100% lightspeed the time of everything not accelerated to lightspeed passes so fast as to occur in an instant. A space trav-eler thus moving at the speed of light can effectively live forever, but the entire Universe would have passed out of existence the moment they reached lightspeed. Oppositely, at 0% lightspeed the time of anything accelerated faster than 0% rela-tive to our point of observation *does not pass at all*, to instead appear frozen in both time and space, and the space traveler which could decelerate to 0% lightspeed would experience a Universe without change, where even light itself does not move, and objects falling would appear suspended in the air, rain held in perpetual stasis, waterfalls and waves and animals all stationary and unchanging as statues.

Of course, it would be physically impossible to achieve deceleration to an absolute, invariant position in the Universe. The Earth, after all, travels at 67,000 miles per hour just around the Sun, and our solar system is traveling through the galaxy at about 514,000 miles per hour, and the galaxy itself is also moving through space, toward the Andromeda galaxy, and our local group is also moving through space as well, so even if we were able to somehow create an experiment which could achieve an absolute, invariant position in the Universe the experiment subject would have a timespan of effectively 0 relative to ours since the Earth is in both continuous absolute and relative motion (so my earlier math concerning Polaris is not quite accurate as we are in fact already moving at a percent of light-speed). But since 100% lightspeed experiences the relative time of the rest of the Universe as infinitely fast, and 0% lightspeed experiences the relative time of the rest of the Universe as infinitely slow, Time itself *arises as a function of travel through space*.

Because of this, time as we experience time occurs in part from the function of literally living on the crust of a moving planet which not only spins about its axis but also transits in orbit of the Sun in turn orbiting the Milky Way in turn attracted by the Andromeda Galaxy and other movement from the local group and supercluster and so on. Indeed it appears that absolute, invariant motion is mostly an impossible reality in our Universe, because even if there is no point of reference close enough for us to observe motion, say for instance if we were out in deep space, we are still in motion due to the Law of Motion (all the molecules and atoms of which we are made are also constantly moving too, even if only vibrating or oscillating). Essentially then, the Earth moves us through time whether we like it or not.

Yet, scientists have in fact already created an experiment proving this without knowing it—because subatomic particles produced in a lab appear to only last a fraction of a moment, this occurs because they have very shortly lived *time* relative to ours because they are occupying a more or less invariant position in *absolute* space, being born into existence from the underlying energy field which supplies all reality but do not themselves have motion in absolute space and therefore their time *relative* to ours is almost infinitely fast and they pass away as fast as they were born. In fact, these experiments likely prove that there is absolute space and not only relative space, because if the former did not exist a particle created here on Earth would inherit our velocity and thus be longer lived. In fact, when these particles are instead given a velocity they also persist for much longer, now having motion in both absolute and relative space, and their relative time is then relatively much slower than when their position was invariant and as such remain with us in our trajectory through space due to the Law of Motion, for a time. These particles would last even longer in our relative time frame if somehow researchers were able to give those particles more lasting inertia which would thus scale their relative time to match that of our own, such as continuing to accelerate new particles through an accelerator and not only those meant for collision.

The principle of what happens when we decelerate is already demonstrated by Relativity and experiments which support it, such as that time passes faster out in space where gravity is less strong but slower in the field of gravity, because there is in fact no difference between gravity and acceleration, and having mass is the same as traveling through space, while *not* traveling through space (if it were possible) is the same as *not* having mass. Because of this, astronauts which go to space are all very, very slightly older than they would otherwise have been if they stayed on Earth, because less gravity means less percent of lightspeed acceleration. But if they were instead constantly accelerating to speeds comparable to the gravity field of Earth they would not experience any change, or if they accelerated to greater than comparable to Earth would then be younger, and could even risk having time on Earth pass by too quickly if they accelerated much faster and risk their entire families and communities growing old and dying out before they returned to Earth a moment later.

This principle of time is likely the real reason and 'paradox' as why we have never met other intelligent, alien life—that the ability to transit the stars at even near lightspeed is a one-way trip into the future equal to the time (or more) it takes that light to transit the emptiness of space, experienced in the blink of an eye, and lightspeed travel does not and will not come without consequences of losing a great deal of time relative to the rest of the Universe (especially our home planet) unless there is some other way to bend space and cheat the laws of time, which may not be possible due to how reality is actually constructed. So visiting a system that is four lightyears from us such as the Proxima system at lightspeed would only cost a reasonable number of (about) four Earth-years passing in but a moment (no cryosleep required), but visiting a distant system that is even a relatively short distance as 1,000 lightyears away means those travelers would reach the location one-thousand Earth-years years in the future and never again see their friends and families they left behind, even though they too would experience the distance in nothing more than a blink of the eye. Other stars and galaxies which are hundreds of thousands, millions, or billions of years away (which are also not even there anymore anyway but move in their orbits or even gone from existence entirely) which, although we would arrive in what felt like a moment, would have left behind our own Earth for that amount or more of lightyears traveled. So, unfortunately, every science fiction film that shows a spaceship zipping back and forth at lightspeed would leave and return to those locations hundreds of thousands or even millions of years into the future each time they did. This does not mean interstellar travel is impossible, just that it won't be achieved by actually traveling through the Universe at lightspeed.

Because space is warped, such as around massive bodies like the Earth, this also affects time (because spacetime is one thing and not in fact two separate properties married together), and gravity and mass can be conceptualized as accelerating while impeded, which sounds contradictory to reality but in fact there is no difference between falling, i.e. moving through space, and gravity (this is Einstein's equivalence principle), except that the presence of the object creating the gravity also then gets in the way of travel. So essentially we are also at all times *accelerating* toward the center of our Earth even though the mass of the Earth also prevents us from passing through. In fact, all of the mater of which Earth is composed is at all times also accelerating to the center, but the reason this matter does not pass through itself and we do not fall through the matter of the Earth is because the *relative* location of the all the matter of which we and the Earth are composed which is decelerated relatively to 0% lightspeed to each other and thus we experience the time of things around us as nearly stopped, because time arises from motion through space we do not fall through the Earth because of our relative 0% lightspeed acceleration (near 0%), which also means that matter appears solid simply because its *relative time* is also frozen at (near) 0% as ours is relative to it, thus maintaining a relative (nearly) invariant position with respect to each other and therefore mass has physical properties as a function of being relatively frozen in time and not because mass has any truly solid properties.

The reason we cannot put our hand through a brick wall or don't fall though the floor is thus because their *time* is impenetrable, not mass, due to our relative invariant position to it, which freezes time as we literally meet what feels like solid matter which in fact is time at the atomic level freezing to a complete stop to prevent the passage of our atoms through those of others near it. As would logically arise out of this principle this time function which creates mass can of course be overcome with adequate acceleration and energy to overcome that relative invariant position, such as occurs when struck by a bullet or a comet, and thusly penetrate the other mass by having sufficiently increased relative acceleration over the impacted object. But this also means that all the great planetary forces which shape the Earth and cause the rise and fall of mountains or subduction of tectonic plates appears in literal slow motion because of our almost constant, invariant relative position to the Earth. Obviously, we do experience time as moving forward which, among other reasons (such as discussed in the upcoming chapter on consciousness), occurs in part due to the gravitational field in which we exist that is constantly accelerating us to the Earth.

Similarly, the mass of the infinite density at the center of black holes accelerates gravity there to lightspeed (at the event horizon) towards its center, but if we were to somehow survive traveling into a black hole time would not actually stop for us, locally, because time dilation does not occur, and we are part of the acceleration and time is relative, but being in the gravity field of a black hole at the event horizon is the same as traveling at lightspeed, which means the rest of the Universe would appear to be as if to be fast-forwarded at the speed of light, wherein entire worlds, stars, and galaxies would appear and pass away in less than moments, and the black of a black hole appears black or empty because what we are seeing is the empty Universe in that spot before matter there even existed, so slow is the apparent time of a back hole relative to our point of observation that no light has yet been emitted from its center (or ever will be). Yet if we could visit a black hole our local time would feel and appear just the same as it does here on Earth, but the Earth we left behind would have aged and died in less time than we can even comprehend.

Because mass is in fact a function of time, and spacetime is a unified feature of existence where our experience of time arises from movement through space, mass too is also a function of movement through space, the energy which arises from the foundations of reality to create mass a consequence of movement through it which then bends spacetime into the energy and mass that create all things we see, feel, and experience. While this may all seem grand and incomprehensible I imagine this function of time and mass as travel through space as comparable to other demonstrable principles of reality we can easily observe all around us rather than using abstract or confusing metaphors or paradoxes (which don't actually exist), since reality is constructed from definite rules which should be reflected in all the reality we see and experience. A good example is the acceleration of water by gravity to turn the turbines of a hydroelectric power station—water accelerated by gravity is used to spin a turbine within a magnetic field, and the resistance of the turbine is what produces energy. This is not a metaphor, but the very demonstration of how movement through space, in this case water accelerated by gravity and spinning of the turbine, creates the very energy which composes everything in the Universe by turning against the resistance of the magnetic coil. Time and space work the same way, where acceleration through space is time but resistance from space then produces mass (energy), which then empowers matter to have the properties it does to thus form atoms, molecules, and all that is created from them, and thus we see reflected in the everyday and common features of reality also the foundations of it.

We also unknowingly do observe time travel on a near daily basis, but because it does not take the form as what we would expect such as commonly shown in science fiction thus fail to recognize it as such, and unfortunately is not the act of visiting previous timelines in of the Earth to visit dinosaurs or ancient human civilizations but is instead the reversal of matter formation such as occurs in the state of *plasma* released by fire and combustion, which is the very energy that creates mass being released backward in a reversal of the time which resulted in that matter in the first place. Mass is energy condensed by the laws of spacetime, so fire, combustion, plasma, and loss of heat (commonly also called entropy) is the reversal of spacetime organizational matter and mass back to the ground state from which reality originates. So unfortunately, time travel (to the past) is the release of energy of which matter is composed, into the fundamental light which is the very beginning or essence of that energy which emerges from the underlying foundation of reality as a consequence of disturbing space as what creates time arising from motion through it.

This also implies that magnetism is resistance of space itself, and can be thought of as time in the opposite direction—not negative time, as that is impossible, but simply the force of time moving toward 0% percent lightspeed instead of toward 100% lightspeed as most matter oppositely accelerates.

It has also been known for some time now that the concept of the 'Big Bang' which was for so long a popular theory of the beginning of the Universe, is not actually correct. This makes sense because if there had been a singular explosion which created the Universe there would then be a center from which all bodies in the Universe accelerate away and share a more or less uniform velocity, but this is not the case and instead all galaxies in the Universe appear to be moving in random directions. But, like all reality, clues to how the Universe forms can be observed from its apparent nature in other observable phenomena, the macro reflecting the micro, because regardless of whether we understand or not all parts of reality are built upon and adherent to universal laws, which is indeed the current problem with quantum physics and Relativity is that there is currently no unifying mathematical model to describe everything, as there should be. There is an interesting phenomenon in the Universe where galaxies and other structures move away from each other at such speeds that the difference between them *exceeds* the speed of light, so if one object moved in a direction at the speed of light and another object moved in the exact opposite direction at the speed of light the gap between them grows at a rate twice the speed of light and light from each object can never actually reach the other, ever, and neither knows of the existence of the other because they can literally never see it or reach it, since the speed of light is also the maximum speed of travel possible in the Universe. This rate of expansion does occur, even between objects as massive as galaxies and clusters of galaxies, so that if there were inhabitants of one which possessed telescopes to view the Universe the expansion between them and other galaxies can be so fast as to exceed the speed of light and we then never know they are actually there. We know the estimated age of the Universe is about 13.7 billion years old because we can only see light that is 13.7 billion years old, and if the Universe was any older there would in fact be more light, because there would have been more time for more light to reach us. *But* if the space between us and other galactic structures can expand at greater the rate of the speed of light the boundary of the Universe in fact could very well extend more than 13.7 years from our position, and we simply cannot see them due to the rate of expansion. In fact, there are very likely galaxies and planets and stars that exist beyond the visible range, and we simply cannot see them because their light has simply never been able to reach us. It is therefore entirely possible that the Universe is infinite in both its age and bounds, but we cannot see the rest of it because that space between the rest of it and our local Universe expands infinitely at a rate *greater* than the speed of light.

Because mass warps spacetime this also means that spacetime between large bodies of matter is stretched and warped, which is part of the reason for the relative disparities in time in the deep of space compared to that on large bodies like the Earth, which is a reduced rate of acceleration and thus also reduced rate of time as a percentage of lightspeed. This warping can also be observed by the phenomenon of *gravitational lensing*, where light bends around a large mass, although it doesn't bend but the space through which light travels bends. But one can imagine these forces like a rubber-band which, when stretched sufficiently, *snaps*. Some people say that reality cannot snap, but it bends such that light itself appears to change its course, and I can't imagine a reality where spacetime can bend or stretch *infinitely* such as is caused by the endless expansion of space between great islands of matter without also similarly snapping, since the laws of physics in the micro should also be reflected in the macro if the Universe is to exist, so that properties we see in other elastic physics such as a rubber band or the magnetic dynamo which is the Sun or other magnetic fields should also occur in the warping, elastic nature of spacetime, otherwise it would be exempt from such laws, and since spacetime underlies the entirety of reality it cannot itself be exempt.

So it occurred to me when contemplating this spacetime stretching that it might be the very catalytic stress which causes the disturbances which originate the energy that crates matter, wherein the tension between great gravitational forces which cause the warping and stretching spacetime eventually snap, to disturb spacetime and initialize the motion which creates time and mass and thus the birth of subatomic particles which then condense into atoms like hydrogen which condense into stars and thus the cycle of the Universe in fact goes on forever, as old matter eventually tears open spacetime to create new matter, and this cycle repeats forever, ad infinitum.

Elastic physics are apparently incredibly difficult to do, and the disconnect between Relativity and Quantum physics likely exists only because nobody has been able to yet create an elastic model of the Universe, because the math is so difficult. Yet we know that elastic properties are in fact a fundamental principle of reality as observed in rubber and solar flares wherein plasma follows looping magnetic fields which eventually snap when the forces become too great and explode energy outward. This math indeed might be impossible to accomplish without the aid of advanced quantum computers such as what we are now seeing birthed upon the world stage, due to their sheer complexity which exceeds human ability which will otherwise elucidate these phenomena. But this theory of spacetime snapping would also better explain the movement of our galaxies and other Universal bodies, because the "Big Bang" did not in fact emanate from one singular explosion but instead a large field in spacetime caught between the gravitational forces of other regions of the Universe warping spacetime far from us where also exists countless stars and galaxies, and this snapping or tearing would naturally create all the many varieties of velocity we observe in the visible Universe, the vortexes, eddies, and other random fractal movement and motion of the first matter local to us which would set them on the fractal pattern of condensation and expansion we see reflected back at us today instead of a central point of eminence.

This model of the Universe would thankfully also mean that it does not, in fact, have a beginning nor an end, which makes far more sense logically (because what comes before the beginning, or after the end?), and it is only that our local space in the Universe began around 14 billion years ago, when the surrounding galaxies and bodies we cannot see disturbed the space where we now reside, as the rest of the infinite Universe moves away from our position at a rate greater than the speed of light, and is the force that causes our local areas to come into existence but for the same reason also prevents us from seeing the rest of it. This seems to me much more plausible because while the nature of the Universe is to be incomprehensible it is still also logical, of which a beginning and end are not (what then came before and what came after?), and because we should see the laws of the Universe reflected everywhere, and while we cannot really comprehend just how long 14 billion years is it still does not seem likely that something as vast and incomprehensible as the entire Universe should only be several times older than our own planet. This is because it is not, and the Universe is infinitely old and infinitely vast, which makes better sense, and is a constant, infinite cycle of spacetime deformation and disturbance due to previous masses which then form new ones to keep the cycle repeating.

Of course all *cycles* do end, though, and the conception of an infinitely large Universe infinitely producing matter does create a paradox, which I said before that I do not like. The black hole could in fact be the ultimate fate of all matter, after billions and billions of years returns through the sheer force of gravity the energy from which reality is made back to the infinite source from which it springs, which is in fact simply the resistance of space eventually and finally overcoming the inertia of matter to stop motion altogether and thus time as well, in a fractal cycle of birth, death, and rebirth reflected as much in the macro as it is in the micro here on Earth in the tiny, mortal lives of humans and other creatures, a pattern not dissimilar in appearance to the surface of the Sun, and a function not unlike the recycling of Earth's crust when tectonic plates plunge back into the mantle from which they arose, the Universe both explaining and justifying itself right before our eyes more plainly than we ever imagined.

As humanity then continues to progress in our science and technology and begin to reach for greater heights of exploring outside our own tiny place in the Universe we will need to understand these principles in order to accomplish such feats as faster-than-light travel or antigravity. If gravity is the compaction of space through acceleration, *antigravity* is therefore the expansion of space through *deceleration*. This already occurs, however, as easily demonstrable in such functions as a hot air balloon or one filled with helium, though our knowledge of seemingly complex realities such as the elements and their atomic nature has obscured so simple a principle while it is yet demonstrated right before our very eyes. The vacuum of space and the relative distance between matter is not coincidentally correlated with low gravity, but is the very function by which the opposite of gravity is achieved. A hot air balloon rising in the early dawn on a chilly fall day is considered a simple trick of physics rather than the quantum principles of antigravity it actually is. To be sure, an airplane speeding through the sky is not a demonstration of antigravity but of friction and forces of resistance, but a hot air or helium balloon is in fact achieving true antigravity, albeit through natural means, which is why it has been taken for granted and thus overlooked as the demonstration of how to achieve antigravity.

Yet as the peril of lightspeed travel is instant time travel into the future while everything else passes in an instant the peril of antigravity is that we and not the Universe around us is the thing which, like the laboratory-made subatomic particles, disappears in an instant, since time and mass are a function of acceleration (gravity) and deceleration (antigravity).

Again, this function of time which is gravity and antigravity and its effects on mortal creatures (not to mention the difficulty of engineering the required technology) is probably why we do not see evidence of extraterrestrial, intelligent life, because space travel is likely just really, really fucking difficult, perhaps even impossible without serious consequences to our relative time which then functions as a substantial deterrence. While space travel would be really fucking cool the desire or preoccupation with leaving Earth more importantly betrays a very human adaptive survival mechanism to leave and explore when things become difficult or stressful where we live. It is the same instinct which caused us to spread over the entire planet (and in the process destroy entire species and ecosystems as we went), which is also caused by trauma and is simply a control mechanism which also causes us to misunderstand that we already live on the kind of planet that we would be searching for out in space. It's RIGHT HERE. WE ALREADY FOUND IT. But if we do not address our trauma we dissociate from what we already know and have and thus it no longer is meaningful to us as what occurs in any relationship, even with an inanimate mass as Earth. Indeed, many fantasy works which depict alien environments are always based on things which already exist here. I mean, have you ever seen a fucking WHALE?????? Those things exist right here, on our Earth, so what if alien versions have four fins instead of two. If our whales had four fins you'd then think it was cool if the alien had six, because the point isn't whether something is unique or amazing but our own unresolved trauma which prevents us from recognizing what we already have, which is everything we could ever want in a planet, and cannot recognize that due to human evolutionary psychological coping mechanisms like dissociation. Resolving that kind of psychological ingratitude is also not an function of willpower and attitude, because we cannot actually change the way our evolutionary psychology functions, but instead of purposeful self-care practices which resolve the trauma from which those behaviors like dissociation emerge, such as through inventory practice.

While such scientific concepts as time and reality are often incomprehensible in their scope and complexity they should also be logical and rationally marry what we observe the Universe to be, not instead paradoxical abstractions that cannot be easily explained without the use of stupid metaphors. When that is the case it is likely that we in fact are not on the right course in scientific inquiry, and should look at other possible solutions observable in the reality around us rather than persisting in sunk-cost fallacies. Because, after all, we are only capable of understanding the natural world, not commanding it.

6. *Alternating Fields of Reality*

What may seem paradoxical or complex about the nature of light and oscillating particles, and indeed phenomena present in experiments such as the double-slit where attempts to observe photons before they exhibit an interference pattern, is that while particles oscillate and may compliment or oppose other phases of oscillation what exactly determines the plane of oscillation and why aren't (or are) particles oscillating in complimentary or oppositional phases oriented in space to avoid each other entirely, regardless of their phase of oscillation? Is the oscillation itself along a linear plane? Is it random and chaotic? Indeed, if a particle is oscillating up and down but the axis of the direction of travel is tilted relative to other particles, the probability of collision would logically be limited to *only* when the center of oscillation is *exact* for two colliding particles and thus greatly diminish the probability and rate of collision to that which is nearly improbable such that reality would not even exist.

This is in fact the purpose of particle accelerators such as the Fermi National Accelerator Laboratory in Illinois and the CERN Large Hadron Collider in Switzerland, as actually causing particle collisions is so complex that miles and miles of magnetic fields are required to direct and cause those collisions because they cannot be observed in nature, so rare do they actually occur (which is another reason the supposed identification of destructive interference in lightwave experiments was also so ridiculous, even it did occur, which it does not). But I am not aware of any experiments or inquires which even address this problem nor explore the orientation of the axes of light particles oscillation. If the axis of oscillation were oriented arbitrarily or randomly in three-dimensional space it would thusly distribute particles in a radial field of probability. Instead, particle collisions are more constant, and indeed reality exists, with matter interfering with other matter and light, so it seems there must be a more constant rate of collision which constrains the motion of particles and their direction of oscillation. Taking a step further back, what even causes a particle to oscillate in the first place? By this I mean, what mechanism other than the catchall "energy" catalyzes oscillation, why do particles oscillate around a center point instead of dissipating into nothing?

We often see the word *dimension* used in different contexts. In my old job it was used to define the planes of virtual space in which I could construct computer graphics and animation. Three planes designated x, y, and z, the same as which are used in plotting mathematics since 3D computer programs function on those same mathematics, using math to create art. Dimensions in creative writing most often refer to fantasy conceptions of metaphysical separation between planes of existence, often holding the promise of exotic and ethereal beings and worlds which can only be visited by transcending the cosmic divide which separates reality from everything else. In more mundane applications of the word it refers to simple measurements of various things, such as a boxes for shipping or how much room a piece of furniture might occupy. In physics, dimensions are a marriage of these concepts, being both a mathematical construct but also the seemingly fantastical, otherworldly foundations which create reality which are both united with reality yet separate from our conscious experience of it, such as the fact that the fourth dimension is time itself which for us seems to move in only one direction and often demonstrated with incomprehensible geometric representations like a *tesseract* which, again, fail to actually help lay persons understand how reality is actually made. In attempting to describe the fourth dimension, scientists or science communicators also describe hypothetical 'fourth-dimensional beings,' as if we are not ourselves a fourth-dimensional being. All dimensions create reality, and so we are also participants in every dimension for the simple fact of our existence, since we do not have the power to separate ourselves from reality, and the kind of fantastical hypothetical this thought experiment posits is simply a science fiction fantasy of an entity which not actually bound by the fourth-dimension and so can travel through time at will. But many people also assume that the space in which we exist, move, and perceive and the objects which populate our reality are the third dimension, which they are not—the third dimension simply gives *depth* to entities, and it certainly does not give something persistence through *time*. That anything exists from one moment to the next is not a property of the third-dimension, but the fourth, and any three-dimensional entity that did not also exist in the fourth dimension would simply cease to exist altogether. A simple box which exists on a table is itself the fabled tesseract, because it persists from one moment to the next.

Interestingly, just like the elements, and in fact the very reason for their existence at all, light frequency is not a constant gradient of values across the spectrum but quantized into whole number multiples of Planck's constant, meaning that there are actual, exact values when going from one wavelength of electromagnetic radiation to the next, implying that different frequencies of particle oscillation simply have additive constants which increase or decrease their frequency just like the addition or subtraction of a whole proton to an element will change it to another. There is no carbon 1.28 or oxygen 0.75. They are whole, finite, and defined, and as such there is likewise much evidence (such as this quantized nature of electromagnetic radiation) that the universe is built upon ordered interval states which make up and define the laws which govern it, which is what provides the stability that allows existence to exist in the first place, and this is one of the reasons why particles can collide, since the field of probably of oscillation is thusly limited to a defined set of quantized values, where otherwise a linear graduation of probability would further decrease probability by such degrees as to effectively and statistically prevent collision entirely and thus matter and existence altogether, as collision *must* occur for reality to exist.

One of the reasons quantum physics, supergravity, and other extremely advanced conceptions of reality can be so confusing and difficult to comprehend is because they are mostly not correct. All truths of the Universe, when solved, will be easy to comprehend by lay persons because the Universe operates on logic, unified and reasonable. For instance, though the theory of a multiverse can be somewhat supported by complex mathematical models and theories of physics, but makes people's brains hurt, I see such models as mathematical evidence not of the actual existence of multiverses but instead the math which demonstrates the *probability* of all and any current possibilities—essentially the math which proves the existence of free will, where any and all futures are possible for our one and only reality, rather than describing an actual multiverse, describing the nature of reality as having a course which is undefined and open to many, indeed infinite, possibilities.

But because the Universe is based on logic and not chaos, which can be described by math, its course can not only be *predicted*, but will in fact only play out in one logical path, as a function of causality which cannot be transgressed, otherwise reality would not have sufficient logic and consistency to exist, which also describes fate, and thus both free will and fate exist together, not exclusively, and seems to me more likely because nature is also highly efficient, and having actual multiverses would be the most inefficient reality possible, in contradiction to the apparent nature of natural laws.

Likewise, the existence of antimatter is a logical distraction from the nature of reality created by mathematical problems, where what appears as antimatter is simply the oppositional reciprocal created mathematically by the laws of nature rather than the existence of something which is separately antimatter, but just as there is only light, not different types of light, there is only matter, and though some of it appears not to exist in our reality it is only because we cannot easily observe it.

Theories of reality such as quantum mechanics and M-Theory identify more dimensions than the basic three of the Standard Model, and when considering the nature of a particle and its oscillation there are several questions which must be answered such as what constrains the particle to its relative position, oscillating across a center rather than simply flinging off into nothing in its own dimension, or dissipating altogether? I suspect this is because the universe operates on complimentary alternating dimensional fields, each of which in turn builds upon the previous which came before it to construct reality while constraining the nature and contents of each previous dimension. I say that each dimensional field alternates meaning that one field is a *constant force* with equal distribution of infinite influence throughout the entirety of that dimension, then the next is a *graduating force* which falls away the further from its point of origination, a pattern which then repeats until the last dimension (if there is even a last), where the constant dimensions provide content or substance while the graduating dimensions give that substance structure and definition.

The first of these fields I see is one which is nothing but infinite energy, a Space and Reality which, rather than being infinitely dark and infinitely void is infinitely light and filled with infinite energy. Indeed, the prevailing theory about what existed before the Big Bang was exactly this, a field of infinite energy where mass and energy were coupled, but this infinite field of energy exists in a dimension we can only observe when it breaks through to the later dimensions in the form of energy and matter, light, etc. There could be dimensions yet earlier, such as that which creates energy itself, or those details follow this first, but nonetheless it is there and the one which follows would then be a graduating, gradient force which then defines that infinite energy into single particles of energy breaking through from the first field into the second, and the second being a gradient force then defines the scope of that energy which breaks through rather than also being infinite and creates the boundaries against which that energy reflects to create the action of oscillation across the center, and because this dimension acts as a sort of boundary (although it may not have an absolute boundary like a wall but the falloff likely exponential or logarithmic) the energy waves reflect from the relative boundaries to produce quantized wave patterns seen in oscillating particles, but because it comes after the first dimension thusly compacts the infinite of the first into the apparent finite of the second. This both gives the particle definition while also powering forces like the strong force or gravity, where their influence is because of the energy of the first field but weaker when further apart due to the graduation of the second.

The third dimension which, as this theory suggests, would be another constant force which might for instance replicate the single particle infinitely, because a constant is also infinite (where a gradient is also technically infinite would be practically or effectively finite at some point due to the falloff its influence) to supply all of reality with the particles which create it, powered by the first dimension of infinite energy and shaped into functional particles by the second, but now because there is a population of infinite particles they will collapse into each other with violence, compelled by the infinite force of the first dimension without the influence of the fourth dimension, which is time, which is a gradient force that emerges out of the sum (I don't mean mathematical sum) of the previous three dimensional forces causing graduation of the infinite energy now also propelled from itself by the action of oscillation which then transforms space into spacetime, and because it is a gradient and not a constant causes instead of linear spacetime one which is graduated and thus curves to create the curvature of spacetime as reflected in the shape of planets, stars, black holes, etc.

The point of this is not that this theory is accurate, because it's so wild and amateur I am likely entirely wrong, but to convey that I suspect the balance of the Universe between forces of *Creation* and *Destruction*, dark and light, the feminine and masculine, etc., to be a result of alternating forces of infinite, constant principles that provide stability and laws with those which are instead graduated with a falloff which then provide the instability and chaos which allows reality to be deformed into matter and all that it creates, the constant forces providing order while the graduated forces provide change, fate and free will together, laws of physics opposing one another directly to enable existence, because if there was only constant forces of order there would be no reality, and if there were only chaos anything which might be made would always be destroyed.

Thus, the collision of particles as we observe does not happen in the third dimension, but in the second (or whatever order they are in), all particles in fact oscillating in the same plane rather than the apparent three-dimensional construct that is our experience of reality. These two primary elemental forces of the Universe, creation and destruction, are thus reflected in all things as what occurs in the microcosm, just like we must know pain to know joy, or hatred to know love, forces of creation and destruction also oppose and compliment each other even in the subatomic universe and these fundamental forces underpin all laws of the Universe.

For us to be here those before us must have died or else human beings would never even evolved in the first place. The Earth came to be from the debris of dead stars and planets which had to pass for the Earth also to come into existence, and so must we also pass if time and reality are to progress as they have always done. Those early stars of the Universe were built from the destruction of whatever was before it (such as the disturbance or destruction of space to create spacetime and matter) in an endless cycle of creation that we do not have the ability to comprehend, yet human culture has throughout its entirety recognized the bipolar nature of existence, with all mythologies and traditions ultimately encompassing two primary opposing forces of creation and destruction, often caricatured as magical deities with agendas and purposes relative to each society's conception of such forces and influence from cultures and peoples who preceded them, but given anthropomorphized nature since that is what we are most familiar. But these oppositional forces of the Universe are very real, and their complimentary function in all things and all physical laws are what fundamentally determine all of existence, from our current occupation upon this tiny planet to the current state of our health to that of our bank account, our relationships, and everything in our past and future as humanity has been shaped by millions of years of the past, even our DNA has been altered and directed by forces which we do not control, such as viruses or exposure to ultraviolet light, and everything that humans are and do is determined by our DNA which was in turn determined by the entire past four billion years of Earth's history and beyond by these very forces of creation and destruction.

Creativity is thus not an arbitrary, invented characteristic of human society or thought, but is instead a fundamental law of the constant forces—as creation is not only limited even to living organisms but occurs in the heart of stars birthing the elements which create planets and moons and asteroids and our very living flesh, where stars themselves are born from dust in stellar nurseries and the nebulous cosmos themselves born from other cosmic chaos and destruction. Oppositely, math, intellectualism, and reason are borne of the graduating forces, which give shape to creation, which also are no invention of humanity but instead fundamental laws of reality that we have happened to discover—I don't think many people understand that math is not something that man made up, but is instead the language of the Universe which describes the forces upon which all reality is built, and people who have revealed and discovered math and science are simply describing the foundations of reality, not creating something for no reason (although some do try do that), but are thus also the qualities of the constant principles that are not only practiced by living creatures but also in the stabilized orbits of planets around their stars and stars within their galaxies and galaxies around each other, which also keep celestial bodies like planets from instantly collapsing into their own singularities that we may here exist.

While those with mental illnesses are often considered unfortunate and insane as was my experience, we instead descend into the realm of the underlying dimensions of existence as discussed in the chapter on metabolic health and powers of perception, because when we become ill our biological faculties which are also constructed from these same laws lose their ability to perceive the higher dimensions and instead discover the terror of gravity's indifference for our wellbeing or the emptiness of absolute darkness, the lesser dimensional forces which make up reality but which then cause us to think the world is flat or we are in constant imminent danger (fear which is further driven by high stress hormones), which sounds fantastical except when the tissues and molecules of the matter which construct our body begin to deteriorate our perception is likewise limited to lower dimensions of existence, which are very real since they underly reality but which are also not where we actually reside, which is instead simply a product of our ability to perceive information, not actually transcend it. Such experiences are frightening and drive us as mad as would any human if not empowered with biological functions able to interpret reality as it truly is and psychologically exist in the present dimension where human consciousness normally resides.

Drug use too interrupts these pathways and filters and gives people new and different ways to experience the fundamental forces and structure of existence, though such experiences are not enlightenment for anything other than the knowledge that we have such biological functions in the first place which allow us to operate as a human animal and sensorially observe the world around us, or the effect of such drugs upon those systems, and stepping into that void which humans are not very well equipped to comprehend are as such forever affected, even being driven mad when venturing there too often.

Thankfully, humans are relatively intelligent and we have been able to recognize patterns in existence which have in turn empowered us to create, build, learn, and explore through the passing on of knowledge from generation to generation, such as the development of geometry or agriculture, and other science which helps us further uncover the mysteries of the Universe. Hopefully this concept of alternating fields of reality helps assist that work.

The Girl and the Bee

After a long, cold winter on the first day of Spring a young girl excitedly ran out from her cottage to play in a field of early blooms. Bounding through the colorful flowers and feeling the warmth on her skin she suddenly stopped, spotting a bright, black and yellow bee, motionless on the ground. "I'm so glad I saw you there!" she exclaimed. "I would not have liked to step on you!"

"I am grateful you saw me too," said the bee. "I would have had to sting had you done so, and my sting is very painful, though we bees have but only one sting and after delivering it, we die."

"That's terrible!" said the young girl. The bee nodded in agreement, but the girl was not satisfied. "Why do you not fly like the other bees?" asked the young girl.

"I could not find enough food to keep me warm," replied the bee, "and now am too cold and tired to fly."

"I'm very sorry," said the young girl. "As am I," replied the bee, "for I cannot return to my hive if I cannot fly."

The young girl picked a fresh flower and placed it on the ground next to the bee, who readily climbed inside and began to sip its sweet nectar. "Thank you!" said the bee.

"If you have but one sting, why use it at all?" asked the curious young girl. "Why not save it, and live forever?"

The bee shook her head as she drank. "Whether I used my stinger or not I would die anyway when my time upon this Earth is ended. It is special, yes, but to not use it would deny my purpose as a bee and the reason I have a stinger at all! Those who would do bees harm would no longer fear us, and raid our hive and take our honey without asking."

"Oh I see," said the girl. "Who would take your honey without asking?"

"Oh, bears, and badgers, and honey buzzards," said the bee.

"That's terrible," said the little girl.

"It is," replied the bee. "But there are thousands of us and only a few of them, and though they are great in size our sting is formidable, and after they have fled in pain we rebuild the hive and replenish our honey."

After a little while the bee began to buzz, then dance, then flitted joyously up into the sky. "Thank you!" she said as she flew away, and the little girl smiled and frolicked the rest of the day in the field, mindful of any more bees which might be in need.

7. What is Spirituality?

In my other works I relate a story of kayaking with a friend in the open ocean around the island of Maui in Hawaii shortly after my suicide attempt, and several humpback whales appeared out of nowhere and swam around our kayaks, which was an experience that helped define my life and reorient me from despair to a journey of self-discovery. Though unaware at the time it was my very first experience feeling true spirituality, and, no, it was not dependent on whales, although they are often used as representations of natural spirituality due to their ethereal calls, languid nature, and mysterious existence, but was instead a product of being in a moment which strongly facilitated or even forced reflection that was, for the first time in my life, not informed by my past experiences of trauma and instead a novel and inspiring new experience. Essentially it was the first time I ever felt compassion for myself, as when nature had thought to give me such an incredible gift of that experience (including the man who so kindly thought to take me there) it made me feel even the smallest iota that God might, in fact, love me. During my short time on the island I continued seeking refuge on the beach and in the ocean, to replicate that experience, swimming with wise, annoyed sea turtles, curious fishes, and flushing wily octopuses from their hiding spots. In the months of the humpback migration the calls of the whales carry through the entire ocean, so sticking your head underwater even in the shallows near the shore it is possible to listen to their songs and whistles as if they were merely feet away, and in such places it is impossible to feel disconnected from the Universe.

Because we typically live such an unnatural existence in our buildings, driving cars on roads, and spending our time buying things or working to make money to buy things it can sometimes seem like spirituality requires such exotic experiences as traveling to far off lands in order to transport us away from the contemporary world, so we take expensive vacations to far-flung destinations in polluting airplanes and sleeping in luxury hotels being waited on by underpaid and exploited workers—but at least there was an expensive yoga session and meditation in an expensive pavilion with an expensive fountain and expensive instructor to help us escape the reality of our choices and lifestyles, right?

In my early forties even daily walks along the small trail outside my apartment that followed a slow, winding river, the Sun on my skin, the chill of a fresh breeze, the calls of the ducks and geese, picking wild mulberries alongside wild squirrels and dragonflies and the sight of the towering, snowcapped mountains was more than ample a space in which to have even deeper walks with the Universe than what I experienced with the whales, to replenish my spiritual reserves in awe of the wonder that is creation, more than any yoga session or meditation class I have ever done.

When going outside wasn't an option, reading a good book in a quiet corner of my couch, wrapped in a heavy blanket accompanied by a nice cup of tea, or viewing pictures of the Universe taken by astronomers with the James Webb or Hubble telescopes were generous substitutes (especially the amazing time-lapse photography which shows our Milky Way galaxy spinning through the sky as we rotate around on the Earth's axis). If necessary I could have the same experience in a windowless, bare room without even a bed on which to sleep because while exotic experiences were formative in my understanding of spirituality they are *not* what creates it, but instead is because we exist in reality and are thus also ourselves as much a function of spirituality as anything else in the Universe as whales, a cup of tea, or snowcapped mountains. We only fail to recognize that fact until life finally pulls us out of our aimless wandering.

Spirituality is regularly subsumed by religion but spirituality is in fact entirely separate from them and, rather than what is typically performed by those institutions, is entirely a personal and inward experience not dependent on other persons, ideology, and certainly not institutions, as was demonstrated by my organic and spontaneous experiences simply being in nature. One of the most ridiculous things I've ever seen concerning religion (and perhaps an undeniable repudiation of it entirely) are online reviews rating how much 'spirit' particular reviewers felt at the churches they attended. That anyone believes a spiritual experience is dependent on other humans or even a building and is not an entirely personal and private experience is a symptom of spiritual *illness* caused in no small part by those very institutions which exploit spirituality and have deceived humanity into believing God is only accessible through them whom have power and influence, mistaking excitement for the spirit and the safety of conformity for peace.

But this problem is made all the more difficult and confusing by misunderstanding what spirituality even is, especially when we have any kinds of unresolved trauma, mental illness, or significant stress which motivates us to find relief and confuse moments of relief for spiritual experiences. When I was a child the absolute truth of my family's religion was certain, and there was no universe or world outside of that reality which existed, let alone subverted my indoctrination, but growing to adulthood my eyes were finally and rudely opened to the mistake simply from being an adult and having eyes to see, ears to hear, a brain to think, and a heart to feel. The natural compassion I experienced from those whom oppositely did *not* subscribe to religion was a startling juxtaposition of the behavior of church people which did not actually align to their ideology with the behavior of irreligious people which very often did, and because compassion was never before shown to me, not even by my own family, it was fairly easy to recognize the incongruity within religious ideology and worldview, and so I happily discarded religion.

But this awakening had the unfortunate effect also of dashing spirituality, for if the deception could be so complete, even by my own flesh and blood, that something which before seemed so absolute could be so completely and utterly untrue what point was there in substituting one set of beliefs with another? Those beliefs certainly had no beneficial effect on the suicidal depression which suffocated me at nearly every moment of every day and provided no steps or plan by which I could find salvation other than vague promises of 'blessings' for practicing even vaguer rituals. If there was anything more I could have done to adhere better to the principles of their religion I am still to this day naive of it because nobody tries with more fervency to be saved than a scared, autistic, gay boy in the middle of a conservative religious family and community (except maybe now transgender kids). Running away from my life was a great step on my journey to enlightenment, and eventually I found myself living in Los Angeles which has as many options for spiritual practices as Hollywood celebrities. In fact, Los Angeles itself has a rich, colorful, and violent religious history, and the reminders of its more traditional, religious, and conservative past are still generously spread throughout the city with old churches in nearly every neighborhood. But I was not about to replace one religion with another and instead sought for meaning through alternative practices that could facilitate spirituality without affiliation to any institution, such as practicing yoga and meditation, or occasionally getting high after I tried marijuana for the first time at the age of twenty-six. While there was much value in these activities—significantly more than anything practiced in my parent's religion—they also did not answer existential problems that still crept in at the margins of my mind like what the fuck was going to happen to me after I die, and what is my purpose on this Earth? Is there a God and does that God love me? If so, what do I do with that information? If not, *then* what the fuck do I do?

Through the efforts of putting my life back together in my mid-thirties, recovering from illness and uncovering the origins of much disease including those of the mind I finally also discovered a useful answer to this dilemma when one day recognizing that there is, in fact, no such thing as *belief*. Belief is really nothing more than desire of want, to have something as true and to make reality work in such a way as to bring us comfort and safety, which are in truth simply human biological survival instincts for avoiding fear and things which stimulate fear, to prevent loss and heartache, and worst of all surprise loss and heartache, meant for the survival of our species even at the expense of the individual. I cringe every time someone says they 'believe' me when I try to explain something scientific to them because science has *nothing* to do with belief—science can and should be tested and replicable, not simply believed, so any person needn't believe in the work or opinions of a scientist but can simply test for themselves whether it is true or not (that's functionally what a 'peer review' is). When scientists say that the Earth is round and that the Earth orbits the Sun nobody needs to take their word for it, but can instead do their own study and experiments to confirm it for themselves. Watching a great oil tanker or other large ship disappear beyond the curvature of the ocean for instance is such evidence, but so also is the lensing refraction of our curved atmosphere as the Moon and Sun also dip closer to the horizon and appear larger than when overhead, and anyone can learn math and physics or look out the window while traveling by plane or send a weather balloon (with permission from regulators so as not to bring down an airplane) high enough into the atmosphere to capture images of the Earth. Belief has nothing to do with it.

It has not even been a full century since subjects like the nervous system and hormones entered the general lexicon, where before our time the understanding of spirituality was instead the primary explanation for many things which are, in fact, not spirituality at all like the nervous system and hormones which are in turn relevant to health and disease and the function of both the endocrine and nervous system, which is why we must take care of our bodies in order to best experience spirituality. But origins of religion and thus most people's understanding of spirituality was also to explain things which were seemingly inexplicable such as mental illness as the influence of spirits, or natural disasters as the anger of divine beings (mind you by opportunistic humans who also sought to exploit people's fear and ignorance for their own gain). Once in conversation with a family member I remarked to them how much I now loved supernatural horror films, having been rehabilitated from the incessant fear and indoctrination of our shared youth. They replied that those kinds of films terrified them and wondered if I wasn't concerned about real demons and mental health. This was especially surprising because that person was otherwise educated and worldly and not normally given to such superstitious thinking, but also because we live in an age where the general concept of disease is understood to be biological in nature, even now by most religious institutions. Man's understanding of literally horrific health phenomena like the plague, "consumption" (tuberculosis), alcoholism, and outright insanity was previously entirely limited to speculation since scientific progress had not yet illuminated causal factors like microbes and neurology, and I asked if they realized that the demonization of mental illness by religious institutions was a coping mechanism meant to understand it but which had the effect not only of stigmatizing mental health but also hurting people who needed medical attention, to which they realized the error in this conception.

In reality many problems experienced by the human condition previously attributed to spiritual forces are still today often mistaken as spiritual function instead likewise mediated by hormones and the nervous system. For instance adrenaline and cortisol are significant stress hormones which are highly sensitive to certain stress stimuli, as adrenaline's primary function is to sustain blood sugar so that cells have a continuous but not excessive supply of glucose which they can consume as fuel, to keep us alive. But adrenaline is also extremely sensitive to environmental stimuli such as threats to our wellbeing, perceived or real, and during interpersonal conflict with other people or frightening or startling experiences the hypothalamus of the brain stimulates an enormous release of adrenaline from the adrenal gland which in turn floods the bloodstream with glucose and increases permeability of cells to sodium which drives up the metabolic rate through increased production of ATP (adenosine triphosphate, our primary energy molecule) by cells so that our body has sufficient energy to help us meet the demands of the stress we perceive.

Dopamine is the emotional balance to adrenaline and instead mediates happiness—not because happiness is a hedonistic pleasure but because dopamine is a tool used by biology to inform organisms as to the absence of stress, to reinforce behaviors and experiences which generally promote our survival such as eating food, having sex, achieving friendship and cooperation, etc. Many people with a bias against certain behaviors like sex, drug use, or even just relaxing will pejoratively describe dopamine release as 'reward seeking behavior' as if reward itself is something to be eschewed, but reward comes even from activities and behavior that such people idealize such as from having family relationships, accomplishing tasks, or achieving success, so if dopamine and reward were really so awful those people should sit in their room, alone, starve themselves, never talk to another person and never do anything good for their wellbeing. Thinking there is any difference between dopamine release for getting a good night's rest and hitting a crack pipe is the primary reason a cure for addiction and alcoholism went for so long without being found, as bias and prejudice prevented even those who were learned and intelligent from investigating the proper biological pathways and systems which mediate such aspects of human biology, which is clearly a neurological disease as I discovered in my work, such as by not properly understanding the role of dopamine as an adaptive tool of biological life and not a hormone of temptation due to thousands of years of religious desecration of the human condition and mistaking biological functions for those of otherworldly spirits that, yes, still contaminates contemporary research in biases that researchers are not even aware of.

One of the most delusional conceptions underlying popular conceptions of spirituality is that congregating with other humans in acts of worship is a spiritual function instead of the potent dopamine stimulant it actually is, because as humans we are evolutionarily dependent on other humans for our survival. When people attend a religious service and feel happy they mistake the act of being included and safe for a spiritual experience due to the subsequent release of dopamine and the feelings of happiness, safety, and fulfillment it mediates, since people are literally safer when belonging to homogenous groups. But this is also why many people who are differently gendered such as gays, lesbians, bisexuals, transgender, etc., do not feel happy in religious institutions because those institutions most often actively *threaten* the safety and wellbeing of anyone who is not expressly allowed in those groups, such as non-believers or those the institution chooses for persecution, so even though we try as children to have spiritual experiences and to fit in and do what is expected we do *not* experience the dopamine release normally felt by other members because we literally feel *endangered* rather than belonging and safe and therefore confusingly do not have the same experiences like our parents or leaders even though we do as they admonish. Growing up the only place I ever felt actually safe was in my art classes, where the other misfits usually tended to congregate, but not brave enough to fully join them as friends, and although I did not look like one of them on the outside with my ironed polo shirts and combed hair they were the only people in my youth I ever felt accepted by and thus safe and thus also the release of dopamine that occurs from a sense of belonging.

But even within homogenous groups as those which belong to religious institutions we can still feel an absence of belonging or safety within other areas of our life, such as our own families, for instance when husbands and wives do not get along and experience chronic conflict and interpersonal turmoil. Because religious adherents congregate to assuage their fears, not because they actually get along, there is also often much conflict within those groups between its members in spite of ostensibly shared worldviews because each member instead serves a self-centered, protectionist function for the others and not true friendship, and so they also often feel ironically alone in spite of being in a large group and then feel they aren't being faithful enough, because of the rise of stress hormones, which they mistake as spirituality (or lack thereof). The feeling of being unsafe raises adrenaline and lowers dopamine to increase unease and dissatisfaction so that we understand, as biological organisms, potential for loss or harm, and the role of stress hormones like adrenaline in anxiety and frustration is why many people stupidly take beta-blockers when they give speeches or have chronically elevated adrenaline, because it helps them feel calmer and more steady—even marijuana functions by inhibiting adrenaline which can and does usually make us feel relaxed, and we mistake that feeling for something transcendental.

But when adrenaline release is impaired the body loses its ability to release glucose from glycogen storage (which is why smoking pot also increases hunger), and this concomitantly stimulates the release of cortisol which, unlike adrenaline, actually tears down lean muscle mass in order to release sugar, which is why those who chronically smoke pot but do not take care of their health or practice dieting behaviors eventually experience paranoia and agitation rather than relaxation (as was my experience), because cortisol is even more destructive on our health than adrenaline. Because hormones are the very emotions we experience, chronic stress behaviors like dieting causes many people's domestic lives to be a constant barrage of stressful stimuli and stressful feelings, and some of the only experiences of calmness or happiness may come from membership in a group or institution such as a religion or work, so we then begin to instinctually seek out those experiences because, remember, the function of dopamine is to reinforce the successful behaviors of animals while stress hormones like cortisol serve to dissuade animals from unsuccessful behaviors, and then people who are unknowingly hacking and destabilizing their endocrine system begin to associate family with stress and religion and work with relief. So the behaviors of both dieting and belonging to a religious group in which our needs are continuously unmet elsewhere, due to lack of effective self-care skills, is one of the most effective ways to cause mania and extreme hormone fluctuations and thus volatile and destructive interpersonal relationships full of histrionic displays of undisciplined emotional venting and constant control behaviors.

What's most interesting about adrenaline is that it is in fact made from dopamine and so has many of the same properties, and a tiny bit of adrenaline (or related amines such as occurs in cheese), can cause us to feel thrill and excitement from the combination of dopamine with a little adrenaline. For this reason many religious groups put on shows, performances, and bombastic personalities to draw in people seeking excitement which they mistake as a spiritual experience. Anger arrises from excessive adrenaline release so many religious institutions, news media, and political opportunists perform hatemongering in order to attract people who enjoy the thrill of anger as a form of entertainment and motivation. This dynamic of happiness and stress stimuli and the endocrine and nervous systems can then easily deceive people into believing they are having spiritual experiences when in fact they are simply experiencing feelings mediated by hormones, and because dopamine can also be hacked by drug abuse there is a high correlation of religious trauma and religious stress with drug abuse because religion specifically deprives people of effective life skills which would emancipate them from dependence on religion which then leaves them without tools that would otherwise raise dopamine if they felt empowered to meet the challenges in their lives and not helplessly interdependent on institutions or other people.

While spiritual experiences can affect hormones, *feelings are hormones* and so feelings themselves are thus *not* spirituality. But there is nothing wrong with wanting to feel happy—in fact happiness is a major factor in feeling peace, and happiness is the primary and most important motivator and reward for all human beings. The experience of happiness is also not lessened by being mediated by hormones, and conceptions or ideology that hormones are banal or corporeal and thus not associated with higher existence also originates from the religious desecration of the human body and the human experience that has been so commonplace for the last several millennia. If we step into a winter snowstorm without clothes the sensory feedback that the air is cold is *not* the reason that we will freeze to death in a short period of time—that will occur whether our sensory functions are working or not, so the mediator of that experience, our ability to sense cold, has no validating effect on the actual experience itself. Likewise, if we receive a gift from someone we have still received it regardless of whether dopamine works or not, and dopamine does not itself validate or invalidate whether we accomplished something of value, or had a wonderful connection with another person, or had a lovely day, it merely *reinforces* the fact of the reality that is true regardless of how we experience it. People who experience sociopathy and psychopathy do not have release of these hormones which is why many of them can become disconnected, dissociated, or even harmful as without the function of hormones informing the mind of the nature of our experiences it is easier to *delude* ourselves to the reality of reality. Oftentimes when helping people overcome health problems they experience a return of hormones and thus intense feelings and sometimes prefer the absence of those hormones, especially if they are inconvenient to our intentions and desires. But a state of sociopathy is often detrimental to people because the absence of feedback sensory information like the release of dopamine and adrenaline can impair our ability to understand reality, and thus sometimes act from a place of misinformation as if our actions will not, in fact, incur consequences or cause ourselves significant loss or inconvenience.

But many people with sociopathy do not harm others and are able to recognize, intellectually, the reality of their lives, because hormones are simply a result of stimuli which help us to understand reality but do not themselves mediate it. Serotonin, the hormone which was long mistaken as the happiness hormone, is in fact the hormone of remorse, guilt, and shame, so people burdened with trauma that causes them to feel constant, unrelenting, and undeserving shame become addicted to serotonin and dopamine modulating drugs like cocaine, amphetamines, and opioids (many drugs like ecstasy, an amphetamine, are described as elevating serotonin but are in fact causing neurons to dump serotonin, liberating the brain from its effects while raising serotonin circulation which is confused as being elevated, when it in fact is only elevated in the bloodstream and not our actual neurology). Using serotonin inhibitors or dopamine elevators can then facilitate delusional behavior, because their modulation patterns then coincide with drugs and not behavior and the environment, and many addicts to those drugs then engage in worse and worse amoral behavior because serotonin does not actually remove guilt or consequences, only the information of their presence, and many are left to deal with the consequences of their behavior when they instead sought to escape it.

It is also quite incredible that our biology works this way in the first place, to give us amazing experiences through feelings and senses as part of the quality of being alive, and base conceptions of human biology like hormones being beneath a higher experience are simply desires to control life and biology by trivializing and marginalizing the physical world, and since the physical world includes the physical body its functions and features are also often denigrated and disparaged. For instance nearly every religion vilifies feces even though we would literally die without an anus and the ability to secrete waste or host a highly valuable gut microbiome which breaks down food for us and produces B vitamins, short chain fatty acids, and amino acids on which we are entirely dependent. It is infinitely strange to me that as the vessel for the soul the body isn't more revered by religious institutions rather than reviled, especially since we did not design or build the body, because in truth this is also the result of fear and control behaviors seeking to control and dominate the very source of mortality, the body, which all religious adherents fear, even though the body is the vessel for what we experience to be the soul. An awe of reverence for how the body works rather than abusive exploitation of its functions such as sex, sleep, and food should predominate religious ideology if they truly understood spirituality, but it doesn't because the point is not to facilitate spirituality, but to control mortality.

Because of these misconceptions of spiritually, I thought the seemingly inescapable despair and hopelessness that dominated my life when I was young and suicidal was an absence of spirituality, when in fact my depression was just a deficiency of dopamine and vitality caused by unrelenting stress in turn caused by both a poor diet, harmful dietary behaviors, and psychological abuse at the hands of my family, religion, and community. Indeed, our religion had indoctrinated us to believe such feelings were an absence of spirituality and thus failure to meet their religious standards, which is a common and particularly vile aspect of religious institution meant, like all abuse, to gain domination and control. This dopamine/adrenaline cycle and its role in spiritual abuse is even easier to see in cult dynamics where reward and punishment are used with abandon to lure and control followers and prevent their escape, and many of which are in fact desirous of those dynamics since they are otherwise unable to feel or experience dopamine release without the confines and control provided by such volatile group inclusion (as discussed in the upcoming chapter on cults).

But this also demonstrates how stress impairs spirituality, when the endocrine system, hormones, and neurological health are in constant tumult and chaos a person's primary motivation is to seek relief from that stress, not to seek spirituality, which we perceive as a quest for spirituality not being possessed of the knowledge of what is happening nor what spirituality actually is, and so simply become preoccupied with relief, which is not the same thing as spirituality.

But since the physical body is the vessel through which we experience life—which includes spiritual experiences—in order to experience spirituality we must seek relief from a volatile endocrine system, rapidly changing hormones, and burdened nervous system by caring for our body and our health to achieve a steadier and more reliable emotional state. This is not to be confused with control behaviors such as religionists telling people their feelings don't matter, are excessive, or invalid because that is simply a control behavior meant to unburden a person from the effects that others have on our own experience. Nor are control behaviors like dieting and excessive exercise care of the body, but discipline, nor behaviors obsessed with preventing aging and disease, which are simply control behaviors borne of fear. But, just as spirituality is an entirely personal experience so is the journey and challenge of being healthy, and the goal is not to become healthy, since that is a control behavior which seeks to conquer the body, which we cannot do, but instead to simply care for it, as we can never determine the outcome of anything, but we can do our best and show up for opportunity such as eating regularly, having as good a diet as is available to us, getting enough rest, etc.

As discussed in *Fuck Portion Control* oxidative stress also literally causes agitation and restlessness and, like myself, many people attempt to learn practices like yoga or meditation to find quiet and peace but are unable to sit still or focus because these factors are instead dependent not on meditation but vitamin C, glutathione, sugar, and sun exposure, so how can we expect to experience spirituality when we won't even eat regularly? When the body is burdened with health problems that disturb the endocrine and nervous system those needs preempt spirituality because the body will die if it is too ill, which sort of entirely negates the point of connecting to spirituality, but even mild physical stress can destabilize the endocrine and nervous systems and therefore must be shown care and supplied with the resources it requires that we may then have space and attention to devote to actual spiritual development, since it is through the body that we experience life, which includes things of a spiritual nature. I was absolutely mystified after twenty years of relentless restlessness and psychological torment when I finally experienced peace and calmness for the first time from simply by consistently keeping my blood sugar up instead of dieting and starving my body which had been promoted as not only a way to get hot (it didn't) but to achieve spiritual discipline (it didn't). But after my experience it made far more sense that abuse of the body would lead to spiritual disquiet, while care of the body—the vessel for the soul—would instead help facilitate spiritual fulfillment.

This epiphany of care versus abuse of the body came during my rehabilitation from alcoholism while participating in Alcoholics Anonymous. I had been laid low by life, gotten cancer and lost a man I loved dearly, and the final realization I had no effective tools by which to live. The prospect of adopting belief in religion to rehabilitate myself was onerous, but I was resigned to do whatever was required to be well, including admit that I was wrong about religion and my tormentors correct. But, thankfully, I found that not only did religion or faith have nothing to do with recovery but that the few trappings of religion which remain in 12-step programs are the same control and abuse that affects religious institutions, and that some of the founders of Alcoholics Anonymous came previously from religious groups which were not successful in rehabilitating alcoholics but still brought with them the prejudiced need to incorporate religious principles for means of control (over their environment, not necessarily the people). Some founders realized the toxicity of religion in catalyzing alcoholism and many had desired not to include any religious language at all, but those who still retained religious ideology insisted and so they reached a compromise to keep some of the religious language in the program, including subtle subversive language meant to relieve religions of responsibility for their abuses. My experience in the program was uncommonly rapid and productive, however, because of the additional nutritional and thera-peutic tools learned from Ray Peat I was using to treat my cancer, and while many friends struggled and relapsed I was having profound spiritual experiences due to better rehabilitation of my nervous system, since addiction is a neurological disorder, that completely upended my conception of myself and life and gave me the greatest hope I had never before experienced. The value of Alcoholics Anony-mous in my recovery was not from religion or any religious perception of God and spirituality but new tools which I could use to better understand life and reality, to live better and more effectively such as the realization that I was not responsible for the actions of other people, or that I could actually care for my relationships instead of trying to make people treat me the way I wanted. Mostly it was the support of my friends and sponsor in the program, which taught me that having support from others is far more important to our ability to thrive than any ideol-ogy, and was night and day different than what I had experienced from others during my childhood.

My spiritual experiences during this process were facilitated by healing my nervous and endocrine system, *not* by practicing any religious practices or ideology, which I did not, and was greatly relieved to find myself not only cured from alco-holism but also having a greater understanding, conception, and spirituality that was *entirely* divorced from *any* religious ideology. The benefit I got from submitting to a higher power—meaning acceptance that I could not myself cheat death or command the Universe—did not come from any religious beliefs or ideology but simply the recognition that I am as subservient to the laws of reality as any rock or other animal. This reality is most plainly reflected in the fact that all living organisms die, and that there are real laws which construct how the Universe works regardless of anyone's interpretation of it. After years of searching for my purpose on this Earth and if there was a God and did God love me the real answer was that *I did not actually need to find those answers*—turning them over to the forces which actually govern the Universe, regardless of what they are, relieved me of the burden of carrying them, and thus I found relief, which is what I had been in search of all along.

The desire to know answers—the *need* to know answers—is a control behavior meant to help us feel empowered by knowing what possible threats we might face in our struggle to survive. Religionists think they have these answers and thus armed to subvert death and misfortune, so when death comes for them or misfortune knocks at their door they are driven mad with rage and resentment because of the incongruity of their beliefs with reality.

But it is also the same for the atheist, because atheism is also nothing more than the same exact desire for control through knowing as what drives the religionist, and the atheist cannot know the purpose of life any more than those with religion, and is likewise denied the peace they so desire since none of us actually possesses the ability to control life which in turn also means not having the ability to *know*, and it is acceptance of that inability to *know*, because we are not Gods, where the key to satisfaction lies.

Indeed the future may hold groups and people who attempt to codify my work into organizations, and these people likely understand my work least of anyone because a fundamental disagreement I have with society and institution is the authority-hierarchy which predominates most institutions and religions. While teachers can impart knowledge and there needs be leaders of organized men such as occurs in cooperative capitalism or socialist systems, such leaders should *always* be chosen by the people over whom the leader has responsibility (not authority), not by the leaders themselves, and anyone attempting to organize a group or authority around my work is likely simply attempting to exploit my work for their own personal gain at the expense of others such as what always occurs in religious institutions or even foundations erected supposedly to promote their namesake's work but in reality distort it and use the integrity and reputation of their founder to instead promulgate their own ideology and agenda. Instead, knowledge, skills, and tools should be shared from person to person, freely, allowing each person to make their own choices and decisions with the information they can themselves glean from their environment and experience and people choosing their leaders from the bottom up and not leaders choosing their followers as what otherwise predominates religion, government, financial institutions, etc.

Much of my admonition to those who consume my work is the practice of *inventory therapy* as discussed in my other works, derived from my experience in the 12-step program of Alcoholics Anonymous because I recognized in the function of inventorying life experiences the most effective cognitive behavioral therapy I've ever found (and repurposed it to be both more neutral and effective), and was effective beyond all conceptions I ever thought possible for my own life based on previous experiences with professional therapy which did little to resolve my abundant trauma and absence of effective life skills. The reason for this is because many control mechanisms born from traumatic experiences prevent us from taking effective steps in rehabilitating the body, which is the first step required to facilitate spiritual experiences. If a person, for instance, believes the goal of being healthy is to be lean and fit so other people admire and respect them they are likely to engage in dieting behaviors which impair their blood sugar (or simply just not keep fed sufficiently often) and thereby destabilize their dopamine, adrenaline, and cortisol and metabolic rate which then causes them to feel restless, agitated, unfocused, and resentful and then make yet other decisions based on those emotions which lead them further and further from effective life choices. The *only* way to resolve such self-destructive life choices is not through willpower and intention but through purposefully addressing the trauma from which they originate as is accomplished by inventory therapy which effectively communicates with the unconscious mind to achieve new learning experiences and thus newfound empowerment to become more effective and thus also an increase in dopamine and peace.

Because many people find health, mortality, and death very stressful they willingly delude themselves into believing their God will simply save them from those problems if they believe sufficiently. One very religious young woman I met early in my research recovered from cancer in part using my work, but then completely abandoned every last dietary recommendation and strategy which had saved her, wanting to believe that she had, in fact, been saved and thus never had to worry again about her health. I was sad to learn a few years later that her cancer had come back, because she had not been doing anything required to prevent its return, and in spite of last-minute attempts to save her she still resisted admonitions like re-reading my book or going outside for sun exposure, and she died not a month later.

But it is not only religious people that are conditioned and indoctrinated by their institutions and social groups to actively ignore things that make us uncomfortable, though religious institutions are the primary propagandists for ignoring inconvenient truths because both their followers and leaders desire above all the power to command mortality, so when faced with challenges such as cancer their toolset is limited to praising God and soliciting prayers, lacking requisite life skills like personal honesty, realistic expectations, and self-compassion to survive such things as cancer. As discussed in my book on psychology, self-compassion is *not* a feeling, but an action, such as staying fed, eating healthy, cleaning up the house, making the bed, asking for help, spending time with loved ones, or sitting down to practice inventory, where the opposite of those behaviors is also the opposite of self-compassion. Because opportunistic pathogens also manipulate the body and derange the endocrine system and nervous system, restoration of at least some physical health is vital to facilitating spirituality, and restoring physical health requires much knowledge, self-compassion, changes in behavior, and very specific steps which cannot occur when there is much unresolved trauma and control mechanisms because those actively impair our ability to be effective.

It is also undeniable that life is mediated by the energy which flows through us as a result of metabolism and biological function, but what most people do not contemplate is that what we experience as the spirit or the soul of our body is the very energy which makes us living, the energy for which the body is a conduit. Yet this energy is *not* created by the body. It is *harvested* from food which originates with plants and other organisms that in turn directly harvest that energy from light emanated by the Sun, and the energy which makes us alive thusly originates from the Sun (our bodies can also directly harvest that energy by being in the Sun, which is why Sun exposure most often feels so good).

This energy which makes us living is the very thing we have always conceptualized as the soul, because the energy of which we are made is also that which underlies all existence. Scientists have recently also confirmed what is called an *ultraweak photon emission* which emanates from living organisms, even plants, but not from their dead forms, which makes plenty of sense considering that all energy sources such as the one which makes us living give off radiation such as demonstrated by lightbulbs or even a hot stove which glows with red light, and this discovery only confirms what all of us already know intuitively when interacting with living organisms versus their dead bodies in that the living creature gives off an energy that we see and experience as life. But because the energy which makes the Universe possible is also the very thing which flows through our bodies to make us living this also means the spiritual is *not* in fact something far off in some supernatural, ethereal, other dimension or realm but is in fact right here in our own mortal one, animating this Earthen body while at the same time connecting us directly to that which is spiritual and existential, the font of all creation, and the great delusion of both religion and philosophy is that the physical and spiritual are separate planes of existence when in fact they exist here *together*, married in the creation of reality and our living bodies that we experience while yet living, and life is itself not possible without the spirit which powers it, energy which is both wholly incomprehensible yet also governed by rational laws which organize all existence.

Like mundane, material conduits for electricity, which is the same kind of energy that makes us living, our body also cannot properly host energy if we are very physically ill, which is why we die when we grow excessively ill, and why care of the body which is the temple of the spirit is so important for the facilitation of spirituality and also why those who grow very ill also become mentally unwell and unable to find peace and fulfillment, because the conduit of the body to host that energy is literally dependent on the ordered state that biology normally achieves, and why death occurs when the body dies as the energy which flowed through us ceases when the body can no longer maintain that requisite order. Though many people become obsessive about health this is not the same thing as care of the body, as obsessive fixation and preoccupation with health and disease is a *control mechanism*, not a self-care behavior, motivated by fear of death and disease not care and appreciation of the mortal form, and the excessive stress and anxiety which motivate those behaviors are detrimental to the physical and mental wellbeing as well.

What we experience as the spirit thus requires a conduit through which to be conducted, and the body is the temple of the spirit, but traditional beliefs which agree with this idea oppositely encourage abuse of the body through fasting, deprivation, and trauma, because the purpose of their ideology is to subjugate reality to their desires, not live in harmony with it, which no person has the power to do anyway, and their conception of the body is as a barrier to things of the spirit rather than its very conduit, which along with resentment of the natural world and limitations of mortality, motivates disdain and contempt for the mortal form rather than appreciation and gratitude, even to the very abuse and destruction of the vessel of the soul.

The truth is the body is *not* weak—it is a gift from the Universe intended that we may experience life as well as that which is spiritual, because the very purpose of the soul in a mortal body is that the soul may experience life, and care and compassion for the physical form is key to perceiving higher planes of existence and all that is spiritual, to which we are literally connected by the very energy of life which animates our body, reflected in our consciousness.

Humans have long believed in the concept of the soul because to early humans the animation of biological life compared to inanimate plants, rocks, or dead bodies was difficult to understand or explain in any other meaningful way. A dead person in fact appears as if abandoned by some former inhabitant, confounding especially our familiarity and love for the person they were, and knowing intimately our own experience within our own bodies have difficulty imagining the absence of consciousness in terms of death without some explanation for what our personal experiences of consciousness are.

Unimaginative, atheist skeptics will complain that life is simply animated by energy without any curiosity about the nature of that energy itself, and that for some amazing, unknown, and incomprehensible reason energy even exists, let alone all the Universe, while yet that energy clearly also contains intelligence as a natural characteristic, else we would not have intelligence. It is true that religionists use the concept of the soul as an excuse to be a fucking bigot and justify even the most heinous of human behaviors, but it still has long been established and accepted by great minds of science such as Albert Einstein that energy can never actually be destroyed, but simply changes from one form into another, which means that the energy which makes any organism living continues to persist without end even after the death of that organism, and while that energy may no longer be congregated into a single form such as was our body the simple fact that it is the very source of life itself and the reason for consciousness implies that energy has properties we do not yet fully understand or comprehend, and it exists in perpetuity, and it is every bit as possible as the existence of life on this planet that the energy which makes us alive is in fact the having of a soul, consciousness, and intelligence that we right now experience in being alive, and because energy can never be destroyed there is likely some aspect to the nature of energy which extends the experience of consciousness *beyond* existence of the physical body, to the realms of underlying quantum dimensions which construct our reality that we cannot possibly comprehend any more than we can the singularity at the center of a black hole.

If the energy which makes us living creatures is also the very same function which creates what we experience to be a soul, which is indeed a simple and uncomplicated conception of spirituality, this confirms there is no real separation of the spiritual from the physical, and they are indeed one unified force which we experience every moment of our waking lives and is not something we must wait for until some afterlife. It is plain from my own personal experience and the observation of many lives that heaven and hell exist now in the present, not in some otherworldly fantasy, with the awe and peace of simple realities being more meaningful to me than I had ever dreamed was possible, and the pain and horror of the worst human atrocities and hate being more awful than any fiery conception of hell dreamed by hateful and fearful evangelicals. Conceptualizing an afterlife as separate from this life is like skipping dinner and going straight to dessert—life is the primary meal and many religionists entirely starve themselves of the experience of being alive in the first place, so undeservedly fixated are they on dessert. Though my parents are quite aged and have many, many life experiences they are still beset by the same personal and interpersonal conflicts they had at the outset of their marriage before I was born nearly half a century ago, with almost no personal development or emotional and psychological growth, even continuing to make the same professional missteps and financial irresponsibility as when I was a child because religion trains people to give up growth and maturity in favor of control and conformity.

Of course, unresolved trauma is the source of this instability and inability to grow, because one of the primary human coping mechanisms by which we handle trauma is to *dissociate* from the present, being too stressful an experience for our mind to cope it is easier to check out and live constantly in fantasy and mythology, and dissociation is the antithesis of learning and growth. This coping mechanism is not only limited to those whom are religious, not by along shot, but since life is the place where we have experiences, grow and learn, and develop in spirituality and wisdom the act of dissociation however necessary for survival also deprives us of those experiences which would enable us to grow as a person in both body, mind, and soul. Thus, to experience spirituality we must also be able to live in the present, not the future or the past as many people do and which religion encourages, which is achieved first by rehabilitating the body and resolving restlessness, agitation, and irritation motivated by stress hormones, pathogens, and a dysregulated nervous system, and then resolve experiences of trauma which otherwise cause us to avoid introspection and being present in the first place.

However, the concept of 'being present' is widespread and popular in the general zeitgeist, but even those who profess it and traffic in the concept very often are not in fact existing in the present (if they were they probably wouldn't even be a content creator) because the act of being present is a form of submission to the reality of existence which is something that most people resent and all people find difficult (at first), and the presentation of groundedness and a spiritual existence is a performance contrived as much for their own coping benefit as for the facade they market to their followers.

What occurs in such people is a genuine desire to be spiritual and present, but still a total absence of skills on how to do so, like me in my past mistaking practices like prayer, religion, yoga, or meditation as practices of being present when in fact they are all practices of *escapism* meant to help the mind flee the present and thus are not at all strategies for learning how to exist presently. Instead, the ability to be present in our lives comes from resolving the very motivations for existing in the future or past, which are those unresolved experiences of trauma and the resultant coping and control mechanisms. Living presently is not a choice of mindset but a choice to sit down and practice inventory therapy, to resolve the trauma which keeps us in a dissociated state, because failing to live presently is in fact fleeing from the present, and why would that occur if not because the present is painful, like a deer fleeing a lion our mind escapes either to the future or the past (or somewhere else entirely on drugs), so the only the way to remain present is to first remove the pain by removing the trauma. Once those are resolved it then becomes such a simple matter of being present that oftentimes no effort is required at all, because there is no more pain the mind enjoys the present and *associates* naturally, and as is the experience of other animals on this planet such as dogs and cats becomes our natural state of existence without requiring complex rituals, structured routines, accoutrements, and certainly not followers.

It is this conflict between control and surrender, illness and spirituality within which we begin to see that spirituality and the physical world are not in fact separate, but exist together, because the greater our trauma the more intense our coping mechanisms which are in essence desires for control, and spirituality is the act of *surrender* to the laws of reality which cannot be experienced when we are desirous of control and trying to do things like forcing our body to produce an outcome we desire, which is incompatible with health anyway, because neither biology nor spirituality is compatible with force. It is thus control—both desires for it and behaviors which act on that desire—which is the primary impediment to spirituality. Even working on health can help facilitate spirituality so long as it is not based in control behaviors because the simple act of relenting control by admitting that we do in fact need to care for our health empowers spirituality through surrender.

Prayer is often considered a spiritual practice, but many people espouse belief in God but then pray for relief from burdens like disease or poverty as if God doesn't already know we have cancer or can't afford a new car, and so clearly do not, in fact, believe in God because their behavior betrays a complete and total unfamiliarity with God's nature since if God does exist God then knows we are not rich, or our boyfriend left us, or people are dying in war, or there is a pandemic, or evil people are currently destroying everything. Praying for things that God is already aware of is like a whiny child complaining about disliking the tomatoes in the dinner we prepared for them—It's annoying, unproductive, and certainly not demonstrative of belief in anything meaningful or spiritual and is simply a desire to be relieved of life's problems, even the consequences of our own actions, and trying to manipulate God into doing our will by way of demonstrable capitulation to authority such as the act of praying to a God.

The problem is that this conception of God has been set forth by man from religious institutions, and does not even a little bit reflect the true nature of God which, like other laws of the Universe, is observable in the Universe around us and not some completely disconnected, irrelevant abstraction. This nature is made apparent by the fact that God most often does not, in fact, ever answer our prayers in the manner we desire, so if our understanding of spirituality is not reflective of reality how can we hope to reliably experience spirituality let alone grow and deepen our spiritual nature? The only useful spiritual advice I have ever received from other people, however, came also from Alcoholics Anonymous, which was to engage in prayer (to any conception of a higher power) but *never* to pray for our own benefit. In the religion of my family I was oppositely taught solely to pray for our own benefit, in the desire for success, happiness, food, safety, blessings, even to be made heterosexual in order to comply with their religious standards. Even thanking God in a prayer for such things as food and company was still an exercise in selfishness because the purpose was not to express gratitude but *compliance* to authority that we may continue receiving such blessings, so religious adherents deliver rote, droll platitudes believing that is what is expected of them and entirely misunderstand the purpose of prayer as something to be used for personal gain and power. Many alcoholics and addicts understandably resent recovery programs because the language they use such as for prayer often appears dangerously similar to that which was used to abuse and harm us as children. But for the alcoholic and addict so much of life is spent worrying and obsessing over our wellbeing because of the abuse which robbed us both of skills to effectively care for our needs and trust in life and people, so the act of praying and praying only for how we can be useful and not what we can receive helps demonstrate our surrender to the realities of our condition and refocus our intentions from the fulfillment of our needs, which will or will not happen according to the laws of fate regardless of our efforts, to fulfilling those of others, thus relieving us of the burden of worry and responsibility of our own fate, which is not up to us anyway.

Indeed, the ultimate folly of religion is to consider God a personal shopper who is unaware of our needs and struggles (even when scripture even states the opposite) unless we entreat God to fulfill our checklist, and only being thankful for the things which directly benefit our life or reinforce our worldview, even if it also comes at the expense of others. Being fundamentally a control mechanism this means such prayer can never facilitate spirituality, being all the while speaking at our conception of God and never actually listening, which we find difficult to do because listening is a surrender behavior which opposes the very purpose of religion as a control behavior. Spirituality is by definition the feeling of connection to the larger Universe, what many people call God or Gods, and which can hardly be achieved without shutting the fuck up. Once when I was a teenager gathered with my family in prayer at the end of a long day the prayer being said by my mother was going on for a significantly longer length of time than usual, even to the point of repeating herself in nonsense. It went on so long some of us began to open our eyes and look around, wondering what the fuck was going on and when we might be freed of this inconsiderate, egoist subjugation. My father caught me looking, because he was too, and there was enough time for us engage in a brief, unspoken conversation about how fucking long this was taking. Even after closing my eyes again he tapped my arm, then pointed to my sister who was so bored she was stretching out her neck in ways that made her appear very silly, which then sent both of us into a an uncontrollable giggle fit. As we struggled to suppress our laughter, occasionally spitting and choking on it, my mother's brow furrowed and she kept going even more as if to spite us, saying nothing of consequence in complete defiance of holding everyone hostage. When she finally finished my father and I burst out laughing, but my mother's following anger further exposed the true purpose of that kind of prayer, which was not to communicate with God, but to use God and prayer as a way to control everyone and everything.

While being gay in a hateful, religious home and community was deeply traumatizing, my life path has also led me to do things and be someone I never even considered as a possibility when I was a child, and I am grateful every day for the person I am, not for the remarkable things I've done or what I have achieved but simply because it is me. Much like before I began sharing my experience and thought incorrectly my story to be uninteresting, what makes us unique or special is not the things we have done or do but simply in *being,* which is also why simply listening in prayer is sufficient, because we do not need to be anything more than what we are, and instead to experience the world around us, including that which is spiritual, rather than wresting control from the forces in charge such as when we spend all of prayer speaking at God. Though there have been billions of beings in the history of earth not one of them was born the day I was, gay and autistic, with my siblings, and grew up in a psycho religious cult only to later escape and find himself through a series of life-shattering events. No matter how many people have lived or will live or what combinations of reality occur only five people in the entire realm of existence and all of history have me as a brother, such are other truths for anyone reading this book, and so is every person that has existed, does exist, or will exist entirely unique in the circumstances of their person and life from every other. This conception of myself, like spirituality, is far more simple than the complex constructions of religious institutions, which seek with the understanding of traumatized children to explain complex realties like our very existence, and one of the reasons the answers I found are so much more useful is because reality is not at all as confusing, nebulous, or restrictive as otherwise characterized not only by religion and mythology, but philosophy and skeptics as well.

The pertinent question posed to spirituality is—does any of those experiences make religious principles true? My family certainly couldn't invoke spiritual experiences on demand as I could in the act of meditation, but the circumstances of my meditation class were also clearly exploitative and not because this man possessed any greater access to God or spirituality than I did. Was there in fact an underlying reason that both Eastern meditation and Western Christianity and other religious practices which can sometimes eke out spiritual experiences? I learned meditation because my life was a near constant and unending stream of stress, after having tried to commit suicide spending years afterward searching for a way to find peace, fulfillment, hope, and enlightenment, but as amazing as meditation was it too only had that effect when I was actively meditating, which I certainly couldn't do at work, on dates, or at the gym when otherwise haunted by merciless insecurity, fear, and anxiousness. It also did not actually solve any of my life's problems, and though my experience of it was profound it in truth burdened me with even more confusion—how could something clearly more spiritual than my childhood religion be at the same time unfulfilling and effectively useless?

My work resolving alcoholism and cancer which led me to understand both human biology and psychology would end up providing the answer to that dilemma, which is that peace, fulfillment, hope, and happiness are all fulfilled by hormones as part of the endocrine system, and oppositely inhibited by stress hormones which rise in response to stresses environmental, nutritional, and psychological (and even worse when we run out of stress hormones), and that my restlessness when sitting in a movie theater was not a problem of a spiritual, mental, or willful nature but a problem of dieting and starving myself in attempts to be fit and attractive. The reason I felt a brief sensation of spirituality at a religious retreat as a child was a brief moment in which I was falling asleep because of the dull, droning voice of the speaker and briefly forgot all my fears and stress which then allowed my nervous system a break from the constant stress and abuse inflicted by my family and their religion for the first time in my life, but which then fled when I became aware of the state because that awareness brought me out of the incidental trance. This is the reason prayer feels spiritual to anyone, because the act of quieting the mind and meditating on abstract conceptions of reality or fantasy allows the mind a quiet moment to stop worrying, to feel empowered, which gives the nervous system a break, if we are not too ill or stressed, from the otherwise obsessive, tumultuous lives most of us lead, *not* because we are speaking to God, and this can also be had from any kind of activity that quiets the mind such as other meditation methods, journaling, practicing inventory, etc., no matter religious or philosophical beliefs, affiliation, background, behavior, gender, sexual orientation, traditions, ethnicity, etc., and is more dependent on the state of the endocrine system and not the least in belief or religion.

Likewise, many of us unknowingly practice spirituality through avoidant behaviors like playing video games, watching streamers online, or concerning ourselves excessively with tasks—the point of which are to escape the reality in which we live that is causing stress to our emotional state and nervous system. Lacking purposeful tools like inventory therapy or meditation we instinctually practice other coping mechanisms which distract us from reality to help achieve calmer minds and slow nervous excitation much the same reason for alcohol abuse and drug abuse, which in my experience also led to feeling like entire years passed in the span of months, without remembering much of them, as alcohol literally helps us check out of life by chemically inhibiting the acetylcholine required to produce consciousness which, if our reality is unpleasant, otherwise feels excruciatingly unbearable.

Yet while coping mechanisms thus help us deal with life, whether healthy like meditation or unhealthy like getting plastered several times a week, they are also *accidental* and thus do not extend their benefits into other areas of our life, since we lack understanding of exactly what it is, why they work, and how to replicate their benefits on purpose in more effective ways. Much like religious zealots, any distraction can become an obsession which actually interferes with our ability to function effectively, maintain relationships, experience spirituality, or even succeed in our professions since we feel unable to actually be present in life and, constantly checking out, function at a disadvantage. Many people who use 'alternative' spiritual practices like meditation, yoga, or other rituals as coping mechanisms to avoid dealing with life continue to have the same problems as those with more traditional behaviors and ideology, or none at all, because all are simply attempts to *escape* reality rather than exist in it.

In order to actually accomplish real spiritual experiences it is important to recognize there is nothing wrong with wanting to be liberated of the stress and pain of life—that is a normal, healthy human desire and emotion which also in fact helps fulfill our desires for survival and success, and it is only when we lack *effective* skills in achieving that effect through healthy and productive strategies that we experience the opposite results of what we desire. Many religious people think they wrestle with God for answers to their problems but in fact they wrestle only with their own ego and fear, while God plentifully answers prayers and we instead simply do not like the answer we have received and so continue pretending God has not answered us and plead for him to give the answer we want. But God does not do that, so we are left in continuous distress if we refuse to accept God's answers, which are always what reality is and not what we want it to be. Once we accept this reality, including the fact that we are mortal and subject to the laws of that condition, including the natural desire to be relieved of it, we can then find compassion for ourselves and our limited power in reality, which is acceptance, and the first step in actually achieving relief from those problems rather than simply trying to escape them. *Surrender*, thus, is the proper path in achieving purposeful spirituality.

As discussed in my book on psychology, so-called gambling 'addicts' (it's not an addiction but simply another coping behavior—addictions always involve exogenous chemicals) suffer acutely from this very problem, not understanding that it is okay to feel unfulfilled, bored, anxious of the future, and desirous of excitement and fulfillment, and the behavior is fueled by psychological dissociation from our lives to such a degree that we don't in turn derive satisfying experiences from mundane realities, feeling stressed by the feeling of being dissatisfied as if it is wrong or not okay to be dissatisfied, and thus act on that feeling by seeking out the therapeutic effect of adrenaline in treating boredom by throwing money away at a small and unrealistic chance of gain. Many of us suffer this same problem but seek that remedy in other behaviors which serve to raise adrenaline and give us a feeling of excitement such as interpersonal conflict, or dieting and drinking diet sodas, all of which we then mistake for fulfillment, which is all the worse if that is the only euphoria we have ever experienced in our lives. Just as was my experience, many people have run on adrenaline nearly their entire childhoods and adult lives and have no idea when they start getting well that the absence of adrenaline is not an absence of 'energy,' but an absence of excitement, and struggle through the lull of low adrenaline which accompanies restoration of blood sugar and healthy diet. It is okay to have a hard time, to be bad at life, to not get it and keep trying to find answers. Recognizing this and changing our behavior to stop trying to control everything (such as outcomes and fate) is the act of surrender required of spiritual-ity. If we cannot do that, especially when engaged in spiritual ritual such as prayer or meditation, then spiritual experiences and peace will always be elusive, because we are unwilling to surrender even to the most basic of parameters and limitations of reality which are plain and apparent even to the most obtuse and trenchant of persons, and it is only denial and refusal to admit reality such as what most of us learn in our youths that this ever even is a problem.

Death is the ultimate surrender, which is why it is required of every organism that lives upon the Earth. Yet though death and loss has accompanied every creature that has ever lived and will ever live religious people whom supposedly believe in God constantly pray to be spared of it, even concocting ridiculous mythologies about eternal life that are not only absurd, but entirely unnecessary—The only reason to desire eternal life is a fear of death in the first place, so if we do not fear death it doesn't matter what may or may not happen after we die, though religionists might claim they do not fear death this is only true because they have concocted a solution to it! And are in reality denying the reality which is that we do not actually know what occurs after death, if anything occurs at all, and must be willing to accept reality whatever it is and not what we tell it to be. But atheists too often insist on the opposite of eternal life, that of being nonexistent, and as such they also have no less fear of death than the religionist, nor more certainty of what happens after death than anyone else, so their position is equally arrogant, contemptuous, and delusional *because we do not know*, and why they too cannot find peace even though they consider their position enlightened and rational, which it is not in the least, and the only rational position about the nature of death is that we simply do not know anything about it, so to make any kind of assured claim is a farce of both spirituality and intel-lect, one making a mockery of God, the other, intelligence.

In order to surrender we must first address *fear*, which is achieved through dutiful practice of inventory as instructed in my other works, for the resolution of unresolved trauma and its resultant control and coping mechanisms which are, by their very nature, the *opposite* of surrender, and are nothing more than the desire to control life, people, the body, cause and consequence, fate, death, etc. We of course control nothing but our own choices and decisions, so attempting to do the opposite is the opposite of surrender, attempting to assume the power of God (control of life, biology, people, outcomes, consequences) to suit our will, and when the consequences of that behavior pile onto our lives the trenchant refusal to recognize reality drives us headlong into insanity as we further dissociate from reality to cope with the disparity between our desires and the consequences of our behavior. While my ego put up a tremendous fight at the outset of my own inventory therapy practice I was also lucky to have been entirely beaten down by life and thus finally able to admit this disparity and recognize that I was not good at life in the least, which made actually doing inventory relatively easy in comparison to what I had been doing before, and had help from an insightful but extremely impatient sponsor while part of the recovery community. Most people, however, do not have such onerous trauma to motivate them to action and so can more easily afford to delude themselves to the need for resolution, especially since doing so is also an affront to the nature of the ego whose sole purpose is for the benefit of our survival and control, and in this case it is even more imperative that we put effort into resolving our past trauma and its self-destructive progeny else we risk spending the entirety of our lives in circles of recrimination, dissatisfaction, and mundanity, only recognizing our failure to wake up and take responsibility for our lives at the very end when there is not much time to correct course.

Yet no matter how aged we are there is always time to correct the past and make a new start, because you're not dead yet. A decade ago standing in the kitchen at my father's house he told me how much he regretted spending so much time at work and not more with my brother and I, and then he left the kitchen to get more work done. There is nothing standing in our way but the abstract, intangible, imaginary ego, and no matter the circumstances of our life it is always possible to grow and develop spiritually since, like the ego, it requires nothing more than our own mind (and a pencil and paper), specifically found in experiences like inventory which subvert the ego.

For many people the idea of deprivation does seems like spirituality, such as fasting, arduous journeys, or sweating in out in a sweat lodge, and oppositely regard indulgence as the opposite of discipline, having too much food or sex seen as behaviors of those lacking in willpower and control over the body. In realty indulgences are instinctual reactions to harm and stress, such as infection with opportunistic microbes which steal our nutrients and feed on human tissue, insufficient nutritional requisites like carbohydrate, or unceasing psychological torment from unresolved childhood trauma, and are likewise a consequence to the absence of self-compassion such as caused by control behaviors and abuse of the body. As discussed in *Fuck Portion Control*, there is no such thing as sex addiction, which is instead an accelerated sex drive which occurs in response to severe illness as a biological mechanism to increase reproduction opportunities before the expiry of the organism, which is resolved by in turn caring for the body and healing those diseases, not abstaining from sex or controlling sexual impulses. Conceptions of 'indulgence' and supposed weaknesses of the body are yet also consequences of religious desecration of the human condition and God's creation and the desire to ignore and avoid inconvenient realities of life like disease, death, and inability to control our bodies which then prevents us from properly understanding the body and reality.

The opposite of harmful discipline is thus not indulgence, but compassion and self-care, and the opposite of starvation and dieting is not reckless overeating but maintaining blood sugar, never going hungry, and eating a healthy diet that provides for all our nutritional needs which then likewise normalizes hunger impulses and provides satisfaction when eating normally. Remember that self-compassion is not an emotion or thought, but behavior, taking care of the body and our personal needs, and practicing inventory therapy is thus also the act of having compassion for oneself, because this is the process through which we care for mental health through the active resolution of trauma. Some it is true do not have enough compassion for themselves to even begin the practice, and persist in control mechanisms and neuroses which give them a sense of control, but the path to spirituality is only found through surrender, and the more self-compassion we perform the more surrender we will achieve and the more room there will be for spirituality by freeing space in the mind that is otherwise occupied by trauma, fear, hyperawareness, stress, and constant rumination on the past or anxiety for the future.

So spirituality is a separate function than the effects of hormones and neurology, but which we cannot experience when such stress gets in the way and distracts us from it, and the upcoming chapter on consciousness and meditation explains both how to practice meditation and why it works both biologically and spiritually. It is understandable that we have long mistaken hormones and neurological health for concepts of spirituality or its absence, but now that we understand better how the natural world works it means we can achieve more consistent, productive, and egalitarian resolution of challenges and problems which otherwise impair access to these wonderful experiences, so first care for your health and start practicing inventory to prepare the body and mind for receipt of spiritual knowledge and experiences. While this may for some seem rather daunting these books lay out (mostly) clear and actionable steps, and the process does not at all require perfection or completion, simply action and doing, and progress is commensurate simply with doing and not necessarily achieving (the desire for which is still a control behavior!). Avoiding such requisite changes because of fears of failure (i.e. believing we cannot do it) is also still simply a control mechanism in which we avoid doing something because of a fear of not being perfect, because how can we possibly fail if we don't start in the first place? Avoiding those changes and refusing to make a start or put in effort is merely the active entertainment of those fears and control mechanisms as discussed in my book on psychology, demonstrating indeed the immense power we do have over our own choices and actions, and choosing instead to make time for reading, learning, and writing practice and sitting down and doing it is the self-care behavior which relents that control and propels us into progress. To achieve this all that is required is effort.

8. The Nature of God

Children's media from my childhood made by adults who came of age in the 70's and the golden age of science fiction art and illustration from masters Boris Vallejo, Frank Frazetta, Rowena Morrill, Richard Corben, and Ken Kelly birthed such unmatched masterpieces as He-Man, She-Ra, Rainbow Brite (yes, it is), Thundercats, SilverHawks, BraveStar, and My Little Pony, and one of the hallmarks of 70's science fiction (and thus the 80's cartoons on which I grew up) was an organic, natural context of space travel and exploration, where advanced technology enabled interstellar travel, exploration, and civilization, but humans and other creatures still existed in more or less natural context and forms surrounded by nature (even if it is alien), barely clothed, exploring the vast potential of space and all the possibilities it might hold. Growing up also in a church which attempted to incorporate a limited understanding of science into its mythology, which then resulted in their own kind of science fiction, my entire young life was thus incidentally filled with an abundance of science fiction influences even before I became a teenager and learned of Asimov, Herbert, or Orwell (or Le Guin!).

So after putting my life back together in my mid-thirties and attempting to contemplate the reasons that humans might instinctually search for higher meaning but offended by the tired conception of God as an old, white man in heaven wondered why the Master of the Universe shouldn't be a She-Ra-like figure with piled hair, gilded armor, a magic sword, astride a magical, winged, talking horse? Why was the only alternative to religion non-religion? Could there not in fact be a God which represented better the actual nature of the Universe? One that creates life and incredible wonders and unfathomable mysteries and phenomena instead of just requiring your money? The embodiment of physical science and the powers which organize reality rather than an anthropomorph?

Before participating in the recovery community I had in fact never considered there could actually be something that can be defined as God which is nothing like what religion typically describes, which are always in truth meant for control and power, that is instead the embodiment of all the laws of reality we see all around us. In fact, things we experience in life like love, pain, togetherness, joy, and loss are not constructs of the human mind or culture arbitrary only to our species but characteristics of *all* reality, born from the fundamental laws of creation and thus in turn reflect the fundamental nature and laws of reality and the Universe. Thinking otherwise is simply a human cynical desire to control life, by defining what is and what isn't and thus wresting control of reality from the Universe itself. But of course we cannot actually do that, and we are pitiful creatures indeed compared to the immensity of existence, and it is human nature to devise coping mechanisms to help us cope with such frightening realities of life as there being no God, or worse, that there is.

But belief, prejudice, and bias also originate from the subconscious as defense mechanisms to help us cope with the hardships of life, and because they are unconscious survival tools we cannot be conscious of them except through cataclysmic life experiences or purposeful therapy such as what inventory accomplishes. Because they too are not aware of their unconscious coping mechanisms, atheists and secularists also only recognize what are in fact superficial differences between their conception of reality and that of religionists, which in practice and effect are not different at all, both telling reality what it is rather than asking, dogma which serves to accomplish nothing but to make us feel control, and defensive posturing which separates us from each other into divisions of believers and nonbelievers whom in reality are all governed by the same laws and conditions, without exception or regard to personal philosophy and belief, perpetrating nihilistic or delusional philosophy of a world and life which is in reality both awesome and incomprehensible.

As such, many people mistake the nature of science as something to believe in but, as I have stated ad nauseam, belief is nothing more than *want*, and belief is the realm of religion and superstition where science is the study and practice of scientific principles which are inherently in conflict with both the function and concept of belief or want. Science is *not* something to be believed, but instead something *practiced* and *known*. Unlike religious beliefs, which are subject to nothing more than a person's desires and fantasy, all science must endure critical inquiry—it is never something we "believe in" but instead something we practice, investigate, and find evidence to support what we may think or understand about the world, life, and existence, to discover what is true and not create truth, since we are not God and absolutely cannot do that. As mentioned earlier, "believing" in science risks adopting flawed data and persisting with incorrect conclusions even when they demonstrate obvious and harmful absence of validity such as the mistaken policy of water fluoridation, which was never actually established by rigorous scientific standards as a *safe* practice, or the incorrect adoption of serotonin as the hormone which mediates happiness, scientists mistaking the torporific effect of serotonin in manic persons as a state of resolution simply because it caused a less manic state and not because it was accurate (and failing to identify dopamine as the hormone of happiness due to remnant religious bias and prejudice against reward and feeling good).

But practicing science is no less true when it comes to existential inquiry like the nature and existence (or nonexistence) of God, the value of life, and the purpose of existence, because we do not yet know why such things as cause and consequence exist or what really are the foundations of the Universe, or even the full expanse of it, and the ultimate question of why are we even here which must all have a rational answer. But instead of accepting the challenge to study and investigate esoteric, existential concepts many supposed scientists and rational thinkers simply dismiss them entirely as something not to study or which cannot be measured, even though they have never tried, simply because they assume it to be superstition and hold (deserved) bias against religious people rather than asking why are there so many religious people?

But this is not how science works, and the absence of proof is not itself proof but simply something to which we do not have an answer. Drawing conclusions about existential questions simply because there is no evidence is exactly what religionists do, so people claiming to have beliefs about the absence of God because they are intellectuals or believe in science do not, in fact, believe in science at all because they would otherwise accept the absolute reality of our predicament, which is that we simply do not know. Many people use for instance also the fact that no divine intervention occurs during even the most heinous of atrocities as evidence there is no God, but this expectation that a God or Gods would do such thing is itself a human construct of contemporary religious ideology which, again, is not what determines reality, and to use any person's personal conception of the organizing forces of reality as the litmus for confirming the existence of a higher power is absurd, because the Universe is whatever it is and we do not have the power or ability to change, alter, or impose our desires upon it, let alone define it. Whatever we observe in all existence *is* the will of the Universe (not meaning an anthropomorphized 'will', but instead the fact that it is entirely separate from our will and control), and thus reflective of the nature of the forces which govern it, because nothing can exist within it the laws of reality that is not also adherent to them. But such conflicts between our higher nature and desires for peace, harmony, and retribution is not similarly in conflict with the nature of cause and consequence, and there has never been a single evil deed done upon the face of the Earth that is not accompanied by equal and immutable consequences, even if only perpetrators live in a hell of their own making and will never know what it is to be at peace, but very often in fact meet their doom as a direct consequence of their deeds, even if the timeline is longer or less specific than what would satisfy our desires for retribution.

It is also abundantly clear that humans and other creatures experience what can be described as spirituality— feelings of calmness, peace, and connectedness to others or the world around us beyond what is immediately apparent, and simply dismissing these phenomena rather than studying them are the actions of skeptics and ideologues, not scientists, and this is all the more egregious because we can, in fact, measure existential data and demonstrate this all the time in scientific practices where seemingly impossible data is gathered *indirectly* when it cannot be gathered directly—for instance we cannot directly image planets which are sufficiently far from Earth, so instead must measure light passing through their atmosphere or the bend of light as it passes by their bodies, or their effect on other nearby objects. Light particles themselves cannot actually be directly observed or measured at all, so instead we devise experiments which infer the nature of light based on the effects of those particles which can then be measured, and in fact there is no more apt analogy to investigating things of a spiritual and existential nature than the scientific investigation of light, which is at the same time apparent yet completely intangible, and the same ingenuity applied to spiritual and existential investigations will similarly yield results as has the exploration of light. In fact, the recent study which confirmed ultraweak emission of light from living organisms is exactly the kind of study which can investigate such metaphysical laws of existence.

Most ironically, those who do not believe in higher concepts of existence, often referred to as *atheists* or *atheism*, suffer the very same delusions possessed of religionists, such as to deny realities of existence which are inconvenient or uncomfortable to our personal identity and worldview, acting like our own personal opinions have any bearing on reality instead of the other way around. Considering the many thousands of years of intergenerational trauma and human history which informs people's opinions and beliefs, such as the fact religion is often used as an excuse to harm others, the position of atheists is both understandable and expected, but because atheists do not ascribe supernatural mythology to existence believe their biases and prejudices are therefore validated by intellectualism, when in reality this is entirely a logical fallacy and they are every bit as invalid and superfluous as religion. Atheism is in truth its own religion, possessed of the same kind of irrational emotionalism born of fear, childhood trauma, human evolutionary psychology, and frustration with our material conditions and conditions of mortality as any organized theology. I once saw a video of a mother who identified as atheist haughtily and contemptibly ask her tiny, young child if they believed in God whom then responded with a pithy and equally contemptible *"no,"* the mother appearing to regard the exchange as if was not she whom in every way, shape, and form had indoctrinated her child in the same way as the very religionists she so apparently despised. While religions and many of its adherents have qualities to be despised, how is this behavior not the very thing she pretends to abhor? How does she not recognize that her God is the denial of one? That her behavior the very same as those for whom she holds such resentment?

Life can be frightening, and it is understandable that any person on this Earth should cling to whatever contraptions should help us survive it. But if we truly desire enlightenment for ourselves and to know the natural world for what it is and not what we want it to be we have no choice but to inquire of the Universe its nature. Practicing and developing inventory therapy resulted in the emptying of much space in my mind which was once crowded by so much unresolved trauma and relentless fear and anxiety, and afterward became immeasurably easier to ponder big questions such as why we are here, what is our purpose, and is there a God? To finally recognize the difference between truth and what I wanted to be true. But when I was young God was used primarily as an excuse to harm others, to take without remorse and to justify oppression of people who were different than us, and indeed the great bulk of religion's history is nothing more than a reason to kill people we don't like or to take their stuff and feel justified in doing so (such incredibly primate behavior), in defiance of our inner voice which demands guilt for such crimes. There was venom in their religion, and the only natural conclusion to the kind of violence done by those who believe is that God does not exist.

But I was not atheist, officially, as I suspected even at my young age there was probably some more nuance to reality than was readily apparent, something I did not fully understand, but having no other frame of reference gladly accepted agnosticism as a useful stopgap between the hatred and loneliness of both religion and atheism. Later in my thirties the experiences of losing my health, partner, and family forcibly cloistered me into a yearslong period of deep and ponderous introspection wherein dealing with the laws of mortality in my illness, research, experimentation, and successes came to finally realize that there was, in fact, God, but entirely unlike anything that resembled the one of religion and in truth more resembled the science fiction animation fantasies of my youth, being itself the very foundations of the Universe rather than some dissociated deity which cannot survive scientific skepticism.

An honestly dispassionate evaluation of religious belief shows it has nothing really to do with the nature of God but is simply how ignorant, unenlightened, and fearful humans attempt to describe the nature of the Universe they fear and respect in the same way our ancient ancestors believed in Odin, Ēostre, and Zeus. I came to this realization first from uncovering the workings of the human psyche in which it was apparent that people operate predominantly on fears and insecurities every bit as traumatizing as the ones I suffered (or more), and their desperation to define God was not religious belief but a limited, biased, and overwrought attempt to cope with life and its immutable, inconvenient laws of causality. Concepts of God and religion were simply insufficient tools used by humans to describe the limitations imposed upon them by reality itself, a sometimes seemingly hostile and merciless world full of death, loss, and despair, and that the concept of God was merely their attempt at describing the omniscient reality of existence in that we do not ourselves control any part of it. God being the shortcut to describe all which has taken me pages and pages.

But I also saw that when atheists hear religionists entreat God they laugh, not because they disrespect religionists but because atheists are driven by the very same fear as those whom profess belief, which is a fear of not having control over life and mortality, and where religionists concoct mythological stories to make themselves feel powerful the atheist attempts to control life through science, or what they claim to be science, and to find control in prediction and surety of knowledge. But the problem is that it is no more possible to control life with science than through religion (we can make better informed decisions but that does not mean we control anything), so the atheist likewise remains as dissatisfied as the religionist, and resents their condition for being exactly the same as those they despise, because in reality it is our shared, humble, mortal condition they both resent, and the entire point of both religion and the religion of atheism is nothing more than a banal sense of control. Both are driven by fear, sharing that supreme commonality, which is also why their conflict (and conflicts between different religious sects) is also so embittered since each recognizes in the other the same fear as an old, rued companion they otherwise long to be rid, and their presence thus unmoors them, completely opposite the purpose of adopting such existential presuppositions as religion and atheism.

It is woefully inadequate to say the existence of God, the supernatural, or a higher purpose for existence has been a significant point of debate and conflict over the last several millennia, with no shortage of God-fearing religionists having made their mark on the course of history, and a smaller yet not insubstantial cohort arguing the opposite. But many influential religionists did not only argue for the existence of God, but also racism, eugenics, exploitation, and violence, to use God as the excuse for their control behaviors, fear, and insecurity. Often those who argued against the existence of God were and are also miserable, in some cases going so far as to commit suicide so nihilistic were their personal philosophies. But both the atheist and the religionist are equally erred in their ideology not because either ideology is flawed in its conception or mistaken in its facts, though they are, but for the very reason that *fear* is the origin of both positions. This is a problem because the human response to fear is always an attempt to control that of which we are afraid, so operating from a position of fear means we are always ineffective because we in fact control nothing about realty but our own choices and behaviors—not the time and place of our birth, the contents and direction of our life, the thoughts and behavior of others, nor the place and time of our death. Emotions too are hormones and neurotransmitters so we do not even control how we feel, and being dispossessed of this knowledge our actions are then ineffective since they are not informed of the reality which constrains them.

Nearly every person I have ever met believes it is not only possible but paramount to control personal feelings, and lives in constant frustration trying to do the impossible yet also entirely failing to recognize that our actions taken in response to feelings are not feelings themselves and which are entirely within our control, because that would then deny us the pleasure of inflicting our pain onto others and thus the satisfaction of control. Belief (imprisonment) to religion or atheism is the consequence of this misunderstanding, believing emotions are in our control but our actions not, to be constantly frustrated and never fulfilled since it is the opposite which is true and the only way we can actually be effective in our lives.

Philosophical giant and architect of Communism, Karl Marx was famously atheist but, like myself, Marx was also an alcoholic which, you will remember, is a neurological condition that arises from inordinate neurological stress during childhood both of environmental and dietary causes which exaggerates the imprintation of negative experiences through dysfunctional excess of the neurochemical *acetylcholine*. Also like myself, Marx was witness to the incredible harms done to people in the name of God by religionists, religious organizations, and religious apologists, and in fact the primary motivation for most atheists is not the ideology of religion but the behavior of theists, which you would think might give religionists more pause than it does, and underlying the works of Marx is ultimately a desire to impose upon the world his egalitarian desire, to make by force his dream of prosperity in which he, personally (or more accurately the child he once was) and thereby also others should no longer suffer.

I saw this in his work because it was also how I felt about the world before barreling headlong into the recovery community, beaten down by predatory businessmen, hateful theists, disloyal family, and scheming social cannibals no longer desired to exist in the world, yet found freedom from it not in forcing the world to change but better understanding of both humanity and reality. Before then nothing but an entire bottle of tequila several times a week made life worth living, but the problem was that created yet other problems, as I could not in fact be drunk every moment of every day, and eventually found liberation from the darkness of ignorance and salvation in my own person, a new conception of my value to myself and the newfound knowledge that I need not conform to *anyone's* expectations of me, not even those of myself. Then freed from those bonds which had held me my entire life I began inquiring of the Universe *who am I* and found the Universe answered not only then but it had been my entire life though I was told by others I trusted and loved that those answers were false and to look the other way, that God did not in fact love me.

It is beyond the scope of this book, but egalitarian societies such as imagined by Karl Marx and other socialists and communists can never come from authoritarian means as what has occurred in some instances as Soviet Russia through the killing and slaughter of other humans, because that very act is in complete contradiction to egalitarian ideals and comes with it the very loss of our own humanity. There exist plenty of alternative ways to achieve equality and prosperity for everyone which can help all mankind toward a more equal and productive society which cherishes the Earth and our natural resources, like *cooperative capitalism* discussed in my book on psychology, or democratic socialism, and the desire to use force in the erecting of our plans and designs merely comes from a desire for control, not equality, where opportunists instead exploit the opportunity for it regardless of the pretense. In the chapter on the duality of character weaknesses I discuss how control and coping mechanisms also never occur singularly but are always conjoined with other complimentary control or coping mechanisms, since control and coping mechanisms are based in fear and insecurity which are in turn borne in ignorance are thus are always ineffective and so require yet more coping and control mechanisms to compensate for their ineffectiveness and consequences. For instance, atheists are also often every bit as arrogant as religionists because *arrogance* is a control mechanism which compensates for the coping mechanism of *insecurity*. Insecurity is a coping mechanism which both anticipates failure so that we may better prepare for when it occurs, and also informs others around us that we believe ourselves unable to handle attention, scrutiny, criticism, responsibility, etc., and having this effect on our environment it has consequences such as rejection or failure, which is also distinctly unpleasant, so arrogance is then adopted in order to compensate for those shortcomings. Arrogance then also causes harm to others and we then adopt yet more control behaviors such as defiance or contempt and the cycle of self-destruction continues. Likewise, neither religionists nor atheists are effectively informed when they construct their philosophy, so that philosophy is itself a compensatory coping mechanism entwined to yet others like insecurity, fear, combativeness, etc., as a bulwark to consequences, not a system which has objective validity.

As such, the only way to accomplish egalitarian goals are through egalitarian behavior. Similarly, *confidence* is not a coping mechanism, but a skill or strength learned from lived experiences, and confidence is more effective than arrogance because, unlike arrogance, confidence comes from real knowledge accrued from real experience and accepts weaknesses and limitations and then acts within those parameters to increase the likelihood of effectiveness, being better informed with useful and actionable information, because weaknesses and limitations are not arbitrary or invented but in fact real factors which constrain the scope of our ability as a human animal, and the feeling or skill of confidence arrives from a reasonable expectation of results based on previous lived experiences within those set of circumstances, rather than outcomes we simply wish or hope for. Arrogance is instead a delusional attempt to deny limits and weaknesses because an arrogant person has not experienced success and is unsure how to achieve it, but still desires to act, and the delusion functions to reduce emotional stress from uncertainty but which also imperils action and increases failure since arrogance is not informed of rules, boundaries, and limitations, much like walking through a new home while blindfolded trying to find the bathroom, where the person with confidence has already been through the home and so can find their way to the bathroom even without sight, but the arrogant person will have no idea where they are going but still undertake the challenge, not knowing the home is also filled with pits of vipers and other boobytraps. Both confidence and arrogance are survival traits available to all humans, but are dependent on our individual life experiences, and there is in fact no choice that can be made between them because confidence is *never* a choice but a natural result of repeated positive experiences, and that is yet another reason why inventory therapy helps us succeed because we can *practice* it, to become *practiced* in the act of resolving our problems and thereby over time grow in confidence of our ability to care for our needs.

Based in fear, all religion and atheism is ultimately founded in the fear of *death*, which is in truth a fear of the *unknown*, because it is what may or may not occur after death that we lack knowledge. Yet while such fears may be consuming and preeminent for an individual it is actually the most common motivating survival mechanism in nature, possessed of most creatures such as tiny puppies when they first encounter a staircase or mice when they see a cat, or humans when others of our species threaten our wellbeing, reflecting our true nature as simply an animal species on this Earth which we also share with other animals which all share a common evolutionary tree and thus sharing many of our traits with other creatures on this Earth, and the fear of the unknown is simply a safety mechanism to prevent creatures from wandering into places which may be perilous to our survival—it is neither rational nor ideological but simply a catchall mechanism which increases the survival of species by simply causing them to flee what they fear, which is often anything we don't understand and thereby incidentally increase chances of avoiding harm.

Because of the steps of inventory therapy I was forced to investigate and contemplated my fear of death and the unknown, and in so doing found, for the first time in my life, that just as possible as the hateful religionist's concept the afterlife could instead be something absolutely wonderful I had not ever yet imagined, so oppressive was the indoctrination trauma of my youth, and in fact most of us never consider that which we fear may in fact be wonderful simply because our default instinct to deal with fear is to avoid what we fear and dissociate from it thus never having the opportunity to explore it and find out that, hey, maybe death is actually a *good thing* which merely transports us onto the next stage of our journey, whatever that stage may be. But it's actually also okay that I don't know because I am not the force which governs death and the laws of existence as I was told my entire life in threats to obey and comply, which instead is the domain of what we experience to be God, and death in fact is not my personal responsibility so I can instead turn it over to the forces which it is.

Being unaware of the nature of fear we simply accept what we are told as truth, imagining a heavenly kingdom floating on clouds singing with angels for all eternity (what a hell, do you have any idea how long eternity is???) or complete nothingness and thus no purpose after we cease to exist (actually there's not much difference there), but because we truly cannot *know* the validity of the things we believe our mind is keenly *aware* of that disparity, so we remain trepidatious and unsure and then attempt to handle that anxiety and doubt by redoubling our own indoctrination in attempts to further drown our doubt and thus snuff out any fear, instead of having compassion for ourselves as a single mortal trying our best and our fear and ignorance as a condition in fact of being an animal species on this planet. The entire point of concocting answers and explanations for things that we do not know the answer to is in truth simple denial, no matter its packaging, and therefore it does not accurately even understand the question let alone give an acceptable answer, and in spite of our constant wrestling never quite extinguish doubt, fear, and anxiety.

Because both the atheist and religionist alike are motivated the same by fear, it has never occurred to them that WHO GIVES A FLYING FUCK???? Reality is what it is, and if there is a God then there is a God, and if there is not a God then there is not a God, and whatever reality is we can do absolutely nothing to alter it. Yet we also exist regardless of our beliefs or the beliefs of others, and the embrace of our loved ones is no less warm, a hot meal no less filling, and the soft rain on our face no less invigorating if one of those realities is true and the other not, and the rain falls on the face of the religionist and atheist alike, demonstrating how the Universe in fact cares for us regardless of our belief or conception of God. Indeed the very purpose of engaging in debate and philosophy about God or not God is still control, so the entirety of one's existence becomes lost in fantasizing about the afterlife or not afterlife in order to control our expectations for future pain, loss, and disappointment that we may be better prepared to meet it, while the entirety of our lives pass beneath our distant gaze and we miss out on the things which really matter like the presence of loved ones, the joy of conversation, or a hot cup of coffee.

The answer to whether there is a God or not is thus *why concern yourself with the question in the first place?* For most of us this is a dilemma borne of trauma and fueled by biological human survival instincts, for if we do not survive then how will the species? Yet our species needs no help, and we are so fearfully concerned with fulfilling our natural instinct for survival and control we do not see it is killing us, nor that we are missing the parts of living which are most wonderful, which do not require questions or philosophy but instead simply being and experiencing, which is something we already do without much effort, the irony of asking these questions is that *we do not actually require an answer*. The cultish religion of my parents arrogantly claimed total universal authority on the validity of all religion and, as many like it have done throughout history, condemned anyone else who did not believe it, even those who share nearly the same exact religious beliefs, without compassion for the condition of mortality and the shared struggle of other humans to find truth and belonging, even to the expulsion of their own family whom are not allowed to attend religious ceremony, made pariahs to be embarrassed in service of authoritarian control, all so they can have the comfort of a contrived answer to things which in really do not matter to the realty of our existence.

Since fear and arrogance obfuscate truth, all we who search for answers to the existence of God have not realized that we are not actually concerned with whether or not God exists, but instead *how can we stop being afraid?* In truth the desire for answers is a desire to stop fearing, which is universal and understandable considering the nature of mortality. The answer to this question, rather than the existence of God, is quite easy to find, which is that all fear is based in ignorance, and searching for answers is the commonest responses to fear because it is also instinctual human nature to seek knowledge, and so the way to stop fearing is indeed to learn, but it must also be valid knowledge and not contrivances meant simply to make us feel better, otherwise they fail to satisfy that fear and we remain frustrated, stressed, and fearful. That's why the answer *who the fuck cares?* is so effective an answer, because it establishes what we know to be true which is that we are not, in fact, Gods with the responsibility of determining what happens after death (if anything) or knowing the answers to all things, and death is what it is regardless of our desire or ignorance, but also entirely outside the boundaries of our responsibility as a single, human mortal.

This is why the practice of inventory helped me stop being afraid, because the structured writing helped my unconscious mind unpack all the complex drama and trauma of my childhood which confused and obfuscated such simple truths, to replace it then with evidence which is plainly all around us that anyone can observe, and one of the reasons why most people are not satisfied with truth because it is often far simpler than our expectations, informed as we are by trauma and culture so, like the answer myself and my family received about my gayness when I was younger, we do not see answers even when they are plain and starting us in the face.

The reality of God is similarly that of the Universe itself. A great, unifying force which is both benevolent and restrained, the same as the reality in which we find ourselves, which did in fact not only exist but was personally interested in my wellbeing and had been active in my life from the very beginning, as most of the circumstances of my life both bad and wonderful were placed in my path separately from my own power. God's nature is also clearly without ego and does not care what anyone's conception of God is, which is why so many differing conceptions have existed throughout time, even as disparate as monotheistic overlords to romantic pantheons of Gods and Goddesses and Demigods. Like fear, ego is a construction of mortal biology which helps creatures survive, to recognize potential dangers to the fulfillment of our human needs and wants, but God has no fear and thus no ego, and we errantly apply this human characteristic to the nature of the Universe in order to more closely identify with forces we desire to control and manipulate rather than any true reflection of its true nature, which is plainly apparent otherwise in the fact that all people have express permission to believe whatever the hell they want. But because belief is nothing more than want, seeking to understand reality is instead the true religion, to accept life and the Universe as it is and not as we want it to be (or what any one person says it is), an entirely personal experience that does not need codification or instruction but instead simply participating in life and taking time to reflect and converse with the Universe.

The true nature of God is thus neither a concept of intellectual or mythological nature, but the universal truth that none of us has the power to command the stars, shape spacetime, or subvert death. Indeed even the smallest attempt to subvert fate is often met with a harsh rebuke by the Universe reminding us of our humble position, such as the time in fourth grade when I thought it was a brilliant idea to tie my hand to a rope when going down a slide so I could get back up more quickly and instead dangled helplessly until the teacher on duty came to my rescue, and we are forcibly made to accept reality in spite of our desire to subvert it. This undeniable principle of existence is what naive religionists call God. The atheist, Science. But both the religionist and atheist alike wish also to be *freed* from subjugation to reality, which is something none of us can do, so the atheist choses to deny God by defining reality, and the religionist to manipulate him, and both are left wanting and whining because at the heart of their position is denial of reality and God, because they are, it turns out, one in the same.

A willing submission to accept the laws of reality is not only then the path to understanding God but also to a more effective life, no longer deceived by personal delusions and desires we instead act within the bounds of reality, refuse responsibility for things not within our power, such as death, relent control of others and the outcomes of life, and instead simply show up for opportunity and do our best (which is really easy to do because everyone is already doing it!).

None of us either can deny the reality of our existence, and simply being is testament enough. Then, since we are here, it is evidence the Universe has laws which make life possible, else we would not, and while we cannot know if we had some design in this fact before the event our birth it is undeniable that we can do nothing about it now (and, no, taking a life does not alter the conditions of existence). In this undeniable reality we can then infer that the Universe has put us here, else we would not be here, since we are not the ones which caused it, and without a single labor or effort on our part we are here to live, and while many argue whether life is a gift or not (on a bad day even I will tell you no), there can be no doubt that life is an *experience*, during which we have experiences, many of which are wonderful and some which are devastating, but all as real as the ground on which we walk, the light which meets our eyes, and the flesh in which we wander, which are, by the way, also constructs of the Universe and not our wiles and power, so not only has the Universe made our life but the entire reality in which it plays out and all the ways in which we experience it. If then the parameters of existence are what we experience to be God, the fact of my existence then also demonstrates that the Universe is invested in my being, even though I am but one simple organism on one little planet in the entirety of the cosmos, otherwise I would not be here, because the Universe made it possible for me to be here, and the intent of my existence in spite of being a tiny, powerless organism on one planet in all of existence is the same as the greatest and most powerful of black holes which eat stars and swallow galaxies because, like myself, they also would not exist if the laws of the Universe did not cause it. The same is true for all things, that existence is sufficient evidence of the Universe's awareness of us, because it is constructed such that we can and do exist, because if we are not the power which constructs reality then some other force than ourselves is the reason for it, which is quite remarkable indeed since this means the Universe itself creates life and thus we as a result of that law of creation evidenced by the mere fact of our existence.

The nature of God is thus the very reality we see around us—a cosmos of indescribable scope, mystery, and power over which we have no control but which is also concerned with even the smallest possible division of itself, even to the well-being and life of microorganisms like tiny tardigrades which can be seen by only the most advanced technology. This is not to say that the Universe is concerned with indulging our desires, hopes, and dreams because that is the coping and control mechanism response to fear. Instead, the Universe simply *knows* what we are and what we are going through, and in that sense we are not alone, even when life becomes unbearably painful, because unlike ourselves God and the Universe do not wish for us to cheat death or avoid misfortune, because those are just as much a part of living as things which bring us peace and joy, without which we would not know the full extent of what it means to *live*.

The forces of creation and destruction govern our lives just as they do the life of stars and galaxies, and those forces are what God is—it need not be more or less than reality, because reality is already beyond our comprehension and sufficiently awesome, and being part of that reality we can thus take comfort in the fact that God—the Universe—cares about us and is concerned with our existence, evidence by nothing more than the simple fact of it.

9. Consciousness and Meditation

In a past tumultuous relationship my partner and I fought constantly. Although I loved him very much he regularly proved to be disloyal, dishonest, and mistook his ambitions for me as care for my wellbeing, so I did not trust him, but being unpossessed of effective life skills to establish healthy boundaries I instead simply erected boundaries, and very quickly both of us withdrew from the relationship even though we still lived together for a time.

I did not realize until after the relationship finally ended that I had been purposefully but unconsciously withdrawing from him as a response to both his behavior and my unresolved trauma and control mechanisms, feeling powerless to get what I wanted simply grew to resent him. I did not have the conscious thought to withdraw from my partner, however, and instead the thoughts were things like "yeah fucking right," when he offered romantic platitudes I felt did not reflect his actions and behavior. I would smile and try to be agreeable and return a similarly rote, 'I love you,' which was ironic because I did love him, but in those moments I was instead mostly disappointed at ending up with someone who treated me the way he did, and completely ignorant about what to do about it.

The truth was I did not end up with him after all, as no matter how long they last all relationships at some point end, and every moment of them both persons actively choose to participate regardless if the relationship is recognized by institutions or not. Long after the dissolution of that relationship, after I learned, developed, and practiced inventory therapy, my thoughts and intentions finally became clear for the first time in my life, newly empowered with tools to understand myself, my needs, and my strengths and weaknesses, and while it was far too late to fix the self-destructive cycle in that relationship I knew based on my new experiences and newfound confidence that my life moving forward would not be so unmoored and chaotic, having finally learned effective life tools by which to better handle life.

Ironically, if I had learned inventory therapy early in my life I would not have gotten in that relationship in the first place, because I would have had skills to provide for the needs I otherwise expected should come from others like him. As discussed in my book on psychology, what many of us mistake as the thrill of attraction to another person is in fact the early conflict dynamics which arise from our complimentary control and coping mechanisms. Sexual attraction is what occurs when we first meet someone, but then as we get to know them and our various worldviews, ideologies, and perspectives are communicated through word, deed, and body language we find boring those whom are, by definition, not triggers for our control behaviors, while those which excite us are the strongest triggers for our control behaviors, and part of the excitement we feel is a prospect of reliving new, emotionally exciting experiences, which is why many of us then repeatedly make the same mistakes in choosing partners even when we absolutely try to stop, because control and coping mechanisms are fully the domain of the unconscious which, by definition, we cannot be conscious except through specific tools like inventory practice. Ironically we are probably far better off choosing partners that don't overly excite us, but after being empowered by the practice of inventory therapy we find different traits attractive in prospective partners after shedding the control and coping mechanisms and the trauma from which they originate, replaced by new and evolved conceptions of ourselves and others and what we enjoy and value in life, and instead get to know people for who they really are and not what we expect them to be or what they can do for us.

When we do find ourselves entwined in dysfunctional relationships most of us humans respond to this disappointment and stress by dissociation—withdrawing or distracting ourselves from being present in order to mentally and emotionally cope with the stress of conflict. Many of us believe that we are entitled to have what we want from a relationship, for instance I thought I was entitled to love and kindness from partners, and when it did not come I would insist and fight to achieve peace, harmony, and fulfillment that I knew I deserved. The problem is that while we all deserve to be treated with respect, love, and kindness, we can never make others give it. It must be freely given, and instead of trying to wrest our needs from another person we are instead obliged to build and construct our lives with those who do rather than trying to force those who don't to do as we please, and because of that orientation and coping mechanism it never occurred to me that the simple fact of someone's presence in our life is evidence of their love for us, and our pouting and anger when they don't adequately demonstrate it is instead an exploitation of their love and vulnerability meant to simply make us feel in control, at their expense.

Not understanding this, when we do not get what we want from others, either from romantic partners or friends and family, we often then resent them for not living up to our expectations, or for making us feel unsafe, hurt, or ignored, even if we aren't aware that's what we are doing (simply being disappointed that people do not do what we want or expect is this resentment). Then, if the conflict continues, over time that resentment grows stronger, very often to the point of predominating the relationship to the point we no longer even recognize any love or affection for people we once thought we would give our everything, and this resentment can and does prejudice our perception of others to the point that we live with a completely imaginary version of a real person, and respond to that delusion as if it is real instead of what the person really is. Once, after my old partner was away on a business trip for several days, I was so excited for him to be home that I cleaned the entire house, including mopping the floor, shaved, gave myself a haircut, put on a nice button-down shirt, put a pot of coq-au-vin on the stove, and went out with our dogs to meet him when he pulled into the driveway, an enormous smile plastered across my face and an insatiable desire to tear off his clothes once we got in the house. As we entered the front door and after asking how his trip was he completely avoided eye contact with me and said, "it was nice to be somewhere I was actually appreciated."

But this chapter is not about romance or personal relationships (and God am I tired of talking about him), and we are all human and have no more or less of the limits of humanity, but as such also consider and relate to God (or the Universe or whatever your concept of a higher power is) in the same way we do other humans when we are disappointed or frustrated when not getting what we want, such as resenting God for not getting us that job that was so important, or making as much money as we feel entitled, or when our partner leaves us and our life falls apart. Then over time consider God the same way we do someone who constantly fails to meet our expectations, especially if we have been raised or indoctrinated to believe God is a personal shopper who will do our will if only we obey, and since God (the Universe) does not do this we are often met with constant disappointment and over time come to no longer even know God's nature or feel a connection to that which is spiritual because so much of our spiritual nature is simply devoted to a completely imaginary representation of God based on our own personal disappointments and sense of entitlement.

Many of us pray earnestly yet the things for which we pray are never magically granted, and because we are traumatized and lack effective life skills we are self-centered and obtuse and resent God as we feel that we are ignored or refused that which we most desire, when the reality is that God has answered our prayer, the answer is just 'no,' because that's not how anything works. Many gay people or transgender people, especially those raised in religious homes, pray to God to be relieved of our differences, and when God fails to grant that prayer we then conclude there is no God, or our family and religion concludes we did not pray hard enough, when in fact God more than anyone knows that we are gay or transgender and loves us the way we are and as such refuses to do our bidding (or that of our hateful religions or families). But then we resent him for it, even though God is clearly interested in us being as we are, and our family resents him for it, and our religion resents him for it, which is by definition then defying the will of God.

Above all God does not force people to act against their will or nature, which is why God has never intervened in any sporting competition, award ceremony, or made a partner stay with us when our hearts were breaking, or even that our parents treat us with the love and kindness deserved of children. If God forced people to do his will, or ours for that matter, there would be no free will, and the laws of reality would be upended. Instead, the problem is not with God but in our conception of a higher power and ability to accept reality, especially the delusion that the Universe can be entreated to do our will.

Many people given to hatred and violence will also do bad things, then claim that it was God's will, when in fact it was only the will of a mortal human using God as an excuse for their self-centeredness, fear, hate, and poor choices. *Technically* it is God's will because God does let bad things happen, and his will is not to intervene, but there are also immutable laws of cause and consequence that are also God's will which are always visited upon those whom offend those laws. The great secret is that those whom offend thus are in fact already conscience of the heinousness of their crimes, which is why they hate themselves and protest with excuses in the first place and fear discovery, because what follows is fear of consequences for their own behavior they then errantly think they can cheat, but such laws are immutable realities of existence that can never be subverted, no matter how much we might protest, plead, pray, or panic, and while victims of such offenders may in fact even lose their lives at the hands of those who do evil, they are not the ones which did the offense, which is indeed the very worse hell which exists.

Yet while God does not make anyone do anything, God does in fact help us, and often too if we understand that God's nature is not to be our personal shopper or authoritarian but, like our own role as parents, is a guide and guardian which assists us in our life path if we listen to the answers given, which are in fact plain and abundant such as when both I and my family prayed for me to be straight and God said 'no,' with my family pretending God had not, in fact, answered them, when he very clearly had.

But one reason God did not grant this prayer is because we were essentially asking God to change me so that I could be loved, and instead his message was that I needed to learn to love me for who I was, not what they wanted me to be. I so often discuss this personal experience because it is such a clear and plain example of people telling God what to do rather than listening to and being humble before God, the reality of existence and the immutable laws of the Universe being his will is plain and apparent, because of human frustrations and fears based in really fucking stupid things like acceptance from people we don't even like (other members of their religion) and harming a family member and not even listening to the God they profess to believe, whom has in fact been very clear about his answer.

My family of course did not listen to God's will because of their own unresolved trauma, which in truth was likely as great or greater than my own, and in the course of writing this book found I still harbored resentment toward the higher powers of the Universe for not giving me things I desired such as acceptance by my family, opportunities to feel productive and successful, or more rapid resolution of my research and illness so I could move on to live a normal life without the insane amount of years and effort it otherwise took from me. In conducting an inventory on this resentment *my part* was to have purposefully chosen to remain infected with oral infectious microbes so that I could study them and understand human health, as it would have required no more than a single trip to a clinic to be eradicated of them and move on with my life, and to remain single in the interim when I could easily have dated. But resolution of my resentment occurred from realizing my desire for acceptance in my family is also a shopping list, and that it wasn't God's fault they treated me that way, and that for God to grant my wish would require forcing them to accept me against their will, which God obviously does not do, nor would I want him to, and my resentment was being directed at the one "person" or entity which actually supported and cared for me, which was in fact God, whom time and again throughout my life put opportunity for healing and fulfillment right in front of my face which I could not recognize because of my yet unresolved trauma and control and coping mechanisms.

This realization immediately made me felt accepted, and I knew that while my family chose not to accept me the Universe did, and I also accepted myself, and realized that I was being cared for in spite of not getting what I wanted and needed, through inspiration, direction, peace, and joy, which made me more appreciative of the things I did have such as a bed to sleep in, incredible progress in my health in spite of having high-mortality diseases like cancer and cystic fibrosis, quite profound scientific accomplishments, and plenty of coffee, great food, and friends.

This interaction also helps to illustrate *how* to understand the nature of God, which is simply to observe the Universe, and that God is not a religious concept but the observable fact that we are entirely and fully subject to the laws and conditions of reality which make it possible that we even exist, without exception, and over which we have no control or mastery and are thus also not responsible for the laws, limitations, and conditions of life (such as by praying to be made straight, rich, young, healthy, etc.) which are instead the domain of the Universe to which we are lowly subjects.

But, like my family, most of us fail to actually communicate with God because we tend instead to talk *at* God, which is not communication at all because communication requires *listening*, and we cannot listen if we are so caught up in what we want for ourselves that we cannot even hear or see plain truths clearly being communicated to us. All animals communicate but, because they do not possess language we understand, humans often consider them incapable of communication even while they are actively engaged in it. A dog will stand in front of us and whine they want dinner and we know exactly what they are *communicating* even though they have not uttered a single word. Animals like dogs and cats do in fact have language but instead of vocalizations they use their tails and other body language as primary communication tools, and the speed, direction, and position of the tail is actually a complex language system we do not understand because we do not take the time to, with some researchers even being able to determine the position, direction, bend, and speed of tail motions to mean specific things. Many other animals like whales, birds, dolphins, elephants, other primates, and even prairie dogs do actually have vocal language and use it constantly to communicate with each other, but because humans are incredibly stupid and myopic we do not consider it language for no other reason than those sounds are not ones we make, even though the sounds which come out of our mouths are no more discernible to other animals which also vocalize than theirs are to us.

Communication is *not* the same as speaking, and while prayer is an acceptable practice in communicating with the God of the Universe the use of words do not effectively or necessarily accomplish it. Instead, the most effective prayer in fact employs no words—not even thinking them. For many people impatience may make this impossible, and many of us yammer on at the Universe like an annoying child. But God is never insecure and it is only us, not God, who is impaired by such self-centered and superficial behavior. Practicing inventory to the point that we can communicate with God without thinking a single word is not a difficult process, merely requiring consistent practice and introspection, and this kind of prayer is much like meditation, spending time in silence and feeling our connection to the Universe, without expecting anything but the experience, to exchange feelings of gratitude while also knowing God understands what we need or want without our needing to speak it, and certainly never with the intention for control.

Although prayer is a type of meditation, actual practiced meditation is the most effective way to communicate with the Universe and higher planes of existence, to experience the peace and tranquility which comes from connecting to the divine and feeling a sense of purpose, peace, belonging, and acceptance. But meditation is effective because it purposefully does not involve discourse, and being repetitively structured can help to achieve reproducible, consistent results unlike religious prayer which may or may not have that effect (though usually not since the fundamental point of prayer is control of God, not surrender).

While there are many methods of meditation to learn, meditation is fundamentally the act of *self-hypnosis*, utilizing hyperfixation on one singular foci which constrains attention to in turn quiet the rest of the mind. But to understand why and how meditation works we must first understand *consciousness* itself, as the act of meditation is still a function of the brain which requires consciousness in order for us to perform any action or behavior. Though science does not yet officially know how consciousness is achieved, the conductivity of nervous tissue is certainly not what consciousness is, as conductivity is instead the rate at which tissue conducts electricity, especially in the stimulation of cells to perform certain functions or send sensory feedback to the brain, as the rate of nerve tissue conduction is not especially fast, only conducting at rate between 0.5 and 150 meters per second, and consciousness must be something more than simple electrical conductivity. In fact, neurological impulses are *not* a function of the rote electrical conductivity that is often used to measure the travel of electrical impulses across neurological tissue, which in fact only measures the rate of electrical impulses across nervous tissue. Copper wiring is even more conductive than human nervous tissue, and even rockets can travel at speeds around 7,000 meters per second.

What people (including scientists) don't understand is that the electrical conductivity of cells is *purposefully* slow—as while nervous tissue is conductive it also *controls* that conductivity so that spontaneous electrical signals do not unintentionally stimulate tissue, for instance an eyelid to spasm or muscles to twitch. This is in fact what happens during motor disease and conditions of seizure, when the electrical conductivity of tissues becomes dysregulated, and if conductive tissue was not in fact insulated and muffled we would even be affected by magnetic fields which surround the planet or ion storms emanating from the Sun. When I was very sick I even had ticks and twitches I was not aware of, and when very ill people rehabilitate their health they will often experience temporary twitches as their energy production rises but nervous tissues have not fully regenerated, and when getting better I had for several months some bothersome spasms of my eyelid and muscles but which eventually went away. Because life and consciousness is a function of energy it cannot exist without the presence of energy, but which must be controlled and directed, so the conduction of nervous tissue is meant to slow and prevent conductivity as much as to also facilitate it, and when these systems fail the conduction of energy in random and unhelpful directions is inevitable since energy is present in the first place to facilitate consciousness.

But just as we can infer the nature of the cosmos simply from examining their nature we can similarly infer the nature of consciousness from our experiences since we are in fact conscious beings, the different aspects of consciousness such as occur during states of meditation, wakefulness, sleep, and aging and metabolic disease (while we are "unconscious" during sleep we still have consciousness, which is why dreams occur and we don't wake up dead). Many scientists (such as Ghaderi, Vatansever, and Le Bihan) have recognized instead that the brain and consciousness does in fact operate on fundamental laws of Relativity. Biology is clearly aware of and based on the laws of the Universe, because biology would not otherwise exist as there is no option but to adhere to these laws, and while the construction of life from atoms, molecules, amino acids, and proteins is very straightforward and easy to comprehend the purpose of *why* life occurs and the nature of consciousness is absolutely not, because we have no idea why or how energy facilitates consciousness, or what consciousness even is. Whatever consciousness is, it in truth *must* be a function of quantum principles, not only physical biology, since it arises from the literal nature of energy itself, which is directly governed by quantum physics. But what most people do not realize is that because of the laws of relativity, and because consciousness is energy governed by the laws of relativity, consciousness *is also a function of Relativity*, which people fail to recognize when they fixate on the physics of physical biology or the physical world around us while ignoring the quantum laws that govern the metaphysical energy of which all things are made.

In the year 1806 the brilliant scientist Theodor Grotthuss proposed a mechanism he observed in water conductivity called *proton jumping* in which protons actually tunnel through the hydrogen bonds of atoms without rearranging those molecules. He observed this in *hydronium* atoms, which is essentially acidified water (H_3O), but this was an especially incredible discovery considering the molecular formula of water was at the time incorrectly thought to have only one hydrogen atom, not two as it actually does. Proton *tunneling* as described in quantum physics is based on Grotthuss's discovery, in which protons do not need to overcome the energy resistance barrier otherwise required for the rearrangement of molecules but instead simply "tunnel" through the center protons of molecules at near instantaneous speed, and Grotthuss unknowingly made one of the earliest discoveries of quantum physics.

But what is *most* interesting about this proton jumping he observed in hydronium is that a great deal of the water in cells such as our brain (depending on the state of our health) *is also hydronium*. Not only this, but a primary purpose of biology is in fact the purposeful organization of intracellular water into a structured and aligned state (often called structured water) much like an organized scaffold, and while most scientists do not know why this is the case, it plainly appears to be the facilitation and exploitation of quantum principles like proton jumping for the facilitation of life and consciousness, and proton jumping is likely THE primary mechanism of consciousness because where the plain electrical conductivity of nervous tissue is not at all fast, proton jumping in hydronium molecules is about 3×10^3 to 3×10^4 centimeters per second, which better approaches the speed of light (3×10^{10}) and so better qualifies as a potential underlying mechanism of consciousness.

We can also confirm that the Universe does indeed have consciousness because we have consciousness and we are also part of the Universe, not separate from it, and any claims that the Universe is not conscious is a completely delusional thing to state as a conscious creature living in the Universe of which we are part which assumes that consciousness is a machination of our doing and not some function of the Universe in which we exist. Humans are not even the only conscious life form, as every single one of all the animals which live on this Earth alongside us also possess consciousness, so the Universe *must* have consciousness else we would not, and it is nothing more than arrogance—and a trick of human evolutionary survival psychology—to think we are greater than the Universe, or that the consciousness which flows through us is somehow exempt from the laws which govern all of reality. But because the purpose of metabolism is to create energy, and that energy is used in part to organize water, which in the body is also often in the state of hydronium, it makes plain sense that consciousness is a fundamental function of quantum mechanics which is achieved through the organization of water and hydronium molecules for the facilitation of proton jumping, and that proton jumping itself is the mechanism of consciousness.

But *how* does this exactly achieve consciousness? In the earlier chapter on the nature of Time I discuss how time is a function which arises from movement through space, so energy such as in the form of protons moving rapidly through space as what makes up our consciousness when they jump from hydronium molecule to hydronium molecule means that *consciousness is a function of time*, in the very movement of protons through spacetime such as occurs in proton jumping it is the movement itself which causes us to then directly experience time relative to the environment around us, and therefore the experience of time is the experience of consciousness. In fact, while medical professionals also do not officially understand how anesthesia functions to achieve complete and total loss of consciousness without causing death it likely does so by simply interrupting the biological organization of water which then blocks proton jumping to then logically cause *unconsciousness,* but since anesthesia does not also stop energy production in cells it does not also kill the body, and when the anesthesia is removed consciousness then resumes. So while our bodies are still living during surgery, our consciousness is effectively gone, and by that measure we could be said to have in fact died during surgery but since the body is still functioning and producing ATP are easily brought back to life once the anesthesia is removed. If anesthesia slows or stops proton jumping in brain tissue then proton jumping is the mechanism of consciousness, and a very easy experiment to confirm the role of proton jumping as the mechanism of consciousness would be to observe the effect of anesthesia on proton jumping in hydronium molecules or other structured water.

But consciousness is also memory which is also learning, which are all the same thing, and as protons in our brain jump from atom to atom moving through spacetime in the space of our brain cells they then pick up the information met as a result of their movement through time, such as probably colliding with yet other protons or electrons which carry information delivered to our brain through the sensory organs, and in this collision the information received from the outside world would collide with those of our consciousness to alter the state of those jumping protons to store that information as memory through imprintation upon them, probably not unlike making a physical impression of our face into some mud or plaster, or the recordings on a vinyl record, the macro reflecting the micro, where subatomic particles carry information from the source and nature of their origination through their literal structure, being altered in turn by what they also interact with on their journey through spacetime, and then likewise change and alter other particles with which they collide according to the nature of their composition to thus distribute that information for recall by literally changing the nature of the particles with which they collide.

The discovery of proton jumping by Grotthuss all the way back in 1806 before the molecular structure of water was even established demonstrates also how readily apparent the mysteries of the Universe are actually observed, and that reality is not quite as complex as we are often led to believe (though it is complex, to be sure), which is instead a problem of human-invented paradoxes and over-wrought complication of simple realities which can instead be observed all around us, which are the same in the macro as they are in the micro regardless of our inability to yet connect them mathematically. Indeed consciousness being a literal function of time is quite simple, probably too simple for more intelligent people to have guessed, but which makes a great deal of sense since time is a literal dimension of reality that constructs the spacetime in which we live.

But this mechanism of consciousness is then also the direct reason why metabolic disease and problems like hyperammonemia disrupt consciousness and cause symptoms like brain fog, forgetfulness, dissociation, senescence, and psychosis because toxic compounds like free ammonia directly *neutralizes* hydronium, to steal its extra hydrogen atom to then form plain, disordered water, inactivating the very conditions necessary to achieve proton jumping and thus consciousness itself which, when severe enough, eventually leads to death, and this is also why it is so difficult to help people who suffer from mental health disorders like substance abuse, depression, psychosis, or even diseases like cancer because their disordered state of consciousness is compromised by ammonia or other factors which disturb the production of energy required to maintain the ordered state of water required to facilitate proton jumping. By the way, as a result of this book and other research into structured water many predatory and opportunistic business people will try to sell special structured water products—but structured water is a function of healthy biology, and structured water can even be made at home simply by boiling water or letting it sit in sunlight, as the addition of heat or other energy to water excites the electrons in H_2O which causes the hydrogen to jump off to other water molecules to form H_3O and hydroxide, which I discuss in *Fuck Portion*

Control as being a more effective vehicle for the delivery of most supplements than simply swallowing the pills and capsules. Adding a little acid like from lemon juice makes this reaction occur even more easily, and it can also be done by applying magnetic fields to water which in turn helps promote the growth of plants when used instead of plain municipal water (since sunlight also causes hydronium water formation that is a method which uses free energy).

But people who do not understand physics and biology will say that our perception of time while meditating is simply a function of our perception and not of the actual laws of Relativity, not realizing that perception is a function of biology and not arbitrary psychology. You must remember that *time is relative,* and measuring by our understanding of time does not also reflect the *experience* of it, and because proton jumping in the brain accelerates between thirty to forty percent the speed of light the relative experience of time by the brain would change depending the rate of proton jumping. This is, in fact, how meditation works—that the state of meditation is a purposeful *acceleration* of the rate of proton jumping due to increased focus and concentration in the attention center of the brain and thus also an increase in the rate of perception which, because there is *more* rate of perception in any given frame of time (i.e. more protons moving through these centers) this in turn achieves a perception of *slowing* of the time around us because of an increase in the perception of time. Because meditation is practiced in peace and quiet, this also thus extends the amount of peace and quiet we experience to take up a greater proportion of our relative life experience than would otherwise occur. When doing the opposite of meditation, such as zoning out, dissociating from our experiences, drinking excessively, or experiencing disease this oppositely *reduces* the rate of proton jumping and thus also the rate of perception to then *accelerate* the perception of time around us since there is decrease of proton jumping in any frame of time, and even entire months and years can go by in what feels like only a few days.

When writing this section of the book I had not mediated for several weeks, and took a break to confirm and estimate by how much exactly this may occur, intending to try meditating for a full twenty minutes I got a little restless (because I was focusing on this goal, impatiently), and decided to end my meditation around what felt to be about fifteen. Looking at my phone after coming out of my meditation I was still shocked to see it had only been a little over nine! But this was always a constant experience when meditating, that what often felt like much longer in fact was a much shorter period of time. Anyone whom meditates will estimate that the feeling of time passing slows by about a perceived 40%, which is not for no reason but instead based on the fundamental laws of physics on which our brain operates, as there is no other reason than the laws of physics because every single thing in the entire Universe is made by them.

This is also the reason for the saying, 'a watched pot never boils,' because the focus on something like staring at a motionless pot of water or paying attention in class when in school is much like the focus which occurs in meditation which thus increases the rate of perception and thus the experience of greater amount of time, while zoning out or being distracted oppositely decreases the rate of perception thereby accelerating the experience of time. This is also why time seems to go by so much slower when we were children, which has nothing to do with our limited time on the Earth as is so often suggested but instead because healthy young people have such an easy time maintaining healthy brain function and thus a very high rate of proton jumping and thus rate of perception that is otherwise impaired by factors like disease, a low metabolic rate, ammonia, etc. In the first edition of *Fuck Portion Control* I talked about how restoration of sodium helps to restore normal time perception, which occurs by increasing the metabolic rate of the brain and thus our rate of perception (but took it out later to make room for more important information), and this is also why resolution of at least some metabolic disease and excess of ammonia is required to achieve meditation because it is impossible for our biology to facilitate an increase in proton jumping if the water in our brain cells is not sufficiently structured as is required to facilitate proton jumping, and symptoms like anxiety, restlessness, agitation, depression, insomnia, and brain fog are all symptoms that indicate reduced proton jumping capacity of brain cells, and when these symptoms are instead improved through the care and nurture of the body and mind then meditation is far more successful through the increased rate of proton jumping in the attention centers of the brain, which can then be used to slow down our relative perception of time and extend moments of quiet and relaxation longer than what they are otherwise experienced, finding great pockets of peace, rehabilitation, and introspection interspersed throughout our daily lives that are in fact relative in length depending on factors like our health, mental state, and practice of behaviors like meditation.

While many meditation disciplines use the breath, thought processes, or external instruction to direct meditation the most effective uses a single thought-sound or *mantra* (the Sanskrit word for thought), in my experience this is because it not only is the most simple but also originates in the mind rather than the body such as when using breath meditation, thus more effectively training our concentration with the center of foci in the mind itself, and certainly not requiring our hearing as occurs in guided meditation which entirely defeats the purpose and effect of meditation in narrowing our focus as locally as possible.

The secret about *mantra meditation,* as I will refer to it hereout, is that this method does not actually require learning special words, sounds, or mantras from other people, institutions, or specific disciplines, and in fact can be any kind of sound or short word and doesn't matter one lick what it actually is but rather it is the *repetition* of the sound or mantra which focuses the mind, not the sound itself, which is why the content does not matter. One caution I was taught, however, was that the mantra should not be a word that is associated with speech, such as 'car' or 'dog' since we associate those words with objects or beings in the outer world, and instead the use of a sound or short word which is made up or invented and has no real association. The word or sound (mantra) I was given during my instruction was something like '*shhring,*' where the 'r' was rolled, a short, complex, and dynamic sound much like the ringing of a bell, the character of which likely helps to focus the mind on this to thus more effectively tune out all other stimuli. Anyone can use this sound, or make up one of your own, or take a common, short word you like and rearranging or replacing some letters until it is nonsense.

After this, the meditation practice itself is extremely simple (and thus also accessible and practical, which are other considerations I make when saying it is the best to use), and requires nothing more than finding a quiet time and place to sit, such as cross-legged against a wall or other vertical surface where you feel comfortable, hands in the lap, head up but eyes closed, then repeat this mantra over and over in the mind at a leisurely pace. That's it. In the first practice it may take up to ten minutes to enter the trance state of meditation, but when it hits it is extremely obvious and is experienced as a state of euphoria where the body suddenly relaxes and the mind's awareness explodes beyond all expectations, and after entering this self-hypnosis the heart rate and breath will also naturally fall, and every subsequent meditation after it will become easier and easier to enter the trance state, now being familiar with what it is, using the mantra as the guide or trigger to enable it by focusing the mind so purposefully.

It is common to become distracted even while meditating, however, which is normal, and when we find our mind wandering away simply come back to the mantra and continue meditating (over time this becomes easier and more natural but never stops entirely since we are just human). As mentioned earlier, the state of meditation also causes the passage of time around us to feel incredibly slow, so practicing for just ten or twenty minutes every day (or even just several times a week) is more than useful, but only so much as we also practice other forms of self-care in all aspects of our lives, such as a good diet, staying fed, caring for the body, and resolving past experiences of trauma through inventory therapy, as meditation is only the act of communicating and connecting with the spiritual and itself does not itself feed the body or resolve trauma.

There is one important caveat to this meditation practice, however, which is to *never* come out of a meditation immediately and resume normal activity or consciousness. If this is done it will actually cause a headache or migraine, and I have experienced that personally when I did not listen to instruction and hastily resumed my day after a deep meditation. I believe this occurs due to the sudden shift in brain function away from intense focus in one area of the brain which then results in actual local metabolic stress and inflammation to those cells, and instead it is required to quietly end meditation over the course of several minutes (at least five is required when doing the full twenty minutes), no longer repeating the mantra but keeping the eyes closed and being quiet, then after several minutes have passed can slowly rise and resume your day.

Understanding how meditation works in terms of physics and Relativity makes it even more amazing than what I experienced in the past, knowing that while the experience is relative to myself in the moment it is also literally a result of the nature of the laws which govern spacetime across the vastness of the Universe. Any why shouldn't it be? We exist in the Universe and are not separate from spacetime, though many people who discuss science speak as if we are literally exempt from the laws of physics, but we are not, and we can tap into our own experience of relativity at will through the act of meditation which requires nothing more than a little patience, effort, and time in the real world, to find and exist in a small, peaceful pocket of reality for longer than what would otherwise be fleeting. Regardless of the how and why, the *what* of meditation—experiencing this state of greater awareness, relaxation, and relativistic reduction of the speed of time—*is* the act of communicating with and connecting to the Universe and what we experience to be God, by *feeling* and experiencing rather than speaking to or demanding from the forces which govern our existence. Words and language must travel *across* time and space, so no matter how well we communicate those forms of communication are always messages from the past, even though it is the imperceptibly recent past, but *feeling* the Universe in the moment we exist occurs as it happens, due to the passage of protons through the atomically small space-time of the molecules which make up our being and create our mind, and thus is more real, meaningful, and useful than both religious or secular rites, rituals, and membership for experiencing spirituality and finding peace and insight.

This is not to say we cannot ask the Universe for help—in fact part of the usefulness of therapy is simply to communicate our struggles, and because the Universe cares for us it also cares to hear us and listen to our problems—the difference however is that it does not grant wishes, but listening is something it does very well. Just remember to primarily listen and receive in turn because there is nothing we can say or do that the Universe (God) does not already know.

When first beginning my academic study of science and biology in attempts to understand and solve my health problems (and those of others) it was often difficult to arrive at solutions to complex problems, especially if they seemed beyond my ability. Having only recently completed much inventory practice and also possessed already of the skill of meditation I was one day struggling with uncovering the underlying biological cause of alcoholism, since I had already realized mine was cured through my efforts to treat my cancer but did not understand why. Frustrated by the seeming intellectual wall which stood between me and the answer I decided to take a break and mediate. After meditating for some time and no longer feeling frustrated, my attention then drifted and came organically to a paper on the neurochemical *acetylcholine* written by Dr. Peat which I had recently read, which discussed how the parasympathetic nervous system contributed to cancer and how antihistamines could block excessive neurological stimulation to help promote recovery from cancer. I had been using antihistamine for this reason, hoping beyond all hope that it would keep my cancer from getting worse (along with other tools learned from Dr. Peat), but from my time in recovery I also learned that antihistamines were used to treat severe alcohol withdrawal, without which the worst alcoholics can actually die from withdrawal. Though the reason for this usefulness of antihistamine in withdrawal treatment are not even known to medicine it suddenly it occurred to me that my recovery in Alcoholics Anonymous had been exceptionally productive and effective *because* of the use of antihistamine to treat my cancer which incidentally blocked the stressful effects of acetylcholine, and that acetylcholine dysregulation was thus likely the underlying cause of alcoholism and addiction disorders. Finishing my meditation I got up and went to my computer to search for studies which discussed any relationship to acetylcholine and alcohol and sure enough it turned out that alcohol *destroys* acetylcholine, though nobody had yet made this connection of neurological function to addiction and alcoholism, and since acetylcholine facilitates learning, memory, and consciousness (which are all the same thing), which are also impaired by alcohol, it was apparent this was indeed correct.

While my intention was to get better I never intended to actually discover cures to any conditions like alcoholism, cancer, depression, or cystic fibrosis, yet the act of meditation took me there when sheer will and persistence had not, because finding a connection between disparate parts of a problem to arrive at a solution not yet known by any other human being in existence requires the divining of knowledge out of the Aether of existence, and while it may be tempting to infer that my own brain put 1 and 1 together to find the answer the mind operates on fundamental principles of quantum physics in which it is powered by and connected to the energy of which all matter and existence are composed, and it is human ego and survival instincts which chooses to elevate human intellect above the incomprehensible nature of reality from which springs everything that we are, including our intellect and our thoughts, which do not come from nothing but instead from the source of *everything*.

I would go on to research alcoholism, both academically and in my own personal experience before fully solving the underlying problems which created it and solutions to cure it, and then to further research and study other yet unsolved problems like depression, thyroid disease, and cancer, and whenever a challenging problem stalled research and progress and was difficult or impossible to overcome, the act of sitting and meditating and then pondering the problem in a conversation with the Universe was every single time granted knowledge, often in a very short period of time but sometimes on occasion requiring longer or several mediations, but always as if it were placed in my mind by some divine carrier (Mercury?), and then I would finish my meditation, rise, and continue my research now empowered with greater understanding than before, and the only credit I can take is in having done the work required (though I really had no other choice). By the time of later research such as finding the cure to cancer and problems with microbes I had done so much meditation and inventory therapy that connecting to the underlying force of spirituality for help and insight no longer always requires actual meditation, but it is fully probable my work and achievements would never have happened without learning and practicing meditation which, in combination with inventory therapy which relieved me of the trauma which prevented learning and progress, and finally caring for my body rather than abusing it through starvation and excessive exercise, was the primary method by which I was able to discern the truths hidden within the endless sea of information which exists at our fingertips.

Finally also learning my place in the Universe through the practices of inventory gave me the confidence to in turn also share my work with others and risk the harassment and hatred that would also come (and the hatred was by orders of *magnitudes* less than I feared) which I never would have done previous to the practice of inventory therapy. Do not misunderstand that confidence as being ignorant or naive to the experience of rejection, but instead the newfound knowledge that I could now handle it when it did. Eventually the cure for cancer would have been discovered, but it was discovered in no small part due to the effectiveness of inventory therapy in resolving trauma and enlightening the mind to the deeper realities of existence, and the ability to connect to those realities through the practice of meditation and its effect upon the mechanism of consciousness and increase in perception that results.

Aging also impairs the rate of perception and proton jumping as our ability to organize cellular water becomes impaired by disease and chronic colonization by opportunistic microbes, which not only interfere with the nutrition we require to be healthy but also actively produce toxic molecules like ammonia which directly interfere with the organization of cellular water. Much aging and metabolic disease is associated with hyperammonemia, the characteristic symptoms of which are brain fog, dissociation, poor memory, and even senility if it gets very severe, which are all fundamental impairment to the state of consciousness. Many people fight aging and disease because of fears of disease, death, losing control, and being dependent on others, and while it is okay to want to be well the experience of aging and helplessness is integral to our experience in this life and inevitable for nearly all animals at some point in our lives, and the point is not to further attempt control over life, which would indeed demonstrate very poor understanding of these principles, but to accept their existence as fundamental laws of nature.

Dissociation is indeed a coping mechanism most of us use to endure stress, and when stress is unceasing such as occurs during abusive childhoods and adulthoods informed by those experiences we instinctually learn to dissociate as a coping tool which helps to better survive trauma but which then also causes us to miss out on life as it speeds by without our even noticing. For those whom adult life continues to be stressful we can prefer to dissociate or get drunk that time does in fact purposefully speed by, but which is why practice of inventory therapy is so crucial so that we may learn new, more effective coping tools that we may instead enjoy life rather than resent it, and feel empowered, to slow down our relative experience of time and sense of being present through the restoration of both body and soul.

This complex interrelationship of trauma, disease, and consciousness is why no person can chose to be present through willpower alone as is often presented, because it is a function of both human biology and human evolutionary biological psychology that functions on fixed rules of physics and nature that must be addressed in order to achieve change and balance. When we try simply to force our will upon the body, mind, or the world around us we typically incur only more stress and disappointment because we do not possess the power to subvert those laws and rules and must instead fulfill them as required.

Many people fixate on the body because we are afraid of rejection and the perception of others, but taking care of the body should come because we understand how life works and have compassion for ourselves and our mortal condition, not resentment, as with compassion comes wisdom and the ability to adhere to the rules and laws of reality rather than defy them. Only after enlightenment to those rules can we then attempt their fulfillment, let alone actually accomplish it.

When used as a tool for distraction and escape even meditation will fail to give us relief beyond the act of it, but when it is employed as a tool for accepting reality and our mortal limitations such as is demonstrated by the act of practicing inventory therapy and caring for the body, meditation can instead become the most powerful conduit to connecting to the Universe in which we live and thus find the peace, enlightenment, and purpose we seek, all because of the incomprehensible laws of quantum physics which govern the reality which God has made.

10. The Comfort of Tradition

In 1984 when I was four years old my family lived on the island of Oahu in the town of Kailua. My father's sister was nannying us and my grandparents visited during the Christmas holiday, and one of my very first memories (if not the very first) is looking up with pure, unbridled joy from a newly opened Christmas present of a large, aquamarine, plush *My Little Pony* with purple hair (yes that was a pun too). It was the most amazing gift, something I couldn't believe existed, never mind that other humans cared so much for me as to bestow such a priceless present. So I was equally horrified when my parents reached out to take it away, "Oh," said my mom, "That was supposed to be for your sister." The scream I loosed in response to the threat of losing such a prize made all the adults recoil in confusion and bewilderment, the puzzled look on my aunt and grandparents' faces sending me into my first existential crisis—could they not see how incredible this was? Why should they take it away? Why are they so confused by my totally normal and not at all unexpected reaction? Obviously I later understood it was meant to be a *feminine* toy, and I was not meant to be feminine, so they were confused as to why their son wanted it and probably their first suspicion their child might not be heterosexual, but it is probably my first memory because of the intensity of the emotional response I experienced that morning, both in the receiving of such a wonderful surprise and the intense stress which immediately followed, but because they let me keep the pony it is also one of my favorite.

Indeed, Christmastime growing up was often filled with cookies, banana bread and, when our family's fortunes were up, mountains and mountains of presents. But Christmas was one of the only times of year when there was actually some peace and joy in our home, everyone taking the time to relax and enjoy the cold, wintry days once we moved to Utah, or carted off to ski in the nearby mountain resorts, drinking hot cocoa and being with friends and seeing our cousins while Christmas music played on an endless loop.

Every Christmas Eve we were also conscripted to reenact the christian nativity scene, narrated directly from the Bible by my father as each of us took turns whining or causing trouble, making shepherd staves or angel halos out of tin foil, and from my parents' bathrobes and ties banded about a towel on our forehead to fake a *keffiyeh*, for wise men, while the youngest played baby Jesus, laying in a laundry hamper manger lined with towels. One Christmas, to the ire of my own parents, we received a *Nintendo Entertainment System* from my maternal grandparents, and the marvel of the age of video games began in our house as it would the entire whole world. At another I received a small parakeet (but whom died three days later when my siblings accidentally left the front door open and he froze to death in his water dish), and at a later one our very first computer. Gift-giving became even more satisfying as we matured and began to understand how just much it means for a human to give up something for another, especially when you're old, tired, and selfish and don't have energy or resources to do things for other people.

As a child my mother also always made a big deal about our birthdays, and would often decorate elaborately and make sure presents were perfectly wrapped and piled on the table, making or buying a cake that was personalized for us atop which the candles were exactly aligned, and everyone encouraged to sing the happy birthday song as is such a ritual for birthdays in the West. But while such traditions were fundamental to our childhood and the way our mother ran our house they were not done out of love for us but instead a need to feel security in her own life, such as what occurred at my fourteenth birthday when I first realized how little she actually cared about me as an individual and instead that everything simply go the way she wanted. While this might sound as a harsh characterization meant to shame her it was in reality borne of the most severe of trauma, wherein my mother was very likely sexually abused as a child, something that I did not myself in all my own trauma have to suffer. As mortal human beings we only have the skills that we have, and being deprived of ones which are more effective (such as how to survive sexual abuse) are limited in our effectiveness in any capacity of life to those skills we do possess which, for my mother, was also found in desperate attempts to establish and maintain reliable and dependable traditions to which she could anchor her identity and thus be relieved of the terror which haunted her as is so common an experience for women betrayed by the men in their lives.

Although I never received confirmation of my suspicion I was horrified during my time coaching people with their health problems to learn just how widespread sexual abuse by their own fathers was of women from her generation, which never in my wildest estimations would have matched the frequency which was revealed to me. Those whom are so abused present with very specific neuroses and coping behaviors, especially the control of their environment in order to avoid feeling helpless and powerless, obsession with sex and the physical body, and the constant absence of particular, prominent male family members from our lives as children whom were the likely predator, which made more sense when I became wise to human behavior considering most abusers do not maintain relationships with their victims which are a constant reminder of their vile behavior.

As I discuss in both of my other books, sexual abuse perpetuates in part because of our collective desire to ignore both causes and solution, which is first rooted in the abuse of children in the first place whom then grow up to be abusers in turn—not necessarily sexual abuse, although that is a common cause, but primarily the torment of children which teaches them to fully indulge emotional volatility and control behaviors, to fear powerlessness and wield control at all costs, even the forfeiting of our humanity, which is then accelerated as adults by both physical and psychological illness and compulsive sexual arousal that accompanies disease.

One of the primary problems which perpetuates such crimes against the innocent is the very insistence that we are not an animal species, which is an understandable opinion of naivety but which quickly falls apart when witnessing basic human behavior, especially the kind of depravity of which all of us are capable such as the rape and murder of children, petty socioeconomic disputes and conflict, feverish adultery, desperate dishonesty, or the animalistic hatred and violence which perpetuates war and genocide. In fact, one tool of genocide is to characterize victims as animals and this fuels violence and persecution towards them because of other humans who want to believe that we are not, in fact, an animal species, though it is plain to even marginally objective observation that we are not elevated above other animals, but part of them, and the reason why we act the way we do.

One reason we prefer to believe (want) we are not an animal species is because of what we, an animal species, do to other animals, and the power we wield over them through our advanced technology and cunning, to systematically kill and consume their bodies as food or take their environment and resources without recourse and so in turn instinctually fear our own fragile and fleeting mortality which can just as easily be extinguished from reality as we do to others. The dread which fills our heart at the idea of being so casually consumed by more powerful entities as we do to those with less than us even inspires science fiction stories of this occurring from invasion by alien species which scoop up humans the same way we do to cattle and chickens, and the thought offends our survival ego defense mechanisms precisely *because* we are animals which possess fear of death a survival mechanism. Defiantly resisting the reality of our position through unrestrained consumption and insisting dominance over all life and impunity from reality then helps us delude ourselves to the obvious reality of our condition, which is every bit as as mortal and fleeting as those animals over which we have contempt, and as a consequence now find ourselves destroying the very planet upon which our very existence is wholly predicated.

But this dilemma is not a *product* of consuming animals, as is evident even in those who refuse to do it whom anyway descend into emotional psychosis, opposing the killing of animal life even to the point of killing of other humans, which by the way are also animals in complete contradiction to their stated beliefs (wants) and revealing instead nonsensical and irrational ideology which is itself desperate to avoid the same fate, because it is the same fear of our own mortality and death which motivates it as the ideology it supposedly disagrees with, simply handled through different strategies the very same way that atheists and theists commit the same logical contradictions and for the same reasons. Some people have even tried to force their dogs to consume a vegetarian diet, and indeed there is a major disconnect between abhorring the killing of animals but owning a cat, whose food is entirely made of dead animals but conveniently packaged into pretty bags with nice labels and marketing.

Likewise, lions, alligators, and wolves kill other animals regularly in order to survive yet nobody condemns them or tells lions they should be vegan, because we understand they are animals which subsist on other animals as part of their biological makeup. Yes, humans are more intelligent (not all, though), but human beings are *also* animals and dependent upon food to survive as much as any other. But other animals are also not exempt from the laws of cause and consequence as ourselves, and other animals also experience the same life events and dynamics as do human beings such as loss, mistakes, joy, surprises, disappointment, love, retribution, altruism, and consequence as does humanity. It is well known, for instance, that male lions will kill young offspring when they take over a pride, which is very similar to human experiences of violence and exploitation, but what is not well known or discussed is how the *female* lions react and handle this situation, and behave in ways in which they attempt to protect their offspring by fighting back or by deceiving the invading males by mating with them to help obscure the parentage of her offspring by making every male which overlapped her territory think he was the father. In this way male lions also suffer the same cause and consequence for their actions as male humans, which is that self-centered violence and exploitation not only deprives us of real fulfillment but also exposes us to danger and liability by depriving us of the support of those we hurt, or even winning their distain and contempt. While it is common for male lions to kill cubs that aren't theirs is it *not* the only reality that occurs, it is only *common*, and because female lions are extremely protective of their cubs I imagine they similarly possess the ability to recognize when an adult male does not kill their cubs and thus not only do not feel threatened by him but protected and will in turn help defend the male against marauding males as has been frequently documented, or otherwise leave an asshole lion they do not like to his fate.

While using terms such as "like" and "asshole" are human terms the major flaw in zoology and biology is assuming that we have, for some fucking god-complex reason, different emotions than other animals even though we are also directly descended from the same animal tree, and emotions are not unique to humans but are the mechanism by which instinct and behavior are achieved in every animal. A honey badger harassing lions is not going through some mindless, unemotional reaction to the presence of lions but likewise feels both fear and bravery as adrenaline and cortisol triggers their flight or fight reaction (though in the case of honey badgers it should be just the fight reaction). Studies in fact have established such emotionally complex behaviors in animals, in fact, such as that lionesses also recognize the reliability of other females and actively take into consideration whether other females can be relied upon for assistance in territorial disputes and base their cooperation on reciprocity, not simply biological instinct or relation, indicating that traits such as integrity, dependability, and helpfulness are intrinsic natural survival skills and *not* characteristics unique to humans, which are even more readily observed in more social, longer-lived species as whales and other cetaceans which also have complex social dynamics, language, and culture just as we do.

Animals therefore also experience the disappointment of rejection, the pain of loss, the emptiness of loneliness, and fear of death, as was documented by the journey of a lone, lost narwhal who eventually paired up with a pod of beluga whales for companionship. While cynics may say things like the narwhal probably just sought safety among a group—YES, that is the same motivation that we as humans have, good job recognizing the similarity! Which is motivated by hormones and instinct which we feel as emotions, because animals are also possessed of the same neurochemicals which are the literal emotions we feels such as dopamine, serotonin, adrenaline, progesterone, etc., because we all come from shared ancestors from which these biological functions first originated, we humans being a direct animal Descendant of other animals just as every other animal is as well.

Unlike other animals we are spared from most conditions of the natural world to instead luxuriate in technology and comparative safety. But living in the modern era full of incredible technology it is amazing that people are bored and unfulfilled while yet living exactly the kind of life our ancient ancestors could not even dream, to be so relatively saved from death and disease as we are, to have an unending supply of food and ability to receive effective medical care when we are injured or sick. Some of us continue to spend our lives lost in dreams of societies like the mythical Atlantis or Camelot even when the society in which we now live has greater technology, convenience, harmony, and advancements than those mythical places, none of which had air travel, scuba diving, MRI machines, or smart phone video calling.

Yet humans are every bit as disinterested and contemptuous as we have ever been, because the dissociation from reality which causes this disinterest is rooted unresolved trauma such as most of us suffer in the ways in which we have been raised and abused, and thus dissociated from reality can only be satisfied by dreams of what we do not have, which will be simply whatever we do not have even when we have everything we could ever want, as a survival instinct of creatures which are as much a product of nature and biology as any other animal which lives on this planet.

Our very apathy of the awesome technology which exists around us, as if we even know how it works (you don't), is blatant evidence of what animalistic creatures we truly are, to mistake such abundance and marvel as mundane or everyday, to take for granted the high survival rate of children and long lifespans. But it is in recognizing this reality that we in turn find resolution of these problems, and it is very reason that we are animals that practices such as traditions bring us comfort, as the value in cherished holidays, rituals, and family traditions is not because of any inherent ideological value they may have (which is why so many disparate types exist) but because they instead fulfill fundamental needs and nature of human animals, things like togetherness, love, cohesion, food, cooperation, and knowledge of the tools by which we find and achieve these things we need not only to survive, but to thrive. As a species we humans do not possess especially effective individual traits—absent of claws, strength, speed, and even fur, no one of us could survive on our own outside of society and other humans. Even hermits whom live in the woods have many possessions made by other humans, and the very woods in which they live remain standing because of other humans who fight to keep them from being destroyed, and the hermit learned earlier in life to cook and prepare food, and what even constitutes food, or how to build and make shelter and clothes, from other human which taught them that knowledge. We are like this because we evolved as a species through the power of the group, which together can accomplish more than any one individual, through division of labor and assistance and cooperation, and our human evolutionary psychological biology is thusly oriented to support that nature, even to the harm of individuals, through desire for inclusion and behaviors which uphold hierarchy, institution, and society such as what are accomplished in tradition.

This is why also many people are disturbed by those who try to tear down even the most harmful of institutions or groups, because our innate survival instincts as human animals compel us through fear to uphold the group and the structures which make it, unless that group finally and undeniably imperils our own personal wellbeing. Failing to understand this because we fail to recognize ourselves as an animal species empowers harm to ourselves and others because our shared evolutionary biology is not concerned with our individual wellbeing, but that of the entire group and survival as a species, but since we are each individuals in turn risk being destroyed by the very things we seek to uphold if they do not in turn protect and care for all members of humanity. To this end, traditions like dining together with family and friends, or having seasonal and reliable holidays and celebrations, and rituals such as marriage or coming of age parties or any other invented institution which brings people together provides us stability of predictability, to satisfy fear and survival instincts for belonging to the group, which can and does then help us feel safe and satisfied and thus more relaxed, productive, and selfless, which never has anything to do with the ideology behind the tradition but instead simply the *act* of doing it.

The Christmas (Yuletide) holiday growing up was never lacking in celebration and gift giving, and the joy of the season in my youth was precisely why Yuletide as an adult was oppositely so frightening and lonely after my family kicked me out on my own for being gay, because I was denied that closeness and love which all humans require which is amplified during the practice of traditions. Many other gay men who belong to equally hateful and dangerous groups of people sometimes choose to hide their identity in order to sustain the privilege of participation in these groups and traditions, because of our animalistic instinct to fear the group, but end up constructing lives built on lies, deceiving women, tormented by an endless internal conflict which often anyway costs them their sanity and happiness. Toward the later end of my childhood my mother suddenly became overly religious about the Christmas holiday, making enormous effort to shift our celebration from one of joy, togetherness, warmth, and good food and gifts to pious and despairing obsession with her religious dogma and ideology and the supposed amoral nature of the world, and turned the holiday into a time of fear and anxiety rather than celebration and happiness. An innocuous looking folio bound in red and green felt held a collection of purported Christmas stories, one or two of which were read every night, but were in truth disturbing tales of authoritarian compliance and rote practices of religious indoctrination hardly anything to do with the holiday and instead meant to sow self-hatred, shame, resentment, and distrust of others.

Indeed this is because shared experiences that we all have are also ideal targets for control. Sex, sexual arousal, sexual relationships, body image, weight gain, disease, love, and death are all universal aspects of humankind and thus become the subjects of ideology and targets of authority to compel adherence to institutions and compliance to ideology. Almost nobody would care one bit about hell and damnation if the requisites for going were a preference for true crime drama, because although there is a substantial audience that likes to hear about the murder of other people for their entertainment it is not an inherent feature of humanity, but personal tastes, and thus not effective fodder for manipulation and exploitation. Membership in the group and our need and desire to belong and participate with family and loved ones is one of our strongest needs as a human and thus the most powerful occasion to shame people for their needs and wants as means of control, by threatening their very ability to fulfill this basic human need of love and belonging, to then turn traditions from exercises in celebration and joy to devious authoritarianism which undermine family ties and ruin relationships in service of power.

Christmas was otherwise very dear to my heart, and after being sent out on my own I would put up a tree and observe the season, making tremendous effort to survive my depression by replicating my mother's cookies and amassing a cherished collection of prized ornaments, cookie cutters, and the biggest Christmas music catalog of anyone in my family. Even though I was no longer religious, Christmas was and still is a time of year that I love, but even though I continued to observe and celebrate the season I was still regarded with animosity from even my own family, recruited into their amoral, hypocritical, and very un-Christmas-like culture war against those who do not share their exact religious beliefs. One year when explaining to my mother how much I loved the Christmas holiday she psychotically began imploring me to return to church and religious worship, completely missing the entire point of what I was saying and failing to have any empathy for me whatsoever. Though my family sought to avoid outright conflict during the holidays the absence of engagement, emotional and physical distance, and the superficial niceties which papered over their general imbibing of hate and fearmongering eviscerated the joy of the season and caused no small amount of tension and conflict during a time which should be about love, family, food, and joy, because to them the practice of traditions and holidays is for the authoritarian subversion of fear, where for those like me it is a celebration of life, love, family, and gratitude for all that we have.

During a relationship with someone who was important to me and before I finally learned how to resolve the trauma of my childhood we both fervently reveled in the Christmas holiday, but the trauma of our youths was so intense that it also became one of the most tumultuous times of the year in our relationship, each of us burdened with fears of abandonment and disappointment in unrealistic expectations of the other to meet the challenge of the season in spite of the amazing amounts of pain and heartache we each carried. My father was kind to save for me all the unopened bottles of wine and whiskey left from his vacation renters, which made visiting with my siblings who were and are raising my nieces and nephews in a religion which teaches them to hate me more bearable. But when that relationship ended I simply didn't even bother going home for the holidays anymore because I no longer had anyone on my side to buffer me from the pain of indifference and emotional isolation at a time when that is the last thing which should occur, especially in a family. Though their abuse of me growing up led to dangerous suicidal depression, I was and am still the one expected to accommodate *their* sensitivities and beliefs, as if there must be a choice between the two rather a spirit of harmony and inclusivity, as if my personal beliefs and needs are invalid or detestable simply because I am not religious and they are, even when their beliefs actively demonize me and imperil my wellbeing both in our family and the larger world outside.

The true irony of the Christmas holiday is that every single fundamental aspect of it is still entirely pagan, the history of which goes back far before Christianity was even born and has its roots in our shared humanity as a human family and is not at all about the religions which have co-opted it for their own purposes, and is the primary reason I love Christmas. One of the most prominent icons of the Christmas holiday is the Christmas tree which, like many Christmas traditions, religious adherents celebrate without any consideration of where it came from or why we do it. Like, seriously please explain to me what the fuck decorating a tree has to do with the baby Jesus? The giving of gifts being explained as a tradition established by the visiting Wise Men makes a little more sense, although also errant, but stringing lights and ornaments on an evergreen tree does not exactly mesh seamlessly into the nativity story about a middle-eastern family living around Jerusalem. Western Christianity of course originated in the Roman Empire in the first century, and before Christianity appeared on the world stage nearly the entirety of our ancestors which would later come to adopt or be coerced into Christianity all had different religious beliefs and traditions local to their own tribes and communities, whose beliefs and practices were more oriented toward the natural world and often consisted of Polytheism (the belief in many gods). In fact, the reason we use the blanket term *pagan* to describe those religions is because before the advent of the major Abrahamic religions Christianity, Judaism, and Islam religions did not have names and categories—they were just simply the traditions and religions of the people to whom they belonged. That is why there is no term for Greek or Roman religious mythology, or those of the Babylonians, Egyptians, or Hittites (Zoroastrianism might be an exception). Christmas too is filled with words and traditions like Yuletide, Yule log, twelve-days, and caroling which, while familiar to us, also do not have any relation to the context of Christianity. Yuletide was in fact the name of the holiday before it was stolen by Christians.

Before more recent human history, human life was very hard—rampant with disease, loss, scarcity of resources, and rife with conflict, and though we had spread across the globe were still very much at the mercy of the natural world, especially winter and changes in seasons and weather or the spread of disease for which we had absolutely no real understanding, and so our destiny as human beings was very intensely focused on such variables and potential liabilities to our survival and explained away by supernatural forces, spirits, and deities because we could not see the causes of those variables in our survival. Because it was such a time of stress, winter was often the subject of ritual activity meant to appease the gods which man believed in control of our destinies but also for comfort our families and societies during times of want. But because life is cyclical, these beliefs incorporated those cycles into their systems of worship and mythology, to explain how the natural world worked and to promote awareness, appreciation, and preparedness for these variables which could potentially cause us much heartache and loss if we were otherwise ignorant, and the celebration of surviving those traditionally harsh times were the point of holidays, to literally celebrate life, survival, and the human spirit, and to express gratitude to fate for the things we have which enable life.

Similarly, the roots of all major religions are far older and deeper than is apparent in our modern era, often lost to history because of the consuming nature of institution and authority and the spread and competition between religions. Christmas was in fact originally called *Yuletide*, or ġēol or ġēohol in Old English, and was a celebration at the Winter Solstice, which lasted for 12 days, which is why we sing about the twelve-days of Christmas, which has nothing do with anything about Christianity. The Yule Log was a tradition of burning a portion of a log each day during the 12-day advent, and more than any of our traditions the Christmas Tree is most important to our shared history as people who come from these traditions because worship of nature was most symbolized by this marvel of the natural world which persisted unchanging through devastating, dangerous, deadly change in seasons—Most animals and nearly every other kind of plant must accommodate winter, but not the mighty evergreen, which stands unchanging against the snow and cold and freezing temperatures to defiantly endure ice and darkness (and then we cut it down and haul it indoors for our amusement where it dries out and dies). Our Celtic ancestors when celebrating Yuletide would also bring food out for overwintering birds and tie them to the evergreens, which is where our decoration of trees began and why we even still put fake birds onto trees to this day. A large pig or boar was also slaughtered for the feast, which is why we get the Christmas Ham, which is actually a Yuletide Ham, and caroling of traditional songs was also another Yuletide tradition which is to this day repeated and practiced naively as if it isn't completely and entirely unrelated to the Christian religion. Even the Christmas star is rooted in our ancestor's past and the use of the heavens, astrology, and celestial events to predict or understand the world through divination, as a new star in the sky (which yes happens all the time when supernovae explode) was long taken by people as a sign of auspicious coming events, not something unique to Christianity, which instead coopted that common belief.

The primary point of celebrating holidays during the dead of winter was to pay respect to the natural world and give thanks that winter was not eternal, the knowledge that spring would come again, and that this time of year was significant in the great pause it demanded of all life subject to it. The successful preparation for winter as well, storing up lots of food and crafting necessities for survival such as clothes and shelter meant that we could also endure, and we celebrated this with great feasts and rituals meant to help pass the time, feed everyone, and promote joy, hope, and comfort. But even some traditions which seem more contemporary like gift-giving and Santa Claus or Christmas lights are also still rooted in other ancient rites and traditions like the Roman holiday

Saturnalia, also celebrated at the Solstice, which promoted unbridled revelry and personal gift-giving as well as the personification of ideals in the form of Father Time, or more accurately *Saturn*, whose Greek name was *Chronos*, the Greek word for time, whose festival *Kroina* worshiped Chronos and preceded Saturnalia which preceded Christmas, showing how Christianity, which originated in and was promoted by the Roman Empire, subsumed and repurposed cherished traditions for political power across millennia. Romans would also line the city streets with wax candles, which must have indeed been a sight to behold and why we still put out lights even today. Hanging Mistletoe was also a Saturnalia tradition—literally nothing about the Christmas tradition has anything to do with Christian ideology which instead co-opted our ancient traditions for political purposes.

You could spend days on the internet and still not reach the end of the trail of the origins of such holidays, because they encompasses many, many lifetimes of tradition passed on from generation to generation and encompasses the rise and fall of entire empires, languages, cultures, and civilizations. *Easter* is another ancient holiday subsumed by Christianity which in truth celebrates the Anglo-Saxon goddess of fertility, Ēostre, whom was celebrated in springtime which (in the Northern Hemisphere) is associated with an explosion of new life. This made much more sense when learning this fact as an adult since even as a child I wondered why the word Easter was not once uttered in any of their religious scripture or used apart from nothing but the title of the day and why suddenly rabbits and eggs held any relevance to their religious institution. As reproduction and birth is a common human experience, the celebration of life and sex and babies and springtime is also understandable and relatable, even by children, who often witness birth of other children and of baby animals and honors the sanctity of life and the promise of spring, hence rabbits and eggs (because rabbits reproduce so prolifically) as opposed to the celebration of some man I never knew and concepts like resurrection which no human on Earth has ever seen or experienced, let alone children.

All words are arbitrary and derived from sociopolitical pressure and the heaving sea of history, and the real value of holidays like Christmas and Easter is not in what they are called or the rote rituals and ideological beliefs but the richness of our past and the true origins of our beloved human traditions and history which, like the evergreen tree, have endured through darkness and despair because they held meaning for each of our ancestors who came before us whom also marveled at the lights, the sounds, the food, joy, and love. The inherent value in tradition simply because it *is* tradition is something which is far more awesome and wonderful than banal religious trappings or personal or religious ideology, which is in its very nature not universally shared by all people as are concepts of togetherness, food, love, birth, survival, and death. It does not actually matter what a holiday is called or what traditions they may encompass, which is why muslim children celebrating Eid feel the same way as Christian children celebrating Yuletide, because it is the togetherness and love which satisfies the human desire to be together, safe, and loved which holidays and traditions fulfill, and not the specific religious context.

Human culture is uniquely influenced by the politics of the past, and all of us are an accumulation of every geopolitical and sociopolitical conflict, history, and tradition of millions and millions of other human beings across thousand and thousands of generations. No tradition is inherently valuable other than what is personally meaningful for each of us, and to adhere to such traditions at the exclusion of others and the fomenting of divisiveness for the sake of dogma and individual belief systems only serves to cheapen the richness of our traditions and the amazing and enduring accumulation of human struggle which has preceded us. Yet, many of us in the West have also lost many of our traditions, our songs, poetry, writings, mythology, even our names have been supplanted by the fruits of war and political and religious subterfuge. I hate my name because it does not have any relevance to my lineage, but was instead chosen by my parents because it was a name in the Bible, not one of someone from whom I was descended or related, from a religion which was forced upon our ancestors as religious oppression and violent colonization destabilized Europe for a thousand years after the fall of the Roman Empire. While much religious belief and oppressive ideology is mistaken for useful tradition, the real value of tradition is in helping us to understand not only ourselves but also the context in which we live and the understanding of what it means to be alive. Though one of the original purposes of holiday celebrations such as during the winter was simply to distract people from their suffering during cold, wintry months when food was scarce and sunlight fleeting, it also served the purpose of teaching children that preparation for winter was not optional, and that if we did prepare adequately for this time each year our survival through the winter was not only likely, but also joyful.

For many of us the desire to restore traditions of the past leads us to adopt our version of old traditions and religious beliefs we think our ancestors may have had, such as by adopting "*pagan*," beliefs, but even calling ourselves *pagan* belies the unfortunate superficiality of that effort because our ancestors did not refer to themselves as pagan, which was instead a word used by Christians to prejudicially group all non-Christian, polytheistic traditions together without regard for their content, relevance, or populations in order to better subsume and subvert local traditions and beliefs wherever they conquered, and none of our ancestors had names for their religious beliefs because they were simply what people locally believed and practiced just as everyone everywhere else had their own beliefs local to their own societies and cultures.

These old traditions also do not in fact have personal meaning to us—adopting ideology is not the same as experiencing tradition, which has meaning simply in the fact of its practice, which we have not practiced for many centuries. This does not mean we cannot find alternatives, which is much easier than trying to replicate the old and unknown, although that is an option, but those of us whom have lost our traditions can simply begin to build new ones by developing those current rites, rituals, standards, and traditions we have for our own daily lives, holidays, and relationships through *whatever* is meaningful to us, personally, because in truth all traditions haves many meanings and have evolved throughout history to suit the needs and trends of the age, to bring commonality, comfort, joy, and togetherness during times of need in the satisfaction of the human condition, not specific to whatever ideology it pertains, and are valuable to children and adults alike not for the trappings of ideology but the mere closeness to our loved ones and emotional intimacy these occasions facilitate, which is also why they can be so devastating for those of us who are rejected from them, the irony of inviting outsiders to join our celebrations in order to appear welcoming and inclusive but kicking out our own flesh and blood whom refuse to conform. Call Christmas Yuletide if the former triggers traumatic experiences, but in truth the word Christmas is as arbitrary to the holiday as is Yuletide, the meaning of words being relevant only to those who use them wherein the important part is what we do, how we behave, what we feel, and how we treat and regard others, which is certainly not limited by any language.

Even Christmas music is not joyful because it's actually about Christmas— songs which touch our hearts are beautiful because they are *music*, written by men and women in states of joy which we in turn feel when listening to them, and regardless of the actual words the celebration is meaningful to the human experience, to make and hear music as a reality of laws of creativity, and even when hateful Christians sing about Jesus in their music they are in truth singing about the nature of the Universe, which they understand as being Jesus in their desire to control their fear of death and loss but which can instead be whatever we choose to identify that better matches our conception of those forces which govern the Universe. There are still some songs which are overly religious and associated specifically with the religious cult of my childhood which I don't listen to at Yuletide, but I realized years ago that the revulsion I felt hearing much of the seasonal music was a desire not to validate the abuse and hatred which had been slung at me by my family and community, which is in truth is also a control behavior which got in the way of my own personal fulfillment, spiting nobody but myself, and was only relevant to me because of my past trauma. Recontextualizing that music in the deep history of human geopolitical conflict, intergenerational trauma, and that music is beautiful because of its inherent mathematical and physical realities as a feature of the living Universe removed the stigma which otherwise prevented me from enjoying what I wanted to enjoy, because in reality it is the music which is beautiful, not the ideology behind it, much like walking into a Catholic cathedral and mistaking the skills of man which constructed such an incredible building and the laws of physics which keep it standing as validation of a violent and hateful church which for two thousand years executed people if they did not believe in it and started wars and cultural genocides caused untold amounts of devastation, rather than the marvel of humanity and reality they truly are. It is understandable that many traditions are associated with pain and trauma, but they are not inherently painful or traumatic themselves, simply associated, and if you want to enjoy beauty created by man though it is still associated with trauma recognize instead that it is beautiful because man is capable of such beauty, with permission from the Universe, which has nothing in fact to do with the ideology that it might be associated but instead the beauty of natural laws which make such things possible.

Food is also a very important part of our traditions, and food often features prominently in nearly every celebration we practice. This is not arbitrary or cultural but because of the reaction we experience after eating, informed by the release of dopamine, oxytocin, and other pleasant hormones in the meditation of instinct toward successful behaviors which increase our chances of survival as an evolutionary human animal species. Yet though I grew up eating meat daily and meat often takes a prominent position in celebrations the last several years I have only consumed meat in moderation, about four to five meals total per week, not because of any purposeful intention but which became less and less appealing the more my health improved. As discussed in my book on health, this did not occur by restricting my diet to healthy foods, but instead getting healthy *caused* overwhelming cravings for fruits and vegetables, which I had never experienced before in my entire life. This occurs because as an animal we do not possess the ability to synthesize the *cellulase* enzyme which is required to break down plant cell walls made of cellulose that is required to liberate nutrients like pectin, vitamin K, magnesium, potassium, manganese, molybdenum, etc., which are otherwise locked away behind the plant cell walls. Instead, this ability belongs solely to our microbiome, and after it becomes restored and more robust by caring for it and our health we then acquire the increased ability to break down plant foods and thereby benefit from them. Indeed, plant foods are far more healthy for us than animal foods, having not only a greater abundance of nutrition but also protective phytochemicals like tannins, carotenoids, and anthocyanins which are entirely missing from foods like meat. But because plants are harder to break down we instead crave easily digested foods like meat and refined grains when lacking a microbiome capable of breaking down plants, and when we oppositely restore a healthy microbiome we oppositely crave plants in preference of meat because they are so much healthier.

Our commensal microbiome also thrive on indigestible plant nutrients from which they produce the short chain fatty acids, B vitamins, and even amino acids, and high protein diets carry increased risk of protein fermentation by opportunistic microbes which produces large quantities of ammonia by parasites and other opportunistic pathogens which then destroys digestion and induces metabolic disease and cancer. The fact that we cannot natively break down plants is a major fallacy of supposedly healthy dietary habits like vegetarianism or veganism, which is that willpower and dedication does not make a healthy diet, as it does not matter how much you want to restrict a diet to plants we cannot necessarily digest them without requisite factors like being replete with vitamin D and thus daily sun exposure, because our body shares vitamin D with commensal *Thaumarchaeota* archaea, as I discovered, which in turn produce vitamin B12 (cobalamin) that is required by our microbes to adhere to plant food. So, forcing ourselves to eat leafy greens and other vegetables when we do not feel like having them is in fact *unhealthy* because the reason we do not crave them is precisely because of times and situations in which we cannot digest them, where restoring instead a healthy gut microbiome restores the ability to break down plant foods so we then can benefit and thereafter feeling even *more* satiated from things like spinach, carrots, or beans than steak and chicken if digestion and our microbiome are working properly.

But in late 2024 as a result of having solved the *Warburg Effect*, which is the underlying biological problem that drives most cancer due to loss of mitochondrial silicon caused by excessive ammonia from parasites and pathogens I also discovered how *dietary silicon* participates in the formation of micelles and liposomes during digestion to promote even better increased absorption of necessary fats and fat soluble nutrients required for the production of *cholesterol* which is in turn required for necessary hormones like pregnenolone, progesterone, testosterone, other androgens, estrogens, mineralocorticoids, and vitamin D. Because hormones are indispensably important for our health the absorption of dietary fats and normal function of cholesterol is thus paramount to being healthy, and successfully restoring these pathways leads to greater increase in wellness, energy, and body functions to the amelioration of conditions like impotence, depression, insomnia, and even cancer since progesterone itself is also antiparasitic. Strangely, as soon as I started leveraging high silicon foods like spinach, dates, and green beans (and a small amount of silicon boiled in water for making tea occasionally taken with dietary fats and fat soluble nutrients), I suddenly and unexpectedly lost *all* desire to eat any meat, at all. At the time I had a wonderful chicken soup in the fridge and the thought of finishing it off (I make really good chicken soup) actually filled me with a slight feeling of revulsion and I was instead overcome by an insatiable craving for *minestrone* (a dense, hearty vegetable soup which, in Italian, just means a big, vegetable soup) so overwhelming I threw out the chicken soup and went to the store for minestrone ingredients, and made a large batch of which I then had *four* bowls and afterward felt more satiated than ever in my life.

At first I supposed my overpowering cravings for vegetables and revulsion to meat might be a temporary reaction to some nutrient deficiency, but to my surprise at the time of this writing it has persisted for longer than seven months, each time I walk past the meat in the grocery store simply feel no desire to eat it whatsoever (I am craving some good crab enchiladas suizas, though). Considering the danger caused by a high protein diet which promotes diabetes and other metabolic illness due to the high rate of ammonia production by proteolytic microbes it does make more sense evolutionarily that we should crave plant foods in preference to meat, because in fact the eating of meat is only a backup, opportunistic behavior that herbivores and omnivores utilize when they are otherwise unable to get sufficient nutrients from their environment. Most people are completely unaware that many animals we colloquially consider obligate herbivores like cattle, squirrels, or deer will happily gobble down baby birds if they happen across a nest or unguarded chick, and the line between herbivore and carnivore is not even remotely as definite as we learn in school, because it's simply easier to consume other creatures rather than plants, especially during times when we may be deficient in requisite nutrients that are not as available during certain times of year or during drought or famine (cattle and deer likely do this to increase their calcium, but squirrels regularly predate on baby birds). While many people hypothesize that eating meat was integral to our evolutionary past, there is in fact no empirical evidence to support that claim. As recently as last year a study on the bones of some of our direct ancestors expected to return isotopes proving a high meat diet instead showed a diet mostly composed of plant foods. And while we certainly ate lots of meat *after* becoming human there is instead an abundance of evidence that *before* becoming human we were primarily dependent on fruit, nuts, tubers, leaves, insects, and mollusks, and we became hunters only after evolving into larger primate humans with longer legs capable of chasing prey and wielding weapons, not before.

The idea that we became human from hunting is especially absurd considering the short legs of our primate ancestors which were entirely ill equipped to even chase down small, slow prey let alone big game, and researchers whom claim we lost our fur to keep cooler while hunting in the African heat have apparently never seen any other wild animal like dogs, cats, and bears which don't even sweat. For some reason these people don't understand insulation such as fur works in both directions, to help keep an animal cool and protect their bodies from harmful UV ray exposure (and abrasion injury), not only to keep warm, just as the insulation in your house keeps the cool air inside during summer just as it does warm air during winter. Hunting likely only occurred *after* we discovered fire and could cook meat, having discovered and scavenged the barbecued remains of animals after wildfire swept the land, because we also have no innate ability to chew raw meat (except maybe more delicate types like in fish), nor even a taste for it the way we do foods like peaches or nuts.

The specific reason we became hunters is because of our distribution away from the original environment where we evolved which, as I discuss in Fuck Portion Control, was likely around ancient Lake Mega Chad in Africa, or the Mediterranean as semi-aquatic apes, which is why we lost our fur and grew longer legs, foraging for shellfish in the shallows and harvesting fruits, nuts, tubers, leaves, and insects on shore. Leaving this homeland for whatever reason our ancestors experienced nutrient stress as they radiated outward to new locations and environments which then motivated the increased dependence on hunting and predation to survive. Hunting, however, was not an evolutionary accomplishment that enabled us to spread cross the Earth but a crutch which prevented us from being long-lived and healthy, and after our ancestors wiped out most large megafauna species and ran out of the primary food on which we had been lazily dependent (a single baby mammoth could have fed hundreds of people), were then forced to farming and cultivating plants as our primary sustenance but which also better promotes wellness and survival of our species anyway since plant foods are simply incomparable in their nutrition to animal products and our commensal microbiome so obligately dependent on indigestible plant substrate.

Another surprise which occurred when restoring my ability to absorb fats and produce adequate hormones due to higher dietary silicon (and resolution of digestion problems and vitamin D dysregulation as discussed in my other book) was that protein suddenly seemed a much less important dietary requirement for my wellbeing which, being six-foot seven-inches tall and two-hundred sixty pounds, is even more insane that I don't feel the need to consume more than about 40 grams of protein per day on average (some days I will eat a full pan of pasta and get up to 60-80 grams), and not only do I not lose muscle, I am the most muscular I've ever been in my life. This is not too surprising considering one of our closest relatives, the gorilla, is the most massive of all primates and eats a diet primarily of plants (and some insects), but what most people don't know is that our microbiome *also* produces protein (amino acids) when they are supplied with sufficient quantities of carbohydrates and other plant nutrients, and in normal sized persons can be as much as 20 grams per day (for me it's probably more like 30 grams). Not only this, but healthy microbes also produce a more healthful *ratio* and composition of amino acids than what is typically found in the diet, such as plenty of lysine and tryptophan which is required to synthesize niacin, so although my diet now has about half the protein intake my body is in fact healthier and has more energy, better posture, better libido, more restful sleep, and more muscle mass because the ratio of amino acids is better and accompanied with far more other important nutrients, which thusly then eliminates even any desire for meat (I still use milk, eggs, and cheese though—dairy is essentially predigested plant foods).

While hunting kept our ancestors alive it also limited our productivity and wellness while promoting disease and shrinking lifespans, due to the sheer amount of protein intake and resultant colonization by ammonia and hydrogen sulfide producing pathogens. It also demanded much time and effort which could have otherwise been spent learning, building, farming, and creating stability as what occurred once we finally did develop agriculture, and its tradition is passed on not as something which benefits our families or health and wellness (there are always people getting parasites and other diseases from eating wild game) but instead feelings of control and domination which make us *feel* in control of the natural world, because if we can kill other creatures on this planet we can then delude ourselves of our own mortal limitations as an animal equally dependent on the world in which they live. Unlike wild animal hunters, however, which take the weak and sick and thus help maintain healthy populations of other animals we humans instead take prized, healthy animals without serious illness (though they still carry parasites) and in so doing weaken populations of other animals by preventing those strong genes and survival behaviors from being passed down to new generations which in turn compromises their populations' long term survival. Anyone who practices hunting as a supposed survival practice is indeed completely delusional—do you really think you will be able to hunt during any supposed calamity while the other *eight billion people* also on this planet also turn to the wild for sustenance? In current geopolitical conflict in areas which lose their supply chains people do take animals from the wild and rapidly deplete those populations in a matter of days, because the sheer number of humans is so high in ratio to other animals, everywhere on the planet. The *only* future in which any of us survives, especially in any emergency scenario, is one which both protects and preserves the natural world, since we are also dependent on intact ecosystems, *and* intelligently produces high quality food from ingenious systems like regenerative farming and holistic greenhouse/indoor farming, and during any natural disaster the *growing* of our own food in controlled environments will be the only available and persistent food source, *not* the wild and animals which instead will be taken in short order by those who did not properly prepare.

Having a diet based on plants is thus not only healthier for us, but also more sustainable for our prosperity and long term benefit and survival, but nowhere is this achieved by forcing ourselves to be restricted to plant foods, and is instead achieved by having compassion for the body and providing for all the requisite environmental and nutritional conditions which supports our microbiome, which then naturally shifts cravings away from meat and toward plants (and God you have no idea how good this tastes too), working within the parameters of biology rather than trying to force it, which only leads instead to more illness since biology functions on defined and finite systems and not arbitrarily to our whims and desires.

Especially since there are billions of human beings now on the planet, we clearly no longer require many of our evolutionary survival instincts and strate-gies, and since many of them are also a function of fear which also causes our own destruction the care and nurture of our own wellbeing through understanding our biology and supplying the conditions and nutrition we require to be healthy is the antidote to those basic survival instincts which are so destructive to our wellbeing. As discussed in my book on psychology, most people also think the act of taking care of our needs means doing things on our own, but this is completely antithetical to taking care of our needs because, you will remember, we are not evolved to be capable of total independence from others, and as a human being we are reliant on others to give us companionship, intimacy, and to help obtain food, shelter, and even knowledge, and caring for our own needs means cultivating those resources which includes behaviors like asking for help or caring for others without concern for our wellbeing—because we cannot survive in systems which are concerned only with the self, and like our ancestors of old which are the reason we are here today through the passing down of knowledge in the form of traditions, love, and understanding we must empower and sustain systems which help all people, through which we then in turn also benefit, and traditions are one of the most effective ways to create and sustain those requirements as the in coming together in celebration, to not only build bonds with others but also reinforce effective survival skills and food traditions, the need of which today is not less than ancient times even with the technology we have, as evidenced by the massive rates of disease, disharmony, and dissatisfaction which plagues many industrialized societies.

There is therefore much value in the practice of tradition regardless of our personal histories, and we can all share in the joys of tradition and celebrations if instead of celebrating our ideals and ideology we celebrate each other, life, humanity, food, the world, and the wonder of creation such as what truly matters and what was originally meant to celebrate. This includes the large, social traditions but also those which are intimate within our own insular social structures or even individually, and the eating and preparing of food because the point of all tradition is the cohesion of humans to better survive the challenge of living which, while less challenging now than in yesteryears is still not guaranteed.

While fear and hatred are also survival tools they are the least effective, and more commonly serve to remove ourselves from the annals of history as we misunderstand the nature of cause and consequence, and sustaining tradition and holidays requires love and togetherness, otherwise fear, ideology, and hate destroys family, relationships, and societies, and since as humans we require tradition, togetherness, and cooperation to thrive, hate, ideology, and authoritarianism not only takes tradition, but society with it. So find tradition in what you do now, integrating your life experience and what is personally meaningful to you, and it will be meaningful simply *because* it is, not what it is.

The Glassy-Eyed Raven

A flock of ravens were hunting for shiny stones in a farmer's field when suddenly a large, glassy-eyed raven not of their flock flew down and perched at the top of an old fence post. "Gather 'round, gather 'round!" it croaked, startling the others, "Have I got a deal for you!"

The other ravens looked quixotically at each other, "we're busy collecting shiny stones for our nests," they replied.

"And a mighty collection you have there!" said the glassy-eyed raven. "But how would you like to have even more shiny stones than you could ever find in this lowly field?"

The ravens weren't quite sure what to make of this. "How many more?" asked one. "Ten?"

"More!" replied the raven.

Twenty more?" asked another.

"More!"

"Thirty more?" asked another in disbelief.

"More! More! More!" screamed the glassy-eyed raven as he danced about the post.

"You have that many stones?" cautiously asked another raven. "Yes," replied the glassy-eyed one, "I have more stones than you could ever count! So many I require many nests and secret places to keep them safe!"

"Many nests?" cried the ravens in disbelief, for most ravens had only one nest, but the glassy-eyed raven nodded his head proudly. "Where do you get these stones?" asked a raven.

"It's so simple," replied the glassy-eyed raven, "I used MY stones to pay other ravens to gather more stones to sell for even more stones. I have been very fortunate to earn so many stones, which I share with the ravens who give me stones, and my method works better than any other!"

"How can *we* get that many stones?" asked several ravens whose curiosity and desire were now well aroused by the large raven's story. But one raven was not convinced, and shook his head while speaking up "that's not possible, to get that many stones. This raven means to swindle us of our stones!"

"Quiet!" screeched the other ravens, angry at the one wise raven for doubting the existence of such riches. But the wise raven did not want his fellows impoverished, and shook his head defiantly. "Do you not see his glassy-eyes, pompous manner, and grandiose flattery? Surely this crow is a swindler and ne'er-do-well, and we'd best wish him on his way than fall victim to his duplicity."

"Don't listen to the doubters," said the glassy-eyed raven, whose eyes got wider as he flitted down to mingle among the believing ravens. "My method is fool-proof, and before long all of you will all have as many stones as myself!" The other ravens marveled and jumped for joy, then gave so many shiny stones to the glassy-eyed raven he could not carry them all, and recruited one of them to help get them back to one of his many, many nests.

"When will we get our stones?" asked one of the other ravens as the glassy-eyed one was about to depart, loaded up with his winnings. "Oh, very soon, very soon." Then the glassy-eyed raven cumbersomely took to the air and disappeared over the trees.

The wise raven who doubted and did not give his stones shook his head and sighed. "We should keep looking for stones," he said. "Why?" replied the other ravens. "Soon we will be rich and never have to search for stones again!" The single raven huffed, and flew away to continue looking for stones on his own. Many days then passed and the glassy-eyed raven never returned, but the ravens of the field heard through friends that the glassy-eyed raven had grown richer than ever, and knew their investment would someday be rewarded. One day when the wise raven heard the others discussing the glassy-eyed raven and his growing wealth, grew indignant. "Of course he has grown wealthier!" He shouted, "you all gave him your stones!" The other ravens looked at him incredulously, then shook their heads, "No," they said, "he's just really good at managing stones."

11. Cults and Conspiracy

I never thought that I was someone who might be targeted for membership into a cult, but a few years into my recovery and research after having first published *Fuck Portion Control* a friend whom I had known for several years was the first (and last) which expressed unsolicited interest in my work, and so I naturally jumped at the invitation to share my experience for the first time with someone known personally with whom I felt a sense of familiarity rather than strangers on the internet (though many of those would later become friends).

After discussing some of my work with him, all the while listening intently he then suggested that a medical professional and influencer he knew personally whom also ran her own group would also be very interested. While I had not yet completed even half the work I would end up achieving I had still been able to solve some significant health problems like depression, addiction and alcoholism, and at the time a functional cure to cancer (different than solving the Warburg Effect as I would later do), and immediately fantasized about finally getting the chance to help people and make a difference as I had been so hoping, now having the potential support of enthusiastic others whom not only understood my work and purpose but could assist me in its implementation and distribution.

So my friend set up a video call with his friend and at the appointed day all three of us met and they sat patiently listening intently to my story and work. The other person, a woman whom appeared in her mid or late thirties and very healthy first caused me some skepticism since people tend not to be interested in the kind of health biology I discuss unless they are sick and have no other choice, as human nature is to do as little work as possible and to deny reality as much as we are able. After finishing with my introduction and story she then began to introduce her background with a group which was led by a charismatic cult leader currently under investigation by the FBI for crimes connected to their organization, but suspiciously avoided the communication of any of her own personal life story or personal experiences. I was alarmed, but being still possessed of some people-pleasing behaviors I did not immediately end the call as I should have (lol). She continued to present her grand vision of personal health and the future, which apparently consisted mostly of a retreat where people paid several thousand dollars to come to her home and learn how to be healthy, according to her. It apparently also was not an invitation for me to help create the scheme using my knowledge or work, which I would have rejected anyway seeing as how she was associated with a notorious, amoral, and likely criminal public figure, but instead offered me a chance to buy a discounted ticket and come learn at hear retreat (about exactly what I was still unclear).

We continued pretending they did not know I was pretending not to have realized they were trying to start their own cult after the failure of the one to which they already belonged, which sought to prey on people's fears and illness, as they sat there insulting me by trying to get me to buy their course with no interest whatsoever in my work which had been the entire pretext of the meeting, and after finally being clear that I would not be buying the course thankfully the call ended and I never again heard from either.

For some time after I questioned how someone (especially someone I knew?) could think I might be vulnerable to recruitment into a cult, being as militantly rational and skeptical as I often am, but since most people refuse to even hear what I have to say, let alone actually read my work, I can see how they would judgmentally assume like so many others that I was just another 'alternative wellness' nut and thus a possible victim for health-anxiety exploitation.

But predatory and harmful institutions like cults and conspiracy theorists and groups are a regular plague of humanity, and have occurred as long as there has been society, often and usually associated with religious or spiritual beliefs because of our shared need for spirituality as well as the fact that *all religious beliefs are manufactured and contrived* in the same way that beliefs of cultists and conspiracy theorists are also manufactured and contrived, the only practical difference between established religions and cults often being the fact that religions have simply been more successful in their longevity.

I see the primary definition and distinction of a cult as simply whether or not the institution attempts to control the personal relationships and associations of its followers, because the primary function of most cults is to isolate members from the larger public or any other potential influence which could jeopardize their dependence on the institution, where if a religion instead encourages and supports the relationships of its members regardless of their membership status there is then no functional cult behavior and therefore is not a cult. But even many large religious organizations such as the one in which I was raised do in fact actively control and manipulate the personal associations of its members through guilt, shame, indoctrination, sex control, and punitive expulsion both explicitly and implicitly not only from the group itself but also member families and other social structures which overlap the religion.

But confusion about why and how cults and extremist beliefs exist belies an ignorance about reality and the fact that all beliefs are made up, that just because a belief is old does not make it any less of a belief than those which are new. That is not to say that all beliefs are bad—human history and culture are crowded with story, tradition, metaphor, and mythology which has helped us survive the conditions of life—but conditioning people to expect and respect religious belief as literal also opens the doors to the same for any of them, including those which are actively harmful such as common to cults and extremists since the practice of religious belief does not train a person to believe in God but instead the denial of reality and reason, to then prime them for membership in structures like cults which thrive on such persons.

Failure to see commonality in all humankind, regardless of religion, ethnicity, heritage, politics, beliefs, etc., is the primary deficiency which impairs our understanding of the why and how of cults, conspiracy theorists, and extremists, as while all people have a desire to belong which underlies membership in any group the organizations like common religion, political movements, or even strange cults and extremist groups are *not* in fact founded on the ideology which creates, supports, and enmeshes the members of an institution, but instead simple association, and the ideology instead acts as a gatekeeping litmus-test that functions to restrict membership only to those willing to prove dedication to the institution and its other members through the transgression of rational thought and general social homogeny. When members are willing to submit to the adoption of uncommon or specific beliefs, practices, rituals they are effectively demonstrating their willingness to uphold the institution above all and thus commitment to the group in its entirety in which the members then feel safe because of the commonly shared dedication and loyalty, not the specifics of the beliefs or practices.

To demonstrate this point, a major piece of healthcare legislation in the United States enacted in 2010 by the (ostensibly) left-wing party was a carbon copy of legislation which had actually been created by the right-wing party which, instead of healthcare being paid for and managed by government as a left-wing policy should be, forced citizens to buy commercial, for-profit insurance products and even fined citizens if they did not. But the members in the left-wing roundly adopted it as their own because it was proposed and pursued by its leaders who were in fact far more conservative and duplicitous than its membership realized, and though the legislation was substantively almost entirely based in right-wing, capitalist, authoritarian ideology it was so unpopular with centrist and right-wing voters who thought it was left-wing policy since it was adopted by those they viewed as left-wing that the left-wing government afterward completely lost the government by the next presidential election cycle. Other state-level, left-wing populations and governments thereafter also continued adopting the forced participation in capitalist insurance markets and punitive fines in the wholehearted embrace of a right-wing authoritarian policy simply because it was their party which did it and not because of the legislation itself.

The religion of my parents similarly sends missionaries out to spread their religion but while the missionaries think the primary goal of their work is to convert others it instead serves only as mechanism for the existing members to self-indoctrinate deeper into their ideology. The dogma and ideology they teach also functions as a filter for potential members by limiting membership only to those willing enough to transgress common social norms and rational thought whom are in turn easier to control and manipulate. As such, membership in any group whether it is a religion, cult, or even political movements have almost nothing to do with the actual beliefs of those groups, where the beliefs instead are the gatekeeping litmus for membership, to purposefully sort them from the Other, which is why belief systems range the entire gamut of slightly unusual to outright batshit insanity and, most importantly, why they can be changed so easily.

This dynamic exists throughout the whole of humanity, and can be observed in as small of social groups as families or friend groups in which individuals are willing to censor their own standards, morals, ideals, and beliefs so we may belong, even to debase ourselves and prove to what lengths we are willing to go to be a trusted and included member with all the privileges of membership. This is not inherently harmful, because each of us wants in turn to be assured that those with whom we associate are as committed to our safety, prosperity, and wellbeing as we are of theirs, and often the price for entry into groups of our choosing are things like ego, pride, selfishness, or emotional walls which impair our wellbeing anyway. But because many of us think that cultists, extremists, fringe groups, or conspiracy theorists are different from ourselves or extreme in their ideology we do not empathize with them (a reminder that empathize does not mean tolerate) and thus remain mystified about why such people behave that way when in fact we do it ourselves all the time in adopting opinions, political beliefs, styles and tastes, or even life goals and larger ideology in order to belong or, at least, to avoid rejection.

Exposure to the worldviews, experiences, and opinions of others and desiring membership in groups can change us for the good and not only for the worse, and many times the friends or lovers I made throughout my life helped me see new viewpoints, tastes, standards, or even political theory, due to my desire also to belong, which even at one point helped me adopt a more minimalist aesthetic and lifestyle which not only saved me money but also time required for cleaning and decluttering, to practically simplify and improve my life and mental health in the process.

Even in groups which are destructive and harmful, membership for those members is by comparison often better than what they had previously (or perceived to have had), which is most fundamentally nothing more than a sense of *belonging*, because the human desire for belonging is so integral to our evolutionary biology. As an animal which also has culture and language our sensitivity to rejection and the harms brought by isolation also means we are not at all designed to naturally resist social indoctrination, especially since indoctrination does most often promote safety for the majority of members while sacrificing comparatively less persons and thus functions as a survival mechanism for our species at the expense of the individual, both those whom are victims of those groups but also those which are members whom in turn never get to experience life outside the confines of institution and indoctrination, most often remaining unfulfilled and burdened with stress, fear, and unrealistic expectations, as the only *inherent* effect of any group membership is a sense of belonging. Failing to recognize these attributes as ones of our innate human animal nature, instead blaming individuals which, while responsible for their individual behavior, did not engineer human biological design, makes us ineffective in both understanding others and solving problems shared by many. Recognizing instead that all human behavior occurs because we are an animal species which evolved over millions of years to acquire traits which promote the survival of the species thus helps us understand that it is the universal instinct to feel safe and included which motivates membership in any group and, contrary to how it might appear, serves to relatively increase a person's sense of wellbeing compared to their past experiences and unresolved and unfulfilled fears and traumas, the entry gate which bars membership to that sense of wellbeing thus being the group's codex of ideology which is unlocked by demonstrable commitment in the willingness to transgress rational thought and subvert social norms in the search for impassioned compatriots to fulfill emotional bonds of fraternity.

Conspiratorial mythology is also exciting to followers, through the sharing of commonly appraised mythology and fantasy much like fans of *The Lord of the Rings* share when excitedly discussing the books or viewing twelve hours of film in one sitting, as the act or experience of enjoying shared interests with likeminded peers serves to deepen and strengthen social bonds through common experience, to further strengthen feelings of safety and belonging in the recognition of shared ideals, goals, dreams, and worldviews as separately from all others and in a world which does not inherently provide such fulfillment. What anyone with membership in any group on which we feel beholden such as cults, religions, political parties, etc., do not understand (due to insufficient rearing) but which is often learned through the practice of inventory is that we can and should provide for such needs *ourselves* as what we often expect from others—for instance feeling vulnerable when lacking friends and groups with whom to associate is a fear we are not interesting or worth loving, but we can learn that we do have inherent value through self-care behaviors like practicing inventory. Friends often take care of our needs, and parents, family, and lovers give us affection, but we can also take care of our needs or give ourselves the affection we normally expect from others and through our own behavior provide for ourselves some things for which we are sometimes willing to sacrifice our wellbeing or best interests, not knowing that we are not quite as dependent on others as our fear often leads us to believe. Otherwise we might find ourselves seeking the same kind of reassurance as what occurs on a date when two persons find commonality, but with shared worldview that is actually harmful to our best interests—we know that in a shared future together it will be easier to experience things we enjoy because the person we are with is also motivated to do the same, but with people whom are racist, bigoted, isolated, delusional, predatory, duplicitous, controlling, hateful, etc., they do not actually have our best interests in mind because we only serve a self-centered survival function for their needs and insecurity, and thus are disposable and replaceable by anyone who can fill that role. This fault arises from the primary motivation for all membership in any group, the most selfish human emotion of all—fear—and when we stop fulfilling that function for them so does our usefulness and we are either discarded or harmed as they in turn seek new methods of control and power to assuage it.

It still may be confusing to understand why anyone would start or lead such apparently insane ideology and behavior, let alone join them and uphold the delusional dysfunction at their core, but in reality this is because of the very simple explanation that human beings instinctually desire the power to subvert those things we fear—loss, rejection, pain, conflict, aging, disease, violence, rejection, loneliness, dependence, etc., and when we are sufficiently traumatized and our mind dissociated from reality and desirous of freedom from suffering are more than willing to believe anyone and anything which simply says we have such power, even when in reality we do not, because belief (want) serves to liberate the mind from the stressful burden of fear by delusionally deceiving it to the reality of reality. This then not only lowers stress hormones but also increases hormones of reward like oxytocin and dopamine due to association with others which further reinforce the delusional and dissociative behavior and membership in the group which is the only source of dopamine reinforcement they likely receive, and because dopamine is the hormone of happiness they truly only feel happy when participating in a group of likeminded people believing they have literal power to subvert reality.

This is also why it is so important to first learn compassion for ourselves, such as through inventory practice (because compassion is not a feeling, but action), because it is impossible to have empathy and understanding of why others behave the way they do when we do not even understand it in ourselves, and anyone can be liberated from such conspiratorial, cultish, extremist ideology by simply having compassion for our mortal, human condition rather than fear and resentment (which again is not achieved through willpower but self-care behaviors like inventory). But the problem of indoctrination becomes even more specific when outsiders confront or debate those within, or their ideology and behavior, because nobody properly understands why and how these groups and people exist in the first place and as such are also not effective. For instance a prominent hate group in the United States is often confronted using arguments based in ideology, errantly assuming (usually implicit) that those whom they are confronting are possessed of rational faculties and motivations to do good which can be reasoned with. But non-members are unaware that the ideology is not even and never will be the purpose of membership—membership is the point of membership—and the ideology is simply the gate which separates members from nonmembers, so arguing ideology in fact does nothing but *reinforce* the gate and bonds of membership within, further cementing the distinction between Them and Us which then further empowers the cult or group or leader rather than rescue or enlighten its members because of the underlying dynamics of human desires for belonging which motivate the behavior in the first place, in the very act of conflict reinforcing their perception that they do not belong with us.

As another example, a famous content creator once made a (very good) documentary on people who believe the Earth to be flat, in which both the believers of this obviously errant ideology and the documentarian engaged in debate about the merits of the ideology and evidence for or against it. Not knowing the ideology has nothing to do with their identity the content creator was mystified when finally providing irrefutable proof that the Earth is spheroid, with footage of the surface of a very large lake in which the water obscures the far shore since water always finds a level and thus seeing the very curvature of the Earth from its surface as irrefutable visual evidence. In spite of this irrefutable and obvious evidence of the falsity of their position he was told by the members to "just pray about it." Dumbfounded, the creator was at a loss of how to explain their intransigence (since nobody has known this concept), but still the admonition to "just pray about it" was also not a refutation of the creator's evidence but instead was an *invitation* to membership with the group, the price being a willingness to subvert rational thought and reject reality to prove the willingness required for membership. The error is to assume the point of such ideologies is to understand reality when in fact the purpose is surviving it, through bonds of association which in this case are determined by actions, not ideology, wherein the ideology is completely and utterly meaningless except in its function as separating members from nonmembers.

So even the act of debating, disproving, or criticizing the ideology in fact reinforces the system of dividing members from non-members by proving we are oppositely *unwilling* to participate, therefore making the status of membership *more* apparent and *more* defined for those within it and thus further entrenching them through dynamics of belonging and not belonging, and thus in spite of our intentions end up participating in the very system which separates them from us. This is why even when cult leaders die or beliefs of the conspiratorial are entirely defied by reality the group don't care in the slightest and simply make up new ideology and change their views accordingly, because all along it was merely a filter to separate members from the rest of the world, the specifics completely unimportant to that function, and was itself always immaterial and thus also malleable and changeable according to their needs and convenience. This is also why groups like this become more insular and more resistant to outside influences, as those influences attempt to rescue or antagonize them, because the entire function of the group and ideology is membership, and leaders instinctively tap into this human psychology because they too have the very same instinctual desire for fraternity and power to handle their own fears. No doubt there will be some people in the future whom attempt to use my work to do this very thing, to establish groups using my ideology and research as the litmus for their own insular and exclusive membership in which they can feel in control and exploit others for their own personal or financial gain at the expense of others (and will no doubt entirely misunderstand my work because the point won't be my work but instead using it for power and control).

Because the existence of extremist and fringe groups is rooted in our common desire for belonging and safety, the catalytic desire for belonging which motivates participation indicates the absence of belonging in the first place, and it is through the satisfaction of belonging and fulfilling of needs, not debate or ideology, that these dynamics are subverted and resolved. In this sense the term 'belonging' does not mean literal belonging, as many people who adopt these behaviors and join such groups already belong to families, friends, or other social organisations, but their experience is of *feeling* disconnected and dissociated in their need to belong, whether due to emotional trauma, fulfilling relationships, or the absence of effective life skills which can provide for these needs, or are in fear that membership is insecure due to social conflict and stress within families and communities. The stress of not belonging also needn't even be expressly targeted or specific such as occurred in my parent's religion toward people like myself who were gay, as it is simply the *absence* of emotional intimacy in families and social groups which stimulates this need, as a common consequence of unresolved trauma and its resultant control and coping mechanisms which can and do leave people feeling lost, afraid, bored, without purpose, conviction, or sense of identity that otherwise comes from strong social bonds, good education, and healthy self-expression which we are each capable of creating ourselves if we are taught or learn how.

Cults, religionists, and conspiracists are thus much like a metaphorical Newtonian fluid—the more you press them the harder and more entrenched they become, while simply ignoring them or refusing to engage their ideology is more productive at liberating people from destructive and isolating behavior than direct pressure. To address their *behavior* directly rather than ideology is more effective. For instance if a family member were becoming preoccupied by flat-Earth conspiracies the approach would be not to discuss evidence for or against that but to state that we see them withdrawing from the family, that we miss them being with us, and following up that expression of love with actual demonstration of it (because love is not a feeling, but action). Those with spiteful political beliefs should not be engaged on their ideology but instead the deficiencies they experience which are the root for their control desire—for instance the recognition that (as discussed in *The Perfect Child*) voters who live far from cities tend to be conservative while those in cities are more progressive humanists because those whom are not in proximity to power feel neglected and ignored and then use spite as a means to get attention and control, which is why direct engagement of the electorate is always the most effective approach for any political party or candidate to gain votes, since even the simple act of showing interest makes people feel included. Making individuals also aware of the *effects* of their behavior both on their own lives and that of others rather than shaming their ideology can actually induce shame, guilt, and remorse for the harm they cause (even if they remain outwardly defiant), because no person is ultimately motivated to harm others, only to gain control, where harm is secondary to that purpose, and thus made to become conscious of things they otherwise try to dissociate, such as the consequences of their actions.

Another example of the potential destruction which can result from not understanding these dynamics and how we play an active role in their genesis were some of the murder cults which arose in the 1980s in the United States in which charismatic figures led groups that actually killed people. These small, violent cults were not directly opposed by any specific outsiders until *after* they committed murder, but were instead formed and emboldened by the Us versus Them dynamic through the larger societal, fanatical obsession with Satanism and devil worship which irrationally gripped the country at the time, creating through its sheer inanity and fervor the requisite antecedent which gives membership in such a cult any meaning in the first place, whom were themselves already emotionally ostracized from the community at large because of hateful, religious obsession and persecution. In obsessing over supposed devil-worship those whom were religious and conservative in fact catalyzed the fanatical, extremist formation of violent cults in complete opposition to what they thought they were doing because of the Us versus Them divisions which result from such hateful behavior and in the process not only emboldened cults, but vindicated them (because they fully embraced the designation placed on them).

Since denying people's needs entrenches them into membership, helping them full those needs can do the opposite and liberate anyone from the shackles of fear and insecurity, to provide for ourselves without dependence on the group. This does not mean coddling the behavior of people who are clearly doing harm as might a Karpman 'Helper' type but instead recognizing what need it is which they are attempting to fulfill and helping them fulfill it in ways which are healthy, productive, and effective and not at the expense of others, and instead addressing behavior, not ideology, which directly addresses that which is substantive and not subjective to ideology. Those who are antisocial such as thieves, liars, rapists, violent, and bigoted did not design human biological psychology and are as much subjected to the mortal condition of humanity as anyone, and we can relieve humanity of their burden by empowering them with skills and knowledge to achieve needs and goals in ways which do not also cause harm, because the reality is that while many of our needs such as the acquisition of food, clothing, and shelter are dependent on others the entirety of our non-material needs such as validation, love, self-esteem, confidence, and even the acquisition of effective life skills and talents can be met through our own efforts without reliance on others or membership in groups or institutions which, although they can help, is not required and which can and should come from growth experiences as occurs simply in the act of caring for our needs (such as by doing inventory). Of course, recognizing this possibility is itself a skill that most of us do not possess unless we are lucky to have life experiences which enlightens us. Above all, the key to liberation is compassion, first for ourselves, then for others.

12. Morality

Growing up in a cultish religion our childhood was predominated by the isolation typical of such insular, fear-based societies, with our entire social network comprised only of members of our religion yet only ever socializing with them at church functions, and even family members that were also in the religion but not especially religious were avoided or marginalized, but being children it was all that we knew of life and the world.

Our family's religious ideology was indeed cult-like. For instance many people who are Jewish, Muslim, or Protestant Christian exist within the world without even considering the ideology or religion of others when they form friendships, relationships, or other associations, because a major part of their identity and worldview as a Jew, Muslim, or Christian comes from their *heritage*, not their religion, such as from family traditions, celebratory holidays, songs and art, clothing, and food culture. But cults and cult-like behavior oppositely control associations such as friends, marriage partners, and even family members in order to restrict and limit potential exposure to other worldviews and ideologies, in fear that exposure will liberate members from dependence on the group, and assumes it is a bad thing if someone does change their religious or ideological philosophy, when in reality a person is still a person regardless of what they believe or who they associate with.

Because of their emotional isolation but lacking skills for developing real relationships my parents replaced them with television personalities and a habit of consuming nightly local and national news but which is also sensational and exploitative and heavily populated toward geopolitical conflict, crime, violence, and other fuel for fear and anxiety loved by religious extremists. Most of my younger siblings during this time were very small and experienced regular nightmares and night terrors from exposure to bombings, shootings, war, and other themes that children should absolutely not be exposed to before their mid-teens. This so disturbed some of my siblings that they now in turn abuse, beat, and scream at their own children regularly and *purposefully* expose them the same horrors of life they were exposed to from a misguided sense that it is somehow normal. One of my siblings during the 2020 coronavirus pandemic sat down her children and told them that grandpa and grandma were going to die because of the pandemic, and all of them cried for most of the day and were entirely traumatized even though my parents did not even get serious symptoms when they caught it, let alone to go the hospital, due to my tireless efforts helping them be healthy, and my nieces and nephews will no doubt have similarly intense fears and obsessions concerning disease and death unless they are lucky enough to find healing in their adult lives (such as through my work).

When I was a teenager the entire country was obsessed by a scandal in the White House of a President whom had extramarital sexual relations with a young staffer, as if this wasn't a common occurrence, and while that president was absolutely amoral it was not because they had sex with someone whom they weren't married as the brouhaha was framed by the right-wing opposition but because it was an abuse of power, transgression of trust, and that President also did many other amoral and dishonorable things including the killing of foreigners to maintain American hegemony and the betrayal of the entire LGBTQ population in signing bigoted hate-legislation. But the controversy enflamed by the opposition party made a mockery of morality because at the same time they were led by a known pedophile (to political insiders) which, if that sounds incredible, was allowed to remain free from prosecution because the implicit threat of prosecution allowed them to control and manipulate him and as a result became the longest serving right-wing house Speaker in US history. The drama of the prosecution of the President for his indiscretion, not the deaths of people overseas or concern for the victims of the pedophile, dominated our religious community and family for a time because the President was a member of the party opposite my family and general community. When later the leader of their party was revealed to have been one of the most heinous pedophiles they didn't care one fucking bit and it was never brought up in any conversation and most people today remain ignorant to it.

Religious groups are often dominated by leaders which complain about the moral state of society, longing for days of old when there were supposedly better morals and the implication that society and people were generally better for it, but what many people do not realize is that even hundreds of years ago when the Pilgrims first came to North America and there was yet no such thing as televisions, cars, or radios the religious leaders of those times were saying the exact same thing, which is the very reason they came to North American in the first place, condemning the place from whence they came as sinful and wicked even though those people were also captured by extreme religious authoritarianism. In fact, the murder of people at Salem, Massachusetts for the supposed crime of witchcraft were of people living in those very communities and whom were also faithful members of their community's religion, and it seems that rather than any time in our history having any acceptable degree of morality for those concerned is that religious leaders are in fact never satisfied with any state of morality, ever, and that the entire thing is a charade meant to control and manipulate followers, which is then why many of them turn out to be the very vile and terrible offenses to humanity as murders, rapists, and child predators, because it never has anything in fact to do with morality but control and power.

This occurs because people who are angry and emotional are easy to control, and it is possible when whipping up followers into rage (whether something is true or not) to get them to do anything, even to abandon their own humanity and take the lives, property, or freedom of others. Both the leaders of such groups and their followers live lives almost entirely based in fear, and because we are in truth nothing more than an animal species on this planet an animal in fear will do anything to survive, even that which is the most amoral and detestable, because the point of religious ideology is not morality, but control of fear, so when morality fails to assuage their fear it is readily and easily abandoned such as most of my family's supposed religious standards in unquestioningly support of political officials which entirely transgressed their professed moral ideology simply because they gained through doing so more power and control, the very point of that ideology to begin with.

But this is not to say there is no such thing as morality. In fact, real morality is simply a direct function of the laws of cause and consequence and our ability to intuitively recognize that law—for instance if we take from someone there will be consequences, even if those consequences are simply to ruin the trust of that person or be forever after burdened with guilt and the shame of being a thief, but more often can be worse such as retribution or even imprisonment. Long ago the first religions attempted to codify morality in order to better organize people and society, naturally recognizing that social discord and transgressing laws of morality makes it difficult for people to live and prosper. The threat of divine retribution, or hell or damnation, was used as a tool of manipulation to motivate self-centered and violent human beings to engage in compliance with expected moral standards—but only those already adherent to morality are those affected by that kind of manipulation, and rather than being an effective tool it ends up creating *more* amoral behavior due to engendering of fear, hatred, resentment, and desperate competition for resources.

But morality is in fact born of biology and our nature as a social animal, because our entire wellbeing as a species is dependent on the survival of other members and we are also hardwired to instinctually recognize moral standards as we age and mature that we may both benefit from and participate in social cooperation to thus increase our chances of survival. Shortsighted philosophers and researchers often debate about whether our biology and natural design is more important in human development and individual morality or whether it is society itself and the rearing of children with principles and knowledge. But both positions are exceptionally myopic and misunderstand that biology and knowledge are the same thing, and morality is inherent to biology if it is not corrupted by experiences of trauma to then trigger basic survival instincts. We also cannot know morality without a mind and body which understands language and human culture, of which we are also biologically wired to do, and in my other works I discuss how when a person takes too many substances which alter their biological function they often stop acting like a person at all, the mere existence of this phenomenon as well as the pantheon of psychological diseases of which we can be possessed demonstrate that biology is both necessary for and actively involved in the formation of individual and collective concepts of morality and social cause and consequence. After all a person cannot be moral if they do not even exist in the first place.

As human beings we are also entirely a product of evolutionary biology, and even the ability to think that we are not is just as much derived from our common ancestors as our lack of fur, or hands and nails in place of paws and claws, instinctually driven to cooperate and associate as a function of our literal evolutionary biology, as there is nothing we are that is not rooted in our evolutionary biology. We would not survive as a species if we all had instincts to kill each other, so there must also be instincts for moral behavior, which are built on and informed by the immutable, natural law of causality in that sharing promotes the whole while selfish myopathy only promotes the individual, and the species itself cannot survive on the individual alone as we are not an organism which reproduces by budding.

The role of nurture in our development, therefore, is more about not *causing* harm to children which impairs those natural instincts, since children are in fact inherently possessed of instincts to participate, learn, and cooperate. This can be more complex than might be readily apparent, such as that those of the Karpman 'helper' and 'persecutor' type will always claim that being too easy on children can predispose them to laziness or selfishness, implying that allowing children to remain undisciplined is wrong, which in truth is designed not to empower children but to incapacitate them and predispose them to indentured dependence on parents for the benefit of the adult at the expense of the child by depriving them of life skills they require to otherwise live successfully. Every child naturally demonstrates instinctual desires to assist their parents in tasks that we, as adults, often take for granted as boring or inconvenient such as doing the dishes or folding laundry, then (as discussed in my book on psychology) dissuade them from helping since they are not capable of doing it right which not only indoctrinates them to believe that cooperation and self-care behaviors are abhorrent but that doing things perfect is also a control behavior which can be conveniently used to manipulate and control others.

This is also why rich people often give to charity, because of the recognition of the disparity between their grossly accumulated wealth and the suffering of those without means causes them to feel guilt and shame, but offering to charity is a way in which they can delude themselves and others as being part of the solution when in fact it merely serves to enrich the charity industry while maintaining systems of dependence by not actually liberating the victims requiring charity from the material conditions which cause their dependence in the first place. If such people were truly concerned with the wellbeing or suffering of others they would instead not chase wealth in the first place and would use their resources to alleviate the material conditions of the people from whom they otherwise take, to build quality and affordable housing, pay good wages, preserve the environment, prevent toxic contamination of the water and air, produce not only sufficient food for all persons but healthy and high-quality food, and facilitate access to effective medical care, education, and opportunity. Because all wealth relies on the exploitation of others rich people can *never* solve fundamental problems of inequity because they in fact rely on those problems to become rich in the first place, and thus then erect charities which do nothing but simply keep the working poor alive and provide cover for the truly amoral behavior which causes these problems in the first place.

Politicians are likewise so commonly corrupt and amoral because they are nearly always a textbook case of believing 'two wrongs *do* make a right' in complete opposition to the common and cherished idiom 'two wrongs do *not* make right.' This occurs because most people who get into politics are motivated not by a desire to do what is right but the fear that others will not, and so their desire is also one of control which then leads them headlong into corrupt and amoral behavior, scheming and plotting and working in secret as they in turn try to wrest control away from those they fear doing the same. Once achieving office politicians are also actively engaged a sociopolitical conflict against literal adversaries (although they needn't be), and losing office would not only mean they are out of a job but would also lose the political and institutional control they believe they have (I say believe, because no one ever really has control). So even those who believe they are the most moral are often in fact simply attempting to justify behavior which is entirely amoral but which they feel justified in doing because their opponents are, they believe, as well, only to find later in sorry, quiet moments after losing power and facing the end of their lives to have lost their very humanity along with it for the things they did while possessed of power.

Racists are in reality preoccupied by a fear of competition, instinctually believing in their fear of others and potential threats to our wellbeing since human beings are naturally competitive, and seek to neutralize that threat by aligning with those they believe have shared culture, language, and worldviews and use physical appearance and differences in culture, language, and worldviews to identify the Us and Them. But the concept of race is entirely subjective, because there is in fact no such thing as race (but yet why there have been countless "scientists" who attempted and failed to establish race which, even if there were, would still be no excuse for racist behavior). Those who exploit racism and racists for power do not care about race, but about power, and then arbitrarily use the false premise of race to divide people and maintain control, and because there is no such thing as race then often ironically (ironic for the supporters) end up victimizing even their own supporters who become victims of their own racism and willingness to hurt others being turned upon their own heads in the most poetic iteration of causality. While there are many vile human behaviors racism is an amalgamation of many, such as dishonesty, fear, selfishness, covetousness, violence, laziness, incuriousness, being judgmental, bias, hatred, self-pity, excessive seriousness, faithlessness, etc., which is why it is especially revolting, and since the bigotry and fear that fuels racism is arbitrary anyone at all is in danger from a racist, not only their apparent targets.

The fact that morality originates from biology, not ideology, is also likewise the reason many aging people get less moral as they age, rather than more, because aging also brings an increase in the intensity and frequency of stress and stress hormones and neurochemicals which in fact mediate the fear response, as fear and insecurity are always the underlying motivation for amoral behavior. While young and healthy our endocrine system is more or less in balance and thus when we encounter fear stimuli we do not feel so incapacitated by it, but growing older and experiencing less vitality and resilience and higher chronic stress and colonization by opportunistic microbes stress then has a much greater effect on our mind and body, and if we have spent much of our life in delusional dissociation that delusion in turn becomes more exaggerated by mortality as it takes us. Because phobias and mental illness are also caused by dysregulation of endocrine hormones and neurochemicals those also with phobias literally experience greater terror and anxiety as they age because of the biochemical nature of the fear response and metabolic deterioration of the endocrine system and nervous system which imbalances the endocrine system in favor of stress hormones, and thus greatly increased feelings of anger, shame, guilt, hatred, revulsion, and fear which in turn motivate greater antisocial behavior than those who are young and healthy.

Of course, many old persons become *more* moral as we age and in this case it is not due to biology but less stimulation of the fear response itself due to life skills and experiences which empower us to feel confidence rather than fearful or insecure when confronted by fear stimuli, which then mitigates or eliminates the fear response that would (and does) otherwise occur in those whom lack such empowerment. This is also why taking care of the body and our needs is so effective in resolving the fear response, because not only do we improve the physical body and health of our endocrine and neurological systems but we also learn that we are also capable of caring for our needs and not so dependent on outside factors which thus reduces or even eliminates the fear response in the first place, knowing we can effectively meet any challenge that comes our way.

In the United States Senate the record for the longest filibuster ever (which is a process that delays enacting of laws) was by a notorious old racist in defiance of anti-segregation legislation, and the fact that someone so amoral in an act so heinous would hold this record appears as a blight on human nature, to commit themselves so tremendously to personal inconvenience for the purpose of nothing more than harming others. But in fact such behavior underscores just how much control fear has over human behavior and how fearful the politician was, as fear is the most motivating of all human emotions, and can even motivate parents to murder their own children. Being as aged, delusional, and absent of wisdom such as what motivates such fear in the first place his response to his intense and overpowering fear of the Other (in this case, black people) was to attempt wrestling control for himself from the reality around him, to then completely debase his reputation (not that he had any anyway) and spend an inconvenient amount of time and energy doing something completely unproductive, unhelpful, unnecessary, and undignified at a time when more moral people were otherwise enjoying their lives, helping and loving others, and working for a more perfect world.

In fact, most people's motivation to enter politics in the first place is the very fear and insecurity which motivates amoral behavior, because contented and wise people do not desire power and control, which is why the majority of politicians are not only vile and reprehensible but lacking in humor and good nature, so politics is forever burdened by the most loathsome cretins if contented, wise, and humble people do not otherwise step up to the responsibility of leadership and instead allow it to be subsumed by the power-hungry, insecure, and despotic. When an entire lifetime is then spent in service of fear and the indulging of control impulses the only possible outcome is an amoral life, which becomes more amoral as we age because of the imminent spectre of death and an increasingly dysfunctional endocrine system which serves only to increase fear rather than assuage it, dispossessed of life skills required to endure such stress if we have not beforehand learned them and resolved past experiences of trauma.

Similarly we are often beset by those with ulterior motives in their control of their environment such as those which harbor racist, bigoted, or authoritarian worldviews. Many of us make the mistake of engaging such persons in good-faith argument and dialogue seeking to be helpful but therein fall into the trap set by this control mechanism, which is that the bigot attains control through attention and emotional manipulation of others and thus receives the very power and control they seek through the sheer depravity of their position and actions in turn fulfilled by the very act of centering their worldview at all in interactions with them which is the very *point* of their antagonism, not the worldview itself which, like the ideology of cults, is completely insubstantial to the purpose of the debate in the first place. As while we may regard these as invalid worldviews the worldview serves as a distraction from their real motives which is to use the worldview as a mechanism for control, so the more outrageous or heinous the bigger the reaction and thus the more control is achieved. Even engaging in the debate therefore grants the narcissist and bigot the very thing they were seeking, which is to affect the world and people around them, as an antidote to the potent fear and insecurity they feel toward life and inherent lack of control as a mortal human animal.

The solution to this is thus not to justify the humanity of others and defense of morality, which needs no justification or defense since it is inherent to reality and biology, but to draw attention to the inherent weakness and emotional turpitude of the person behaving in such a way in the first place, because this behavior inherently originates from psychological and emotional weakness and insecurity of their position in the world which, when laid bare by pointing it out instead of addressing the ideology which oppositely gives them control, makes them feel especially vulnerable in complete opposition to their goals and thus more likely to abandon their destructive and anti-social behavior. Larger populations of people can similarly be rehabilitated from such violent and mean worldviews by providing for their wellbeing, material conditions, and empowerment to self-care which thusly eradicates the fear and insecurity which stimulates such desires for control in the first place, to help people realize there is no real reason to be afraid and that we are each of us more capable of providing for our needs that we might realize, dependence on others for those needs ironically being one of the mistaken traumas which motivates their behavior.

The primary solution most people intuitively seek to cope with such dilemmas which cause fear and anxiety such as concerning death and mortality is more *time*, to delusionally put off that which is inevitable, to procrastinate judgement and, like the amoral Spanish Conquistadors in search of the Fountain of Youth, desperately chase anti-aging schemes, dieting behaviors, excessive exercise, money, harmful pharmaceutical use, and even surgery rather than fix what we have broken, care for what we have, and learn how to truly live with the time that is given us, and in the process end up ironically hastening both moral and physical demise as the body becomes destroyed by such abuse and the mind insane, addled by a lifetime trying unsuccessfully to bend fate and the will of God. But those instead which accept reality and live better within the bounds of morality as what is even apparently evident in the nature of causality in our behavior within human society instead relents control of the body and fate to the forces which actually govern them, and thus relieves us not only of a great and impossible responsibility but also the price of time, energy, and humanity such behavior otherwise abstracts.

Similarly, much effort has been made to indoctrinate people against antisocial and amoral behavior and crimes, such as racist violence or political violence such as through education about world events as the American Revolution, the Civil War, slavery, or the World Wars and the Holocaust. Yet one reality of life that many are not aware is that things we think happened a long time ago did not, in fact, happen a long time ago at all—for instance, if the oldest people in any generation live to see 100 years, end to end there have not been even three full generations since the founding of the United States. Geronimo, the famous Native American resistance fighter, was still alive when Bob Hope, the comedian and talk-show host famous in my grandparents' generation, was born. Harriet Tubman, famous emancipator of people enslaved in the United States died only the year before the start of World War 1.

When I was a child (around six or seven) before I knew how cameras worked and saw old movies or TV shows in black and white I literally thought the world only existed in black and white and only in our modern life was everything in color. Seeing old footage of World War I and World War II in black and white gives them a sense of ancientness that is unfamiliar to our contemporary senti- ments, but the fact that they are filmed *at all* is a testament to their recentness, as it's only been a little more than 100 years since both the invention of film and the beginning of World War I, and at the time of this book we still have fifteen years to go before the 100 year anniversary of the *start* of World War II. By any measure, if there are still people alive today which witnessed something it was not long ago at all, and there are not only still people alive who were born during World War II but who actually served in the war! If not even a full generation has passed, the events are not part of history but are still, in fact, contemporary.

The World Wars were also only separated by about twenty-one years, and children born at the end of the first war were recruited to fight and die in the second just as they reached drinking age, and since both wars had tightly related causes I am of the opinion they were not at all separate conflicts but one single conflict bisected by a short pause, and the conflict was a major catalyst for the following dysfunction of generations since and massive unresolved trauma which set the stage for our current state of affairs as discussed in *The Perfect Child*. Growing up we were then taught about such world conflicts and the atrocities which occurred as if they were history, ostensibly to prepare new generations to avoid such awful tragedies, even while those whom perpetuated and participated it were still alive in Germany, America, and Argentina and even as our own country actively and continually committed genocide after genocide and conflict after conflict even to the present day and publishing of this book, distracting from the fact that those things are not history but contemporary, and that all mankind is capable of committing such heinous acts.

Authors of policies and ideology do not understand that it was never and has never been ideology which drives such conflict, which is why all their efforts to abate racism, geopolitical conflict, and genocide over the intervening years has utterly failed but, like the circumstances which precipitated the World Wars, is instead caused by the material conditions of people whom then become desperate for relief from their constant struggle and use racism or nationalism simply as an excuse, and since much effort has been undertaken to suppress and exploit the working class we yet again find ourselves in the very same condition in spite of endless propaganda and indoctrination. No person on Earth wants to engage in conflict as a primary means of resolving problems, so natural is it for humans to loathe conflict and confrontation, so the stakes must have less or comparable risk than what will occur without that conflict in the first place, and when people are deprived, impoverished, and disenfranchised there isn't much to lose.

A practical and very specific example of this problem is the entire anti-smoking campaign in which the United States government spent literal billions of dollars in taxes derived from tobacco companies attempting to brainwash people that smoking is bad for them instead of using that money to relieve people of the stresses and nutritional factors which lead to dependence on smoking. Indeed I don't think the government is even interested in solving addiction (I ended up doing it instead), and so when alternative products like e-cigarettes and nicotine gums came on the market their use exploded, even among children, and the government didn't accomplish shit. The failure to resolve nicotine addiction occurred because the material conditions of people who smoke never changed, not because they weren't indoctrinated enough—Well, that's not entirely true, because there *was* in fact a major decline in smoking from about 1995 to the current day, and what happened in 1995? It wasn't an especially clever anti-smoking campaign (the D.A.R.E. program famously worsened drug addiction), but the advent of the fucking *internet*, which served to connect people to each other, and because social inclusion is a major stress relief for humans we see an enormous decline in stress associated behaviors like smoking, crime, teen pregnancy, etc., since the advent of the internet, not anti-smoking campaigns, because it helped alleviate some material stresses by providing access to social inclusion, information, opportunity, etc., which are otherwise impaired in a car-centric, capitalist economy and lack of third-spaces.

The far majority of human beings are good people and desire nothing more than social cohesion, peace, and stability, and refusing to facilitate that will always, always result in cataclysmic civic upheaval no matter the time, place, or ideological indoctrination, as has throughout the entirety of human history, because no person can eat ideas or sleep with principles, and human animals are designed by nature to seek relief when in pain, resources when we are wanting, and spite when we are harmed. Morality is thus built into biology because without it we would not survive as a species, requiring cooperation and empathy as a tool to our very survival and not only those which do harm. Morality is therefore a product of nature and a law of the natural world, as a function of causality, which is also why it cannot be transgressed without consequences, which are also the domain of nature, and without these tools we thusly lose relationships, security, and resources like integrity, trust, cohesion, cooperation, and wisdom we require to survive, worst of all the enmity of others we may win in doing harm and thus invite our own destruction.

But because biology is achieved through hormones, neurology, and biochemistry we can and do also lose these characteristics even through no fault of our own. For instance many people treat narcissists as if narcissistic people *chose* to be the way they are when in fact no human has control over our biology nor even the characteristics of human evolutionary biology such as what causes the condition of narcissism as a result of childhood abuse and neglect. That does not mean narcissists should be excused of their *behavior*, only that they did not cause the condition in which they find themselves. A psychologist I know frequently uses their position to further stigmatize narcissists and blame them for their condition rather than empathically providing insight and solutions to resolution of the condition (to be fair, nobody but myself actually had a solution for it until now as discussed in my other works). All people are responsible for our choices and decisions no matter how we may feel or think, but feeling or thinking the way we do which motivates both good and bad behavior is instead the domain of factors like dopamine, serotonin, oxytocin, progesterone, etc. Progesterone for instance is the hormone of *empathy*, which is why women have heightened emotional experiences during their estrus cycle and pregnancy, but which can also simply be experienced by supplementing high doses of (natural) progesterone, and when helping middle-aged or older women recover their health (or any men who use it too) there is also a return of intense feelings due to the resurgence in progesterone, which many of them understandably resent since life is often a lot easier without feelings. But the desire to avoid feelings is also a control behavior borne entirely from the absence of effective skills and self-care tools, and the best state of existence I have ever achieved is both *having* feelings and the knowledge and tools to understand them, not control them. Denying morality is created by biology means there is never any cure for those who suffer psychopathy and sociopathy, but recognizing instead that progesterone functions in part by providing emotional feedback to us that we may better feel our experiences and better know what moral behavior is means that impairment to our endocrine function and the production and regulation of hormones and neurotransmitters by diet, trauma, or opportunistic microbes, can recover emotions for those with severe emotional apathy as sociopaths and psychopaths.

Keeping in mind always the separation between accountability and biology so that Karpman types cannot opportunistically sustain conflict by duplicitously aiding those which are amoral, recovering good dietary behaviors and addressing disease and pathogens or even simply the supplementation of important hormones like progesterone thus can help direct and promote morality though our natural human instincts for cooperation and harmony. These hormones and feelings do nothing to alter the past or amend harmful behavior, however, which instead requires action and changes in behavior such as in taking responsibility for our actions and rectifying that harm as much as possible without concern for our own benefit, with care not to cause further harm in that process. Removing either choice or biology is the domain of the Karpman dysfunction types and the various control and coping mechanisms each of us may use to survive our own experiences of trauma, but both choice and biology together make morality and which can thus be integrated to better understand and achieve morality in both a personal and societal context.

Amoral behavior occurs at all to give the offender a sense of control, and we intuitively understand its effects which is why we often do things which are not right, because we feel justified in doing them because of the control and power they in turn make us feel without even recognizing that very feeling IS of control and power. This is also not relevant only to obviously amoral behavior as murder and theft but even in small, everyday offenses we cause others which are meant to assert our power over them. For instance not showing up on time to arrangements, appointments, and meetings, because doing that is the act of surrendering control of the relationship with others to whom we have committed our time, in the act of being responsible in fact show them that we do not mean to assert control over the relationship and regard them and their time and convenience equally to ours, while refusing to do that instead communicates that we have the power and are unwilling to relent control. Controlling people refuse to be on time, even to the point that it is absurd and transparent such as an aunt of mine notorious for showing up an hour or more late to every single fucking family function we had growing up and famously had explosive conflicts full of hysterical drama with her husband and children whom demanded I not speak to her gay son when he came out.

Many times we feel that problems like being punctual are unintentional, but though they often are unconscious they are a function of the subconscious ego which asserts power and control by failing to take into consideration the feelings and wellbeing of others as equal to that of our own. Doctors often famously require their patients to set appointments then do not even show up for an hour or more, requiring patients to sit and wait to be seen which has nothing to do with doctoring but instead is a control behavior meant to impress upon the patients that the doctor is more important than you are. One of my family members once refused to take me up on my offer to help coach them with their stress and psychological problems as I had done another quite successfully simply because of the requirement they show up on time, because doing that would mean giving up a control behavior which otherwise makes them feel powerful through abuse and neglect such as by *not* taking responsibility for our behavior concerning others.

Other very common, usual, amoral control behaviors which are highly destructive (to ourselves, not to others, which is the entire point, that we are self-destructive and destroy relationships and opportunities that would otherwise be of *benefit* to us!) are things like not paying attention when speaking to someone we know personally (especially on the phone!), getting mad at someone for not paying attention, not taking the time to call friends and family members, not maintaining other relationships when we get into a romantic partnership, frequently talking over others, getting angry when someone talks over us, exploiting people's mistakes as opportunity to embarrass, shame, or lecture them, dominating conversations, refusing to share our personal thoughts or experiences with close relationships, quietly resenting people for not doing things we expect of them (like taking out the trash), frequently rescheduling on people, only calling people we know when we are bored (such as in the car on the way home from work), frequently returning gifts for something else (just ask people to stop giving them), being upset when someone doesn't like a gift or the food you made (because it was all about you anyway), not apologizing when we make mistakes or inconvenience others, sharing hurtful information with someone who doesn't need to know about it, fighting over the check, lecturing people when they ask for help (you can just say no if that's your answer), lecturing people when they offer help (you can just say 'thank you'), doing chores or work instead of being with guests and visitors, talking politics at social events, nagging partners and family members, evangelizing, triangulation (sharing grievances with a third party knowing it will be communicated back to the subject), refusing to ask for help but leading others to offer it, and generally every heated argument we ever get into with a partner (the point is to feel in control, not to solve problems, which is why they never get solved).

When we have deranged endocrine systems and stressed by disease and abuse of the body it can be very difficult to be moral because our instincts then are those of self-preservation which takes precedent to that of the group, since we cannot be useful if we are sick, starving, or dead, but many people's instincts for self preservation are turned on indefinitely by unresolved childhood trauma and systemic metabolic disease, and instinctually chase money and other antidotes to the things they fear, but are nonetheless required by causality to behave morally in every aspect of our lives. Even religious people who attempted to codify moral-ity also sought primarily to assuage their own fears and this is why religion has failed to achieve even its own goals because acting on fear is always insufficient to acting on wisdom, fear being the instinctual coping behavior of all animals which does nothing more than function to spare us from death as long as possi-ble in this mortal life. Morality cannot be based on fear, because then it seeks to control others and parameters of morality, which is itself amoral and which we cannot do since we are no less mortal than any other creature on this planet, but when morality is based on wisdom and knowledge it can and does then function to control the only thing we can which is our own behavior (not even our own emotions!). One of the most formative experiences of my childhood was from a variation on a famous quote by Mary Ann Evans (a.k.a. George Eliot) placed by my mother in the front of the scrapbook which kept my baby pictures and documents which read, *"For what purpose are we here if not to make life less difficult for each other?"* which more than any moralizing sayings deftly describes not only the definition of morality but the very function of human evolutionary biology—that it is not enough only to refrain from causing harm to others but that we must also be engaged in their support, because none of us is truly capable to provide all our needs alone and must do for others as we expect and require in turn.

Similarly, as discussed in my book on psychology, *tolerance* has no paradox-ical function and is in fact a social contract which, when transgressed by one party then nullifies that contract, and being moral does not mean standing by as deranged and terrible people harm others, which is in fact also an amoral position and the *facade* of tolerance used as a mask for indifference or even complicity by equally hateful and destructive people that wish instead to be spared the humiliation of doing the harm they truly desire by allowing others to do it for them (the Karpman 'helper' type). Making life less difficult for others means *actively* standing up against those who would do harm, but very often we shortsighted humans feel so justified in our moral anger and outrage that we also in turn do things that are, objectively, also immortal to also transgress the age old idiom, *'two wrongs do not make a right.'* For purposes of clarity I should explain this as meaning that when we witness or expect others doing something amoral we then in turn also do something amoral believing we are excused of the consequences of our behavior by the actions of the other person which justified it.

This is an especially perilous moral fallacy which endangers the integrity, humanity, and even freedom of many ostensibly well-meaning people as we in turn try to control those we fear being controlled by through in fact doing the very thing we fear of them. Failure to understand this simple application of cause and consequence then leads many supposedly well-intentioned people to do things which actually harm themselves, others, or their intended cause, for instance during some elections when people believe the other party is cheating will in turn cheat as well, in order to make up for the supposed harm or wrong that is occurring, but very often in contemporary societies the actual rate of voter fraud is so marginal as to immaterially affect the outcome at all, and many politicians will lie and claim fraud even when there clearly is none so the people who end up breaking laws because they believe others are breaking them will still have broken the law and often end up being prosecuted and fined or even jailed for it.

In the investigation of crime many law enforcement officers, prosecutors, and politicians also feel absolutely convinced that someone (or a group of someones) is guilty of crime and very famously law enforcement routinely fabricates evidence because they "know" the person is guilty, and when in turn their subversive deeds are found out ruins the prosecution and guarantees the freedom of the person they thought guilty or, worse, to in fact condemn someone in fact not guilty as we might have initially thought and thus rob innocent people of their person and property, and though the persons involved thought they were justified in their amoral fabrication of evidence end up achieving the exact opposite ends because of their transgression of one of the most simple and long-lived truths known by all that two wrongs *do not* fucking make it right.

In fact, there is never any moral justification for being amoral in response to the amorality of others, and this is simply a control and coping construction intended for us to feel power and control, justified in indulging our desires for control at times when we may in fact feel powerless or frightened of losing that power, and it fails because it is every bit as amoral and illogical as those whom we are opposed, a survival mechanism meant to achieve nothing more than our own personal security, not justice or anything that is right or good.

Very famously the United States dropped nuclear bombs on the cities of Hiroshima and Nagasaki at the end of World War II, and what most people are not aware (especially in the U.S.) is that the end of the War and Japanese surrender had *already been secured* when the order was given to bomb the cities and kill their inhabitants. This heinous act was committed not to end the war as we are often told but to sacrifice the innocent on the altar of U.S. power and dominance, to threaten the rest of the world of what we were capable should they not do as we demanded, justifying our desire for power and control as a pretext for preventing war, which we did not even do afterward in the repeated and frequent incitement of conflict and slaughter of peoples in Vietnam, Korea, Chile, Cuba, Haiti, Puerto Rico, etc., committing in fact the transgression of two wrongs do not make a right and our leaders condemning the United States to yet more blood upon its crimson history and the shame of its people who cry out for peace and good will.

There are plenty of ways to resist those who intend to cause harm that is not also amoral in turn. Simply vocalizing opposition and exposing their deeds is often more than sufficient because there is nothing an evil person hates more than being recognized for what they are. But often the amoral route is the more controlling route, attractive to those with unresolved trauma who desire control, not morality, motivated by fear, not wisdom. Considering that all amoral behavior is rooted in delusion and trauma it is also often possible to outmaneuver those whom are amoral when we understand the origins of their motivations and the underlying human evolutionary biological psychology which constructs the human animal and its driving psychological factors, to be more *effective* in our behavior than what occurs when we are not enlightened, and help fearful and insecure people find the security they desire and skillset to themselves be more effective before they feel limited and desperate enough to resort to amoral means.

Indeed this is why all amoral people and movements ultimately fail, because such behavior itself originates from ignorance to laws of cause and consequence and the nature both of humans and life, and thus are not even empowered to achieve their intended ends in the first place because success requires effective understanding of reality, and we cannot be effective when we are uniformed, delusional, and willful. Amorality is born in the absence of knowledge, and because delusion is often a coping tool used by those who behave amorally they are not even *capable* of effective action and often end up causing their own demise simply through their own behavior, which can be accelerated in turn by remaining strictly moral. The tiny, old Mohandas Karamchand Gandhi liberated all of India from the powerful and ruthless British Empire with nothing but his person and a shawl and loincloth, such is the force of morality against immorality.

Because morality is based on the fundamental and immutable laws of cause and consequence morality is therefore built into reality itself, as nothing in reality exists without the express function of reality, where morality is simply the laws of cause and consequence applied in metaphysical existence we observe through our personal, human experience. While morality is good, productive, and of self-interest to be moral it is not only because moral behavior is good that it is good but because it is inherently productive, useful, and rational, not only to us but to others and the world around us as well, in harmony and cooperation with the laws of causality rather than their opposition (which we cannot do anyway).

The Zodiac

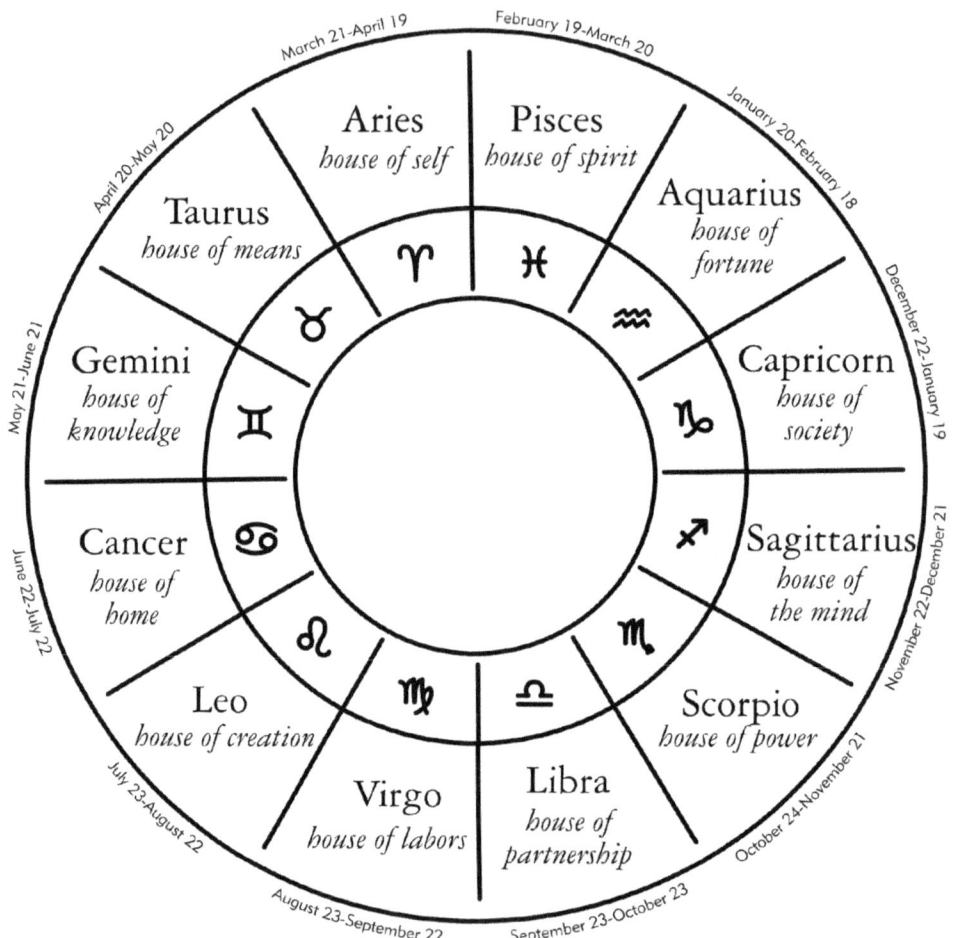

13. Rational Astrology

It is plainly observable that we do not control the Universe. But when it comes to understanding how the Universe operates most people are limited either by superstitious ideology or scientific myopathy and fail to recognize the plainly obvious truth that *we do not fucking control the Universe*, which, yes, also includes life's content as such seemingly mundane behaviors as choosing what shirt to wear for the day or more important things like whether we land that job or when and where we meet the love of our life.

Recognizing that we do not control when we are born nor the time and place of our death (except in suicide) at the end of our existence, and also the years of dependency in which we are too undeveloped or old and invalid in the book-ends of life, or the life of others which affect ours in ways that we are not even aware such as the city workers who keep the lights on and the trash moving, or the immune resistance which allows us to even live which has been passed down not only by our human ancestors but those which were not human, or that we constantly breathe air from a thin and improbable layer of atmosphere protected by a magnetic field surrounding our planet which would otherwise have long ago blown away in solar wind, most of us naively insist we in fact control our lives inbetween those very specific exceptions as if by virtue of having working legs, a mouth, and a brain (never mind that those too are a product of millions of years of evolution) we suddenly have control over the very composition of reality and the laws of causality.

This is obviously nothing more than a delusional function of human evolutionary psychological biology in which our psychology, which is entirely contrived from our biological evolution as a human animal and every bit as much a product of our DNA as any other part of our body, deigns to wield arrogant dominance over laws of nature in harmony with our evolutionary functions of group survival and codependence and instinctual control of nature and systems to conveniently forget that we are also a biological animal with a fleshy brain and DNA and human instincts and limitations, not to even mention the thousands of invisible types of microbes which can in fact alter the function of our digestive system, endocrine system, nervous system, and brain which govern our biological response to life and stimuli, even our ability to learn and act and form the mind as a function of the brain. Obviously, someone who has been in a terrible accident and lost control of their motor skills has very limited ability to act, and are they too in control of their lives? If not, are we who retain the entirety of our biological functions any less subject to the laws of the Universe?

The only possible answer to this question is that of course we are all subject to the laws of reality and do not exist apart from them. But what decides when, how, and what happens to us in the course of our existence? It certainly is not us, and the laws of the Universe do not stop working the moment we decide to steal someone's wallet or help an unsuspecting stranger, or choose to eat cereal for breakfast or spontaneously bump into someone who will then become the love of our life whom only exists because of the structure of reality and not because we desire them (even the desire is also a product of reality and not our wiles).

Most people's conception of reality is that its laws build the Earth, the Sun and Moon, the tides and DNA and evolution and our physical body which can walk upright and the hands that give or take, but the moment we decide to use these gifts the laws of reality break down entirely and have absolutely nothing to do with our experiences, our relationships, personal choices, unexpected surprises, or those patterns and developments in greater society and the history of man which then are all entirely left up to chance for us to do as we wish. Man's ego is truly ludicrous, and serves to comfort us in the face of febrile helplessness and forces we do not understand in a world that is often frightening and merciless, even in spite of our great technological advancements, when our very lives can be stamped out before we have even done anything worthwhile or had any fulfilling experiences simply because other monkey humans want your land for their real estate holdings or don't like you because you believe in a different religious ideology.

In fact, this delusion of human belief that our lives are our own and not subject to underlying laws of reality just as real as those which create gravity and molecules is borne entirely of fear and our desire to dominate life in anticipation of death that we may by our own hand avoid it. But most fishes born on this earth never see adulthood, never get the chance to mate, and many young of other species like baby birds and baby rabbits mostly serve as feed for other animals or succumb to disease as what also used to be the plight of human children before Alexander Fleming discovered penicillin. Livestock raised by humans are mostly alive for the sole purpose of being our food, but the prospect of a similar, sorry fate for ourselves fills us with dread, not because we are special but because fear is a universal survival instinct mechanism designed by nature to keep creatures alive.

Pain too is a survival tool, as demonstrated by human children born without a pain reflex which then chew off their own tongues, or those so high or drunk they do not feel pain or fear and meet fatal injury. But, being deluded to our nature we then concoct fantastic delusions and excuses such as divine ordination because our fear is supposed to be more special than animal fear, and so dominate and destroy everything on the planet, including ourselves, in our quest to be rid of it.

But all aspects of existence are governed by immutable and definite laws of the Universe, including every aspect of our waking existence, as instead of the supposed chaos and disorder often attributed to the biggest of intergalactic structures and the tiniest of atoms, all things are governed by laws of physics and reality and not simply left to random chaos and disorder. If this were not the case we would not be here, because laws give the Universe structure without which there would be none, and without structure there would be no galaxies, stars, or planets upon which life can exist, and a constant law of the Universe is the balance of creation and destruction which cause starts to explode and create the elements from which we are so spontaneously organized.

Discovery of amino acids on asteroids should have put to rest the debate of whether life is a constant law of the Universe, yet it is clear from our own existence that it is because without consistent laws of reality we would not be here, and those laws apply to the entirety of existence, not just our tiny pocket of it.

It is very clear that life organizes itself due to principles like the strong and weak atomic forces, but the same forces which make amino acids or construct solar systems and black holes also govern when and with whom we have children and who those children turn out to be and even things as small as deciding what to have for dinner or whether we even have anything for to eat for dinner in the first place. The absence of scientific explanation for that which is existential and meta-physical is simply an *absence* of scientific progress and unwillingness of research to investigate taboo subjects, not the absence of evidence itself, of the laws which so clearly organize who and what we are and when and how things happen on this planet throughout the entirety and fullness of reality.

Though science has not yet defined these laws they are in fact observable in our daily lives, through the patterns of existence we can see all around us. One beautiful pattern often seen in nature are *fractals,* which are repeating mathematic models that helps reality design complex structures too complex to individually code into DNA. For instance a tree cannot code genetics for every single separate stem and leaf—its DNA chain would simply be too long to fit into every cell in its organism—so instead the tree uses fundamental principles of mathematical fractals to automatically generate stems and leaves where they should go and in what form. The vasculature which also transports nutrients throughout not only trees and other plants but also our bodies, and the formation of bark and roots and inflorescence (flowers), and the genesis of fruit and seeds to disperse for propaga-tion are all based on fractal mathematical models.

Fractals are also not only visible in biology, but also appear in the waves and ripples of the ocean, the rocks of the Earth, and the patterns made by flowing water as is erodes mountains and valleys because fractals are a fundamental product of the mathematical physics which cause order and structure to the Universe. By their nature, fractals are also infinite, and graphical simulations of fractal equations repeat forever, indefinitely, no matter the scale at which it is observed, but this is not only a function of mathematical models but those models reflect reality and describe the nature of fractals in physics, but as such they are convenient formulae for biology to form the bloodstream of humans, the shell of a sea snail, or towering mountains.

But fractals are not also limited to the physical construction of reality, and fractal mathematics are also observable in *time*, throughout the formation, duration, and destruction of planets, solar systems, moons, asteroids, galaxies, life, and events, and govern the ebb and flow of when and why things occur, which causes patterned but organic clustering of occurrences (or their absence) rather than evenly distributed, linear, and dilute time which, if it weren't, would be impossible for anything to exist at all in the first place since the dynamic forces of creation and destruction are a result of the bunching and clustering of spacetime and thus physical fractals we see in the physical world must also be properties of time being that space and time are one and the same. Because of the work of Albert Einstein but also science fiction, which has popularized the concept, it is well known that venturing out into space where there is little mass results in faster relative time than here on the Earth where mass oppositely decreases the relative rate of time in proportion to mass as a percentage of lightspeed acceleration as discussed in the earlier chapter on time. Time is experienced entirely different in the massive bodies of black holes relative to our position here on Earth due to the effect of their sheer mass on spacetime, but if these forces which govern the Universe affect something seemingly so fundamental as time, and time is one of the greatest factors in events and developments in our life and reality as humans, how are the universal forces of the Universe not in charge of when we experience things in our lives from the very inconsequential to the most profound, and our psyche and experience not a result of and subject to these forces, and our delusional perception of control simply part of the evolutionary human psyche meant to promote our survival as a species?

In fact, the practice of inventory as discussed in my other works can and does endow its practitioner, should sufficient quantity of resolution be achieved, with the distinct perception of ourselves as the primal primates we truly are which, in spite of our technological progress, are no less simplistic an organism than our other primate relatives, and the delusion that we are anything but an animal species on this Earth because we live in skillfully constructed homes or drive cars to work becomes so apparent as to be comical. But it is this very primal, animal nature which imperils our understanding of reality and thus our effectiveness in our own lives, in spite of our best efforts to be effective, which can and does also lead us to self-destructive or unhelpful behaviors even when we think something is good, true, and useful because, again, we do not control the laws of cause and consequence or the nature of reality.

Just like fractal patterns, another mathematical pattern we see all throughout the Universe are *cycles,* which are in fact just fractal patterns of cause and effect in every single entity in the entire Universe which connect back to themselves in never ending systems that govern change and growth in all entities which occupy the Universe from stars and galaxies to human lives and the world around us.

The existence of cycles demonstrates that nothing in all of existence has a simple or finite beginning and end but is in fact a continuation and extension of cycles which came before it as well as those which come after. For instance while a single human life may seem to have a beginning and an end we are in fact made from other humans which came before us, both in our physical body as well as our mind, personality, and worldview which construct a life, and which was also made from others which yet came before the ones that came before us, then we make new life from ourselves to also affect and interact those that come after, and the choices we make and the events which affect us influence those which in turn come after whom cannot escape that influence no matter how much they might try (even myself which has recovered from the abuse and trauma I experienced as a child still was affected by it). Even at death our bodies simply change form and become composted nutrients which are then recycled into other organisms after we are gone, as the elements from which we were made, both corporeal and ethereal, still exist and can in fact never be destroyed, only also changing form just as our bodies. Even the patterns of long-lived structures like planets and stars, solar systems and galaxies are all cycles, never really beginning and never really ending as they go round and round and round in their course through space and time, by which we can literally observe the principles of time and cause and effect, and some people will readily accept the manifestation of these principles when conditions are very obvious such as the seasons which occur on Earth, but outright deny any pattern or influence in more existential context but no less obvious like the cycles of relationships, as if the relationship between us and our lovers are any different than those of planets and their moons or stars with their galaxy, that the events in our lives are somehow separate from the laws which govern everything else in the Universe, and thinking somehow such a worldview is not in fact rooted in bias and superstition simply because we tell people that we 'believe' in science.

Dismissing the correlation of observable phenomena like the patterns in stars and cycles on Earth with disciplines or traditions like practices like astrology, or the majority of humankind's sense that there is more to existence than what we can see such as what religion attempts to exploit, is no less a control mechanism than uncritical belief and is the same conjoined problem of control desire as what exists equally within religion and atheism, one side desiring control through fantasy and the other through pretended reason. In fact there *are* very real natural phenomena which underlies the discipline of astrology which are reflected back at us from the heavens, those very laws of causality which govern it which also thusly determine the nature and behavior of Jupiter, Pluto, Saturn, Mars, Venus, the Sun, Moon, etc., relative to our position and thus also correlated with our individual lives since we are equally subject to the very same laws as they.

Not surprisingly, contemporary science also ignores and even obfuscates the fact that nearly all famous, ancient greats of science, mathematics, and astronomy such as Galileo, Pythagoras, and Hipparchus *were all astrologers!* But not only where they astrologers, the primary purpose of their mathematics and science was not for the secular study of astronomy but to understand the metaphysical laws of reality which govern *all* existence and which are reflected back at us in the patterns observable in the heavens such as is attempted by the practice of astrology. Famous Aristotle was not an astrologer but he was also misogynistic, homophobic, racist, believed in slavery, and is famous because his ideas and philosophy were convenient for the later purposes of Christian despots, as Aristotle also thought, wrongly, that the Earth was the center of the Universe and, like many others, tried desperately to use logic and patriarchy as a control mechanism to manage his fear, insecurity, and desire to wield control over life and people.

All these shortcomings does not logically mean Aristotle was wrong about astrology, as that is a type of logical fallacy but, I mean...*come on*. Pythagoras (of Pythagoras' Theorem) preceded Aristotle yet he was an astrologer who correctly recognized the Earth was not, in fact, the center of the Universe, nor was it flat, and he was generally a vegetarian who forbade blood sacrifices (which was still a thing at the time). Pythagoras is sometimes criticized for technically stating that the Earth also didn't orbit the Sun, but some point in space, but *this is technically true* and he was intuiting the existence of the *barycenter* which is the actual point around which two objects orbit, which is never exactly in the center of either, though which in the case of the Sun is still inside the fucking enormous diameter of the Sun which was not known until a few hundred years later with the work of Aristarchus and Eratosthenes.

Galileo Galilei is considered the father of modern science and the scientific method, condemned to house arrest by the Catholic Church for the final years of his life for his research proving heliocentrism, but whom also not only also used astrology but was a well-regarded astrologer who provided astrology services for patrons and *actively taught astrology* to his students. Anyone who similarly believes Galileo wrong about astrology should ask themselves if they are truly smart enough and knowledgeable enough to challenge the father of the fucking scientific method about something they probably don't even know anything about except having an opinion given them by popular consensus, not scientific research, nor that in siding against Galileo and aligning with the religious church which persecuted him is supposed to somehow be a logical, rational position? In truth, ancient masters of science like Galileo and Pythagorus (and countless others) did not believe in astrology because they were from an ancient time and had old belief systems but because astrology is *not* something to believe in but which can be seen and observed with empirical evidence as required by all scientific discipline, and the function and existence of phenomena which underlies the practice of astrology is a very real and very evident function of reality, which is why such intelligent, accomplished, and observational persons thought nothing of its practice and found harmony both within their study of astrology and astronomy together.

In fact, it is also not widely known that there was no distinction between astrology and astronomy until the recent 17th and 18th centuries, when they were finally separated as distinct disciplines by religious Catholic and Protestant theologians, because to early astronomers astrology was not at all a separate religious belief but a science practiced one in the same with the scientific study of the heavens, because they rightly recognized that fractal patterns and cycles of reality underly *all* things, not only physical sciences, and that we are not separate from the laws which govern the Universe just because we are human.

But many prominent Christian theologians since even the early days of their religion like Augustine of Hippo to Martin Luther opposed astrology because, they claimed, it robbed men of free will, because principles of punishment and reward are fundamental to religious control and power, which doesn't make much sense if fate is a property of the Universe, and the subjective bias and prejudice is in fact that against the natural world and phenomena such as what astrology describes, not cynical skepticism of it, which in truth has its origins in theism, not science as we are often misleadingly told, where in fact it was the *scientists* which practiced astrology.

Born in ancient Babylon (Mesopotamia), astrology was also a primary originator of mathematics and science (which occurred also in Egypt, being neighbors), and because astrology was the study of natural phenomena it is also the very first *science* practiced by man, and after conquering Babylon the Greeks absorbed Babylonian science and technology, including the practice of astrology and study of the stars, planets, and geometry. The word 'geometry' originally meant "earth measurement" and its original function to measure and understand the heavens for the purpose of divining the past, future, and understanding the present by understanding the forces of nature on the organization of matter and time is usually omitted from contemporary records by people who have been indoctrinated with contempt for astrology and ancient practices, and describe such use of mathematics and geometry for the purposes of astronomy as if it were a strictly contemporary scientific discipline of astronomy, or remembering when to plant crops (fucking serious, are you for real?) and not the reality that it was entirely in service of astrology, to understand the nature of reality, purpose, and existence.

Many religionists are, however, more interested in astrology than people who are entirely secular since their ideology also naturally lends itself to belief in things they maybe don't fully understand, but some Christians I know have no problem believing God impregnated a mortal woman to birth a man with the power to grant everlasting life but tell them the orbit of Saturn correlates with events of structure and discipline in our lives and they laugh unironically, even though order and structure are natural and observable phenomena in reality but resurrection is not. Worse, secular people also often engage in cultural erasure of ancient peoples' practice of astrology, for instance it is very common for archaeologists to describe Mesoamerican astrology and Mayan study of the stars and heavens as astronomy, *not* the astrology it truly was. The Mayans were absolutely NOT studying astronomy just because they were interested in science and space, but in service to their form of astrology, to understand the past, present, and future though understanding the cosmos reflected back at us, not for the sake of scientific discovery, but existential. Puebloan peoples in the Americas too had detailed study of astrological phenomena integrated into their society, buildings, and mythology that for some reason scientists (racistly) retcon as Western standard physical science rather than being spiritual and mythological in nature. Effectively, so-called rational thinkers and scientists usurp for their own sterilized, prejudiced, professional benefit that which has in fact mostly been the widespread, commonplace, and repeated recognition that celestial phenomena correlates also with our individual lives here on Earth and the societies in which we live and used actively in the practice of most human culture in some form or another in ways which are also useful and practical, and this erasure not only serves to actively deceive people to history and the origins of science but also to very realities of existence which do in fact exist but which are inconvenient to those whom, like the atheist and religionist alike, desire control by imposing onto reality our presuppositions.

Astrologers for sure have not done themselves any favors, however, often using astrology in ways which are destructive, superfluous, deceptive, exploitative, superstitious, duplicitous, and often just completely fucking incorrect. But just as with religion and unscrupulous science these are all problems concerning *human* behavior, not the actual composition of the Universe and the laws which organize it, where the correlation of astrological events and metaphysical characteristics of the cosmos with our own personal existence is a completely separate phenomena from the *practice* of astrology, which is still worthy of proper inspection, research, and consideration regardless of the behaviors, beliefs, and practice of astrologers. Considering this reality, scientists who refuse to acknowledge the existence of phenomena which underlie the practice of astrology effectively refuse to study reality and laws of nature simply because of the *behavior* of other people, which is most ridiculous, very human, and completely unscientific.

But like many Christian theologians, many of us wish also to deny concepts as fate and destiny because we believe deciding the course of our lives then empowers us to prevent painful experiences that life can and does bring, even to the subversion of death, so we try to construct a reality in which we have power and control over the things like fate, then the sterilized logic of math, or belief in religion serves only to embolden and empower this defensive and deceptive survival function of the ego through which we delude ourselves as having control over reality. But there has never been even a single human being whom has successfully voided the laws of cause and consequence or subverted destiny, and the inherent nature of fate as a principle law of the Universe can be easily demonstrated by the mere fact that none of us played any part in our being here, cannot prevent our exit from the mortal plane, and on a daily basis meet over and over again events and influences entirely outside our control. In fact, most people thinking thoughts, including supposedly intelligent and rational minded skeptics and scientists, fully believe that we ourselves, as a function of our innate individuality and autonomy in the Universe is the source of the thoughts we have, springing forth from intellectual spontaneous generation, out of nowhere and yet also to our personal credit as the creature who had absolutely nothing to do with what we are and why we are here. The delusional coping response to a pandemic, also for instance, which repeats in all of them no matter the time period such as we all recently witnessed is even outright denial of disease and our shared mortality which then unfortunately in turn claims many more lives. Growing up my father often crowed about how fit and healthy he was, and indeed he always did have a six-pack and never had to go on a diet, but he also suffered from life-long stomachaches every time he ate any meal, and drank bismuth the way Midwesterners drink soda, and also fell from a scaffolding in his twenties and dislodged a disc in his back which then never fully healed because of the role of opportunistic bacteria colonizing the gut which also caused his constant stomachaches fueled by mass iron fortification of the common wheat flour he ingested on a daily basis because, as humans, we require food not to fucking die. The relativistic dismissal of physics, metaphysics, evolution, and biochemistry for arbitrary personal beliefs such as they apply to our microexistence whether they are religious or secular is not a triumph of thought, but the abject fucking failure of it.

To further demonstrate the absurdity of human coping behavior and sheer inability to properly conceive of reality there have been literal billions and billions of other humans which have existed before us, along with literally innumerable other organisms, every single one of which has died or will die, and yet most of us treat death as an optional inconvenience. The two simple anchor points of birth and death demonstrate immutable laws of reality which are in fact governed by forces we do not fully understand or even begin to comprehend, but science avoids investigation into these more esoteric phenomena of existence simply because how to scientifically investigate such phenomena is not yet within our comprehension, likely beyond even the scope of quantum physics, to investigate the very source of dimensions which in turn enable known physics to construct reality in the first place. Yet this does not mean we can't try, or even find evidence for their existence through indirect observations just as done when studying light. Removing the very same subjective conjecture such as what occurs within religion we must instead recognize that there are laws of the Universe which decide all things, even those which appear arbitrary, subjective, personal, individual, or random to our limited powers of human observation and intelligence, for *everything* is subject to the laws of reality, without exception, and the idea that the physical forces of the Universe stop affecting us when it comes to daily life is nothing more than anthropomorphisation of reality same as what is caused or enabled by religion and mythology.

Although they lived thousands of years ago, those great scientists who came before us who used astrology to understand the natural world and the nature of fate and time were still more educated than the far majority of all people today, where many of us think that because we can turn on a smart phone we also know how it works, yet can't even do basic math nor point out most other countries on map. The Dark Ages which occurred between the time of the first scientists and today happened because of religious zealotry which led to the abandonment of learning by the masses and caused incredible suffering and ignorance for most of recent human civilization. As I mention in my other work, Cyrus the Great of ancient Persia was the first recorded ruler to declare freedom of religion and to outlaw slavery, and that was in the year 500 B.C.! Indeed what progress we might have made since that time if the Dark Ages had not intervened in the epistemological progress of mankind, having now achieved even spaceflight and particle accelerators within a mere several hundred years from the dawn of the scientific revolution? The first (analog) computer was also not a product of the modern world, but from *ancient Rome*, known as *the Antikythera Mechanism*, and if not for the violence, oppression, and greed of humanity in the intervening two-thousand years, a thousand generations might also have had cars, electricity, medicine, plumbing, video games, and smart phones.

But though we do live in a future full of incredible technology that would have made any ancient person breathless with awe there is still an incredible amount of religious bias which corrupts and misleads even the most brilliant secular researchers and scientific minds today whom do not even realize their prejudices against such things as dopamine, reward, genetics, heritage, race, and real physical phenomena such as what underlies astrology originates not from their own enlightened intellect and exacting, rational worldview but from religious prejudice and oppression caused by theocratic zealots which many years ago found it in conflict with their ideology of divinely appointed individualism just as the left-wing voters believe the Affordable Care Act was not a right-wing invention, such is the predominant view by secularists and scientists that astrology is a pseudo-religious, occult practice (the word occult itself even being the language of religion), and therefore failing entirely to recognize that the foundations of astrology are in fact real, observable, and provable natural phenomena and laws which construct reality because of ancient religious oppression which sought to eliminate any challenge to its supremacy.

The entire discourse also surrounding the existence of extraterrestrial life is another perfect demonstration of this problem, wherein scientists and lay persons alike debate whether aliens even exist because we have not personally been visited by them, which is one of the most pure demonstrations of human egoism—making everything about us—instead of the plain and simple truth that we are not any more important than any other form of life, and that interstellar space travel is probably just really, really fucking hard, even perhaps (I hate to say) to the point of being *impossible*. Those who may be offended by the proposition that we are no more important than bacteria, or birds, fish, or alien life feel revulsion not because it degrades our humanity but because we harbor contempt for lower creatures as a coping mechanism for the fear of our own precarious mortality, and debates about extraterrestrial intelligent life is even more ridiculous and absurd than anything in astrology, for instance the assumption that aliens might invade us for our water when there is literally an infinite supply of water throughout the Universe, even floating freely in space far away from the heavy gravity on a planet such as ours. The technology required for interstellar space travel is also way more advanced than technology required to literally just synthesize water without requiring expensive, wasteful, destructive interstellar colonization. Such advanced species would also be able to spontaneously grow tissue and biological mass as needed, so they don't need to travel across space and time to abduct humans for biomass or research (lol). Hell we can now grow hamburgers in the lab and we still can't even travel to Mars, but members of the scientific community will in one moment denigrate the idea of Universal forces governing the course of our existence and happily indulge complete science fiction simply because they were raised with Buck Rogers, Star Trek, Dune, and Star Wars, for no more than our egoist desire to feel control, which we fail also to recognize as simply a function of human evolutionary biological psychology which constructs everything that we are, think, say, and do.

Especially in relationships there are easily observable patterns of behavior which occur to all humans, and stepping back from our personal, insular lives and viewing humanity as a whole the ebb and flow of partnering, separating, birth, loss, etc., is an enormous fractal model which encompasses the entirety of humanity together, each of the many billions of us having the same kinds of experiences as every other yet delusionally thinking we, and not the Universe, will it. Yet even within individual lives these experiences are also cyclical and repetitious characteristic of fractal cycles, such as seeing a partner becoming emotionally sensitive at predictable and repeating monthly intervals, which reveal underlying cycles of extrapersonal influence and our shared biological history which has nothing to do with individuality or free will, yet uncurious about what laws and forces are responsible for such realities of existence. The only reason astrology is not part of rigorous scientific discipline today is simply because of the *bias* caused by religious despots whom imposed monotheistic authoritarian worldviews onto the course of human history in opposition to the rational scientific practice of astrology (different than entertainment astrology that is so common today) which was actually based in observational corollary of the natural world, the ebb and flow of all existence which can be seen reflected down upon us when the sky is clear and the night free of light pollution.

But the truth about the Universe is that there is both fate *and* free will together (as discussed in the chapter on time), and the fallacy of this debate was assuming they are incompatible one with the other. In reality fate and free will both exist in harmony and not in opposition, and thinking there must be one or the other is a failure of human intellect and imagination and not a logical contradiction, which happens because fate cannot occur until the person or being at issue makes the decision for it to happen, so there is free will, but what decision or choice we make is also in turn influenced by *everything* which constructs the entirety of reality, and I mean literally *everything*—the millions of years of evolution which birthed us as a species, the formation of the planet on which we live, the temperature of the climate and absence of large predators which terrorized our ancestors, even the toxic lead in gasoline which made previous generations insane and the other myriad and countless and unfathomable factors like the one time that our parent's friends made fun of them so they then raised us to be insecure about rejection and now we are paralyzed at the thought of having a relationship. All the circumstances and the lives of every other person on this planet which affect and constrain even the occurrence of an opportunity of which a choice can be made in any one moment, and in the end there is only one choice which can be made, but we are the ones which must make it, therefore we are in fact an *agent* of the law of fate, which we achieve by exercising our free will to choose the only choice in front of us.

When realizing fate exists a person may have two responses—one, to give up and stop trying since there's no point anyway or, two, to take comfort in the reality that our responsibility in life is limited to very little which thus makes living a far easier prospect, and in making the first decision a person delusionally thinks this is their own thought rather than the effect of trauma from the past or ideological instruction by parents, society, and culture which shaped us into the pessimist who then makes that choice because. Or another person who could choose to share their wealth willingly with their society, to help build nice roads, good schools, and programs for medical treatment that the society will stabilize and become peaceful, but growing up were taught to love money and fear poverty and so instead chooses to keep their money and blame everyone else for their circumstances, only to be shot and robbed later in their lives by desperate, starving people who must have food in order to survive because of the nature of human biology and the vicious survival instincts of our evolutionary past which drive panic and violence. We *can* choose, when choices are presented to us, which is free will, but what choice we make is predicted by fate, and we become the agent of fate within making that choice of free will, thus fate and free will exist together—the choice you will make is always the choice you were going to make, but you were always the one to make it. If this were not true you would would not be reading this book and have no idea who I am because I cannot express to you just how much I never wanted to do any of this.

There are also many easily observable and easily provable evidences for the validity of astrology, and planets do not of course have actual magic or mystical powers which determine the course of our lives but are instead manifestations of the underlying natural laws which do. The specific reason that correlations can be seen between the movement of the planets or other astrological phenomenon and the events in our lives is the very quantized nature of reality, such as of electrons which do not occupy graduated positions surrounding atomic nucleus but instead occupy exact valences. This means that all things can and do naturally fall into resonant frequencies with each other, to then establish relationships between their behavior and that of others. Even all the planets in orbit and their moons which orbit them have *orbital resonance* which causes them to have quantized relationships with each other, such as a 2:3 resonance between Neptune and Pluto, or the 13:8 orbital resonance of Venus and Earth. If energy wasn't quantized then this would not happen and there would not be observational corollary between things, but there is and so there are observable corollary between all things, and so planets and other celestial phenomena behave as they appear do (from not only our relative position physically on the Earth but time as well) for the same reasons that we are suddenly fired from a job, begin a devastating new romance, or must soldier through a pandemic, which is that all things, including those of a metaphysical nature, are ruled by the same, immutable laws which construct reality.

Any astrologer who speaks in terms of mysticism is a poor astrologer indeed, and likely opportunistic charlatan seeking to prey on the desires of their victims, consciously or not, and part of the justification for bias against astrology is the tendency of astrologers to also use astrology as a control device of exploitation, which in doing is no less harmful than any religious institution and, just like other more established religious institutions which try to control fate and delude us to our place in the Universe, is equally unhelpful, incorrect, and harmful. While understanding of astrology can provide insight into our existence, purpose, and fate it cannot be likewise used to control life and fate any more than a religious person pleading with God to give them money or a car as many unimaginative astrologers practice in order to attract followers. In rational application of astrology its only purpose can be and only ever will be to gain insight into the workings of the Universe and understanding of life, and astrologers who use astrology to promise money or success, or any specific future outcome rather than advise people on their growth, development, understanding, and spirituality should be dismissed as opportunists with insufficient wisdom to be useful.

In opposition to rational astrology, nearly every daily astrology column in newspapers and websites are mostly nothing more than entertainment and click-bait which should not be taken seriously, such as the popular Western astrologer who also happens to be a New York socialite whom typically only sees in the stars whether or not it is a good time for someone to undertake renovation of their home. Most astrologers are, in fact, useless, which I say not to discourage or shame astrologers but to warn the public of the need for scrutiny and skepticism when listening to anyone who claims to know the future, especially when they charge money for it, and it took several years before I finally found sources which were not silly, untalented, or exploitative. I once even asked a question about astrology in the forum of a prominent and respectable astrology website populated by lots of professional astrologers and got a dozen stupid answers that all entirely failed not only to give an even remotely accurate, useful, or pertinent response but to even understand the question in the first place, since most of them operated on prior training and do not fundamentally understand how natural astrology works, but in the process realized I had discovered aspects of astrology never before known, as discussed in upcoming chapters, which resolve longstanding conflicts and inconsistencies in Western astrology.

Many scientists are also duplicitous, cheat, or have conflicts of interest, yet it does not invalidate the practice of science (mostly because we have been indoctrinated to 'believe' in science as propaganda for capitalism, but that's for another book), and likewise astrology cannot be invalidated simply because humans are bad astrologers. After all, what astrology is attempting to do is describe real phenomena we see and experience, which is indeed a herculean task, and of course there will be failures, dead ends, and dead weight, and the point of all science is not to blindly accept what is or what we believe or want but to continue studying and seeking to understand, which should be the guiding principle also of all metaphysical science. All of astrology, even this entire book, are people, writing, ideas, and human observation and interpretation of what we observe and *are not* the actual laws of nature which underly this discipline, but merely our attempt to describe and characterize something that actually does exist, and make no mistake—*it does exist*. Patterns of life follow obvious and irrefutable cycles in which the macro and micro are inextricably and undeniably connected and related, yet there has been absolutely no science which seeks to explain, quantify, and prove what governs these patterns, yet we know that what connects them are immutable physical laws that bear reality into existence because nothing in the Universe can be excepted from those laws, and which also cannot be subverted simply because it opposes our conception of reality and desire for stuff such as is often the case with astrologers and users of astrology alike. Though it is often presented as mystical, spiritual, or religious, astrology is in truth merely man's way of describing merely the *observational corollary* of celestial events and mathematical principles with those which also occur here on Earth.

One of the most profound and obvious of such cycles, for instance, is that unless affected by aging, disease, or medication, women's menstruation cycles are generally the same length and frequency as the lunar orbital period (different than lunar cycles), which is 27.3 days. This is not predestination but a likely evolutionary biological strategy which probably utilized some feature related to the Moon to form specific biological functions across the millions of years of our evolution. Or it could be that our evolutionary ancestors tended to mate more often at night under illumination by the full Moon and thus the menstrual cycle started to align to that, and people may say 'well the moon caused that function'—and, *yes that is the very point*, that the physical laws of the Universe affect our lives directly through laws of physics, which is not a function of the Moon but a function of forces which put the Moon where it is and determine how it behaves, in the patterns and organization of such astrological bodies *which similarly also affect the patterns in our lives.*

Then an argument may be such that, well, planets like Neptune and Uranus are so far away they can't possibly affect our lives, except that Earth's orbit is *not* circular but instead an *ellipse* because of the gravitational influence of other planetary bodies like Neptune and Uranus, and the presence and position of every single other planet in the solar system (including maybe planet X if there really is a yet undiscovered planet) is also affected by every single other planet, influences even which resulted in our even being here at all in the first place, which is no accident but a direct result of the way the Universe is constructed.

Even tiny Pluto, way off at the edge of our solar system, affects the motion and position of the giants Neptune and Uranus, and the Sun itself has regular cycles too, one for instance which is eleven years in which its activity increases or decreases and its magnetic field reverses, and these cycles affect life on Earth even affect technological equipment now that we have such technology. So the same laws which produce gravity and put the planets where they are also govern when and who we are and what happens to us and why and how and when, and being ignorant to those forces is not the same as their being absent, mankind has simply heretofore sought to explain seemingly inexplicable laws of reality over which we have no control, such as death, love, and the unknown through religion and mythology, terrified at the idea of having no control over life, but the Universe operates on laws and that does not change just because we got fired from our job, feel badly about our appearance, because our partner left us, or because we cannot comprehend how life first sprung into existence in the first place.

Yet also are such concepts as justice, violence, cooperation, creation, destruction, love, betrayal, recrimination, or forgiveness created and governed by laws of the Universe, for *all within the Universe is subject to its immutable laws*, not only planets and human bodies but philosophy and hatred, oppressors and the oppressed, secrets and knowledge, for reality is mind-bending with black holes and neutron stars, the curvature of spacetime and a probably infinite expanse to all existence we cannot possibly begin to comprehend, so why do we feign to conceive of things like marriage, love, loss, and the cycle of life as pitiable conceptions of mortality and not fundamental laws of reality? These too are thus also correlated with things such as planets and cycles of the moon, since the same laws which govern anything also govern everything.

Whether conscious or unconscious (usually the latter), the desire to thus control life leads us headlong into more pain and frustration than what would otherwise be the case as we attempt to impose our vision of reality onto the one which actually exists, and because reality and our perception of it are often incompatible or contradictory we then experience greater consequences antagonistic to our expectations, and without the practice of inventory it is impossible to reveal to the unconscious mind the real nature of reality such as what is also observed in astrological phenomena, and those whom dismiss even the opportunity to observe reality are fully deceived by the evolutionary animal mind that such secrets are not worth investigation, because to them the ego is more valuable, and then persist in unending self-hatred, interpersonal conflict, and personal frustration as if we are not entirely run by evolutionary biological instincts which sacrifice the individual for the sake of the whole.

By its very nature the unconscious is the unconscious—it is *not* connected directly to our conscious mind and we therefore cannot actually communicate to it consciously or willfully, which is why simply thinking through our existential problems never resolves them, because we cannot be consciously conscious of that which is explicitly unconscious. The unconscious does take in all the same information and stimulus as our conscious mind, however, and the act of practicing structured writing as what is accomplished through inventory communicates to the unconscious mind, as when we write the unconscious mind reads, and by removing our preconceived perceptions of life and reality and the trauma which motivates fearful, destructive, self-preservation behavior through this practice we can then be more rational animals and thus uncover the reality which exists all around us, not what we desire to see, and so then understand our purpose and journey in the process.

For example I often experienced the very common anxiety for the next day, as the approaching bedtime and promise of sleep also portending every and all possible horror which could come along with the coming day. This anxiety on a Sunday night, at the end of a long weekend for the coming Monday was even worse and my anxiety for the start of the week would often be so great as to require more alcohol in order to placate my fear and calm my nerves, or to call in sick or not take jobs for a week or two when I worked as a freelancer.

Yet no matter how much I tried to talk myself into calming down or rationalizing my fears, or to speak about it with therapists, it not only never improved but worsened as I aged and accumulated more and more negative experiences. Once I got sober and could no longer drink to soften that fear I one day had the thought when new at the practice of inventory to inventory this fear of starting a new day or week and the resentment for days and weeks or structured societal schedules, conflicts, and systems which exploit workers and people in the manner which leads to such traumas as I was forced to endure.

In the middle of writing these entries and pondering this anxiety and what might be its origins I realized the entire concept of days was nothing more than a facade, based on the cycles of day and night and seasons on our Earth, yes, but in reality time is nothing but an uninterrupted constant that does not change and does not delineate into day or night just because we go to bed and wake up the next day, and the cycle of day and night is only a trick caused by our position on the crust of Earth spinning about its axis without regard for human conception of time in the construct of days, weeks, months, and years, and if we stay up all night and watch the galaxy pass overhead there is in fact no real distinction between a Sunday or Monday or Tuesday or Friday other than what we both impose upon reality and deign to accept through our active participation.

My anxiety concerning Mondays thus was caused in part because I did actively *choose* to participate in this construct, but was denying my participation as an act of free will which at any time I could also choose not to, and was not, as I felt, forced at all to participate. If I wanted it was possible to refuse entirely to recognize Mondays, not go to work and not earn money, and treat every day equally and do what I pleased to satisfy nothing but my own fulfillment.

Of course I would be poor, and that choice would then cause yet other consequences of its own, but now realizing that the choice was in fact entirely mine to participate in society it eradicated my anxiety, feeling then suddenly empowered by this new knowledge. However, I also realized that my anxiety was also caused by fears that I could not handle challenges or conflict, and also mistook many people for enemies but whom in reality were just other people trying to get by as best they can too and likely possessed of the same anxieties as I which caused their antisocial behavior as it had my own (and thus also inventoried those things for even better resolution).

Because time is a constant and unending reality every moment is thus as potentially good or awful as any other, which is not itself a function of time but of other forces and factors with which we contend such as other humans, institutions, disease, responsibilities, etc., which have nothing to do with our conception of time, dates, schedules, etc. If it were somehow possible to completely wipe my brain (which did in fact occur from getting plastered several days a week, and the very reason alcoholics drink to excess) and completely start over from total absence of knowledge there would still, however, be sunrises and sunsets, rain on my face and wind in my hair, but no Mondays, or Sundays, or even birthdays, and certainly not a time to show up for work, yet I would not have realized any of this if I had not taken the time to show myself some compassion by doing self-care work like inventory in taking the time to address my problems instead of callously press on through them without regard to my wellbeing, and such is the problem with the study and understanding of astrology, in that it also requires previous knowledge and skills in order to be effective.

Mondays do not exist, and instead we actively choose to participate in the concept of Monday just as any contrived, human institution like marriages, teams, workplaces, and so on. Such also is the study of the natural world, especially existential themes as what astrology attempts to achieve, in which we risk being ineffective or even destructive if our behavior and efforts are informed by bias, prejudice, expectations, and unresolved trauma.

We often have far more free will than we realize, because usually we direct our will to things we do not control such as consequences, biology, or other people, and part of these kinds of fears and anxieties is because we fail to recognize where we do or do not actually have control. Many people study astrology with dreams of avoiding the dangers of the future but that knowledge is found in reflection on our life experience and resolving the control and coping mechanisms which otherwise compel us to transgress causality, not an inability to predict the future. Of course, could we wipe our brain and start over there would still be also hunger, disease, and mortality, over which I have no control (meaning for instance I will starve if I don't eat food) so I would still need to find food, shelter, and education, to understand how to be healthy, and we thus construct artificial concepts of time, space, and relationships between things which do not actually exist so that we may more effectively survive and meet the challenges and requirements for living, but which are still constructs that do not actually exist in reality, and recognizing and understanding what control we do have ironically empowers us with more control because we direct our control behaviors to those things like our choices and behavior and stop wasting it on things like other people, consequences, or the future, over which we in fact have none.

This is the irony behind skeptics of such phenomena as what underlies the practice of astrology, desiring that we are not in fact beholden to such real laws of reality, and choose instead to recognize the ones that we have contrived like Mondays or marriages in order to feel a sense of control over life which we do not actually have. Recognizing that I actually had a choice whether to participate in Mondays, because they are entirely fake, eliminated my anxiety because it was entirely within my power whether or not I would participate which, because I wanted to earn money, be productive and helpful, and stay fed and sheltered, of course I would.

Money is also entirely fake, but the contract between two persons which is implied by the exchange of money is not, but we construct a fake institution like finances and monetary value in order to simplify and thus better control what is in reality uncertainty, to find delusional security in what is really insecure, then in the exchange of money with another it is implied that we will receive something in return and when this expectation is subverted such as getting swindled or exploited we are all the more rudely affronted by the experience if we exist within a conception that fails to recognize the falsity of money or the delusion that bad things won't happen to us just because we did our best to prevent it. This is why the entertainment of self-pity is counterproductive in the practice of inventory, because self-pity is rooted in delusion—but where those with self-pity hear such a condemnation they do not realize the delusion is of our very self-worth! That we are much more capable than we realize and the self-pity is instead an *underestimation* of who we are, and likewise the entertainment of self-pity in the practice of astrology is what also fuels delusional astrology, unrealistic expectations, chasing fortune, or even attempting to avoid fate which we feel unable to handle even if, as was my experience, we have a lifetime proving ourselves capable and resilient in the face of adversity. Becoming thusly enlightened to the reality of life through the practice of inventory which helps enlighten us to knowledge such as that we are not as incapable as we have previously thought we then become more empowered through knowledge to effectively meet the challenges of reality and find comfort in the constancy, rules, and dependability of the Universe, which is actually in charge of such things, not our own limited, pitiable contrivances. Understanding reality we can then also effectively use observational astrological corollary to achieve greater insight into the workings of the Universe and thus our lives, the lives of others, and the purposes and functions of existence ruled by the same laws and physics which put Jupiter where it is or the Sun in its galactic orbit, now liberated from the onerous burden of mastering the Universe and humanity, which was never our charge anyway.

The work of astrologers Robert Pelletier, Robert Hand, and Liz Green and others associated with *Astro Dienst* is the most useful in the entire world, and their website astro.com is in fact used by astrologers worldwide because it sheds a great deal of the entertainment fantasy commonly associated with astrology, focusing not so much on predicting the future but simply to describe the quality of time as what can be expected by any person during any transit. Some of them refer to this type of astrology as *psychological astrology,* meaning that any of our experiences can be arbitrary except for how we experience the quality of time from the point of our psyche, since it is through the mind that we experience life around us. While this approach is laudable there are however many influences which are not only related to our perception, so I prefer to simply refer to designate my system of astrology as *rational astrology* which is founded on study and observation of the natural world and its statistical correlation with our experiences and uncontaminated (as much as possible) by our prejudicial desires, insecurities, and fear. While their site still contains some hullaballoo it is absent of superficial, idiotic applications of astrology such as winning the lottery, getting rich, or meeting your one, true love, which are usually meant simply to exploit human vulnerabilities and desire by talentless and amoral astrologers which have no insight into the very real association of our lives with that of the rest of the cosmos, which these astrologers have recognized and as such produce instead very rational and high quality assessment of probability and a great library of works by many talented and insightful astrologers.

Indeed it is extremely important when indulging astrological practice or studying its phenomena that a *variety* of sources are consulted, and never to consider interpretations of astrology as absolute since ALL of it is all filtered through human interpretation, remembering that all astrology is merely interpretation of celestial and universal events by human beings whom at best barely understand it, and there are many whom are lazy, duplicitous, opportunistic, or outright wrong, and even the best astrology can never be wholly regarded as absolute truth but merely the closest and best approximation we can so far accomplish. Ultimately, the only thing for which astrology is really useful is to better understand life, our experiences, our place in the Universe, and how to live more effectively.

If it was not yet clear, practice of inventory is first *required* to properly understand astrology in a way that is useful, otherwise it will do nothing but entertain delusional coping mechanisms and fear, to obfuscate reality so we can exist in our personal, dissociated perception of it, becoming trapped in a prison of fantasy every bit as any religion, missing out on the present and its content, shrinking from reality, and resenting life. But when empowered by inventory practice which helps us shed delusion, fear, self-will, and control behaviors the information which is then gleaned from the study of astrology, derived from the work of astute astrologists with reputable and rational understanding of astrological phenomena (which can even be you if you apply yourself with dedication, consistency, and self-honesty), can help us to understand the root of all things and thus find wisdom, comfort, and joy as we grow closer to the power of the Universe, which is what we experience to be God, through knowledge.

14. Evidence For Astrology

Being based on geometry, trigonometry, and planetary bodies it is even more ridiculous that learned people dismiss astrology with more contempt than they do religion, since it is the only major spiritual discipline that actually contains math.

But there are broadly apparent astrological correlations which can be easily recognized in human experience and the natural world to serve as anchor points of orientation to recognize, identify, understand, and prove metaphysical astrology phenomena, and astrology is as macro or micro as any aspect of reality or scientific discipline—A quantum computer could compute astrological correlations down to the smallest minutiae of moment to moment life but we as humans cannot perceive events in such complex capacity and it is easier to see broader, more generalized patterns which can be the starting place of investigation. In reality, however, every life has *many* aspects which can be correlated to many factors, and there are many planetary and heavenly actors which correlate with specific aspects of individual lives which, when learned, make far more rational sense than what normally occurs in entertainment astrology.

There have, in fact, been some 'scientific' attempts to investigate the validity of astrology, but nearly all of them are dripping with contempt and lazily constructed by scientists who have no familiarity with the discipline whatsoever, let alone its long history of exploitation and poor practices, and as a result design very poorly conceived trials, controls, and fail to account properly for variables, including their own biases.

For instance many studies claim to measure the accuracy of astrological personality evaluations using a person's zodiac *Sun* sign. But the Sun sign is a colloquially oversimplified aspect of astrology often used for nothing more than entertainment for the casually use of astrology, and researchers using such criteria are doing nothing more than disproving the validity of an opinion column. The Sun signs are one of the broadest aspect of astrology, and large groups of humans share Sun signs because there are only twelve in total, and Sun sign attributes are also those which are most outward and obvious, which is why astrologers determined them to be attributes related to the Sun, which emanates energy outward and is obvious and seen. But the Sun signs are *not* even remotely complete descriptors of *anyone's* personality, they are only the *outward* manifestation of some parts of the personality, the way *other* people experience us (and some aspects of the physical body), and investigations or skepticism for the validity of Sun signs does not determine the illegitimacy of astrology but the ignorance of the critic, for Sun signs are only but one aspect of the entire system (which is itself, remember, nothing more than human attempts to describe and categorize real phenomena, not the phenomena itself).

I also doubt, for instance, any scientists designing these contemptible studies know that a person's Sun sign is more pronounced in their personality at the beginning of their lives, and over the course of a life a person becomes more aligned to their *Moon* sign, which represents our inner nature, which demonstrates how someone conducting a study on any subject must also be familiar with it, and such studies are the equivalent of a researcher conducting a murine study on squamous cell carcinoma without knowing even what 'carcinoma' means and only completed Secondary School biology courses.

A more accurate composition of astrological personality requires instead the combination at the very least of the Sun, Moon, and the *Ascendant* (also called the rising sign), and beyond that there are yet more than a dozen other factors of the personality which are determined by all the other influences. For instance the placement of the planet Venus determines a person's conception of and approach to love, romance, and partnership (including both plutonic and romantic), which would also ostensibly affect a person's personality in ways that are not directly determined by the Sun sign. Some studies on astrology also used accepted standard personality systems as controls, but for instance one commonly used standard has only *five*, highly generic personality categories (even more generic than Sun sign generalizations) so when compared to even the most minimal astrological dataset of twelve 'personalities' determined only by Sun signs, which isn't even how it works anyway, hardly seems a comparable control.

In terms of astrology people are likely to identify more with their Moon sign, not their Sun sign, because the Moon sign represents our inner self while the Sun sign represents the outer self, which is what others see, not the study participants. Another study which did investigate astrological "personalities," whatever that means, used the metrics of anxiety, optimism, or death-anxiety as a measure of personality and I don't know about you but I don't go around categorizing person-alities based on their death-anxiety, and I don't think any astrological discipline does either, nor is there any generic astrological factor which rules death-anxiety (one of the obscure comets probably does), and studies are only ever as effective as they are in their design, and nobody would expect someone unfamiliar with the biology of a cell to design a well constructed and executed study on mitochondrion, the equivalent of which is occurring in nearly every study I've seen which purports to investigate astrology.

Another such study sought to determine if astrological influences correlated with marriage success as defined by divorce status using dates of birth to, yet again, base such inquiry only on the Sun sign, which is not even close to how that works, and while they did actually put some effort into understanding some aspects of astrology they entirely failed to understand that Sun sign marriage compatibility *is a fucking entertainment*, and not how real astrology works, and even used some of the most pulp astrology websites possible as their criteria. What's even more amusing about this study, however, is that they assumed without any sense of irony throughout the entire paper the presumption of astrology concerning marriage only for its *success*, and not instead the real purpose which is greater understanding of *why* or *what* something is, not in its use to control realities of life we cannot control like someone's behavior in a marriage. In fact, most marriages and partnerships take place between *incompatible* partners in exercises of control, wherein our complimentary control and coping mechanisms are the primary impetus for pairing, because the other excites those vulnerabilities in ourselves, which is why most marriages in fact do end in divorce. The study also had absolutely no idea that the *Venus* sign is the primary descriptor of the quality of romantic pairings, also, and again, NOT the Sun sign (which they would have known had they spent time studying any legitimate astrology, like, at all). Yet this does not mean that studies could not objectively measure the validity of astrological influences, and they can do just that when they are designed *properly*, which will not only help prove or disprove astrological data but also progress research into the metaphysical sciences which have for so long been neglected.

One very well designed and simple study without room for bias by the study authors or participants investigated the association of winning a Nobel Prize in medicine and physiology with an astrological sign. This is entirely non-biased because all participants will have an astrological sign and the study authors cannot manipulate or alter the conditions of the study, the controls, nor the resulting data, and the point of it was not to prove or disprove anything (which is itself a bias), but simply to see if there are were any correlations. There was in fact a very, very clear association of winning a Nobel Prize with the sign of Gemini, with a 1.9 odds ratio compared to controls (the distribution of signs working in the sample field). While it could be described perhaps that this is just a random distribution it so happens that *Gemini is literally the sign associated with learning and knowledge*, while the lowest association was the sign of Leo with just 0.35 odds (sorry, Leos) as while Leo is the embodiment of light, joy, and fun they tend to be least interested in intellectual disciplines.

This study alone, being one of the only ever done about astrology, is enough to prove without a doubt the validity of astrological correlations with our life experiences. Yet it also seems like standards of investigation with astrology must meet *much higher* thresholds for legitimacy than other scientific investigation, even though some truly very small margins of difference in supposedly legitimate study subjects will still be described as statistically significant. Astrology is an interpre-tational practice, the content of which is dependent upon not only personal ideas and human interpretation of significant natural phenomena we still do not fully understand, so such variables *must* be controlled or eliminated from any study investigating astrology, to limit human bias, conflicts of interest, and ignorance as a factor.

What's even more amazing about the Nobel study is that Libra is in the smack-dab median position as would be expected of the middle sign of the Zodiac and the representative of balance, a fact that was probably lost even on these study authors, and the results of this study alone should be, if scientists at all understood what astrology actually is, a complete vindication of the entire basis for it. If a sign like Leo or Libra had instead achieved the top position it would have invalidated it, but instead it validated some of astrology's most fundamental and basic principles.

But there are also many major astrological disciplines, from traditional Vedic astrology to Western Astrology to Chinese Astrology, and there will naturally be major imperfections, inaccuracies, and inconsistencies in any interpretation of anything let alone something as non-standardized as astrology, adding in the thou-sands and millions of unregulated and non-standardized practitioners, each with their own opinions, biases, agendas, and experiences, and establishing any baseline without serious, academic understanding of astrology is just entirely ridiculous and is like trying to prove whether individual philosophies are more effective and accurate than others—very difficult and esoteric and so must also take those into account in the controls and methods rather than contemptuously disregarding such complexity.

Additionally, when people conduct cynical or specious studies on astrology they also forget that the purpose of doing studies is not to prove something to be *true* or not but to investigate hypothesis in order to create knew knowledge, to *find* out what is true, not simply that something is untrue, and the purpose and func-tion of studies into astrology should be not to invalidate the opinions of astrologers (because even this book is nothing but my own personal opinion), but to help illuminate the underlying forces of reality which compose it.

For this reason I also find that Western astrology is the discipline most likely to accurately reflect underlying metaphysical laws which cause the phenomena astrology seeks to describe. For while there are things of value in other disciplines, Western astrology has adapted and changed the most over the intervening centu-ries in attempting to incorporate new information and discard old and outdated practices, which is itself a scientific behavior that helps to increase the validity and effectiveness of any system, although other systems can be investigated too, this book is concerned only with common Western rational astrology.

Every Sun sign does have some dominant traits which can be observed, although they could not be categorized into an entire personality, but are still demonstrably more prevalent in those signs. Having the Sun in Scorpio, for instance, makes for people who are often very dominating (in a conquest type of way), often also in terms of sexuality, which may not be obvious to observers since sexuality is often a private matter, but Scorpios are especially interested in seduc-tion and conquest, and a subjective study asking participants whether a sexual partner should fulfill their desires might return the very highest percentage from

Scorpios (or anyone with Venus, which rules love, in Scorpio).

Similarly, Leos are always, always the center of attention, and one of the most obvious and reliable constants in the zodiac is that simply asking whether or not a person likes attention will return more results from Leos (or those with a Leo Moon). This is so reliable that once in a chat on a livestream in which the conversation turned to zodiac signs I guessed a random chatter I had never interacted with as Leo simply because of their unbridled use of exclamation marks, which are a method by which to draw attention to oneself without being antisocial (because Leos also abhor antisocial behavior).

Aquarians are *always* eccentric, Cancers *always* love cooking (once they've had the opportunity to learn it), Geminis talk, and talk, both Taurus and Aries are *always* strong-willed, Virgos are always complainers, Sagittarius are always adventurous, Pisces are always overwhelmed, Capricorns insular, etc., etc. (obviously I'm using the term 'always' hyperbolically in this context and not literally).

One particularly interesting constant among signs is that, unless there is an aspect which undoes this effect (which would require understanding such exceptions), nearly all Libra men and Libra women present with some deviation from stereotypical gender maximums of their gender than those from other signs, meaning that Libra men are always more soft or feminine than other men and Libra women always more masculine. This occurs because Libra embodies *balance*, though most people think of balance as its opposite meaning (to be well developed, whole, or well-practiced) when balance instead means more equal distribution of all parts—equal play as much as work, equal chaos as much as peace, equal give as well as take, and God are Libras indecisive because every choice has balanced pros and cons between which they feel equally compelled. In fact the time it takes various zodiac signs to choose between something like flavors of ice cream could also have similar usefulness as the Nobel study free from bias and influence of authors and participants.

But because personality is subjective it might instead be possible to measure literal physical gender characteristics like the amount, density, and distribution of body hair on male Libras compared to other signs, and since high dihydrotestosterone shifts body hair patterns to be centralized to the sagittal plane and forearms. Or the increased forearm circumference in ratio to wrist size in males since dihydrotestosterone also increases forearms size. Or the pitch of the voice in Libra women compared to other signs could be studied too.

Being on the cusp of a sign would also affect results, however, so studies would also need to control for that, but none do because the people investigating astrology have such contempt for it they don't bother to even understand what it is they are actually studying, and the purpose of their work, unlike studies into things like mitochondria or the nervous system, is to spit on astrology rather than understand it, which is itself a significant and inexcusable bias for which researchers should be ashamed to call themselves scientists. Like the Nobel study, investigators can use measurable, non-subjective data and statistics in order to reduce the introduction of invalidating biases, prejudices, and inaccuracies of subjective data, which will also become more informed as the study and exploration of astrology advances such as what I attempt to accomplish in this book, due to greater understanding of the underlying realities that construct reality, much the same as the often absurd, early reconstruction of dinosaurs by early, amateur paleontologists did not invalidate paleontology.

To this end there are other plain, apparent, and measurable transits and aspects in astrological discipline which can be measured empirically which are not given to such subjective and insubstantial evaluation as a 'personality summary,' and even on relatively short time scales, although long term studies would also be advisable. One such reliable point of observation is the transit of *Mars*, which moves fast and represents action, ambition, and initiative, and thus has correlative effects in people's lives that are nearly always very obvious and very strong. Instead of measuring large, generic, group trends based on popular entertainment astrology, recording whether individuals experience the common effects of prominent Mars aspects to their natal chart during its transit can return more obvious correlative data. For instance, Mars transit opposite a person's natal *Mercury* correlates most often with aggressive and confrontational *communication* such as angry emails, social media fights, or in-person, vociferous interpersonal argument. This is true because the nature of transits take on the nature of the planet at issue—for instance Mercury is the planet of knowledge and communication, but it is also fast, fleeting, and not very powerful, so most transits by Mercury are short, fleeting, and without very obvious impact except during its retrograde (the effects are obvious just not so much as to cause pause or disruption). Mars however is aggressive and assertive, and its influences very strong, and while it transits the zodiac rapidly it is still slow enough that its influence is more obvious than Mercury or the Moon, so the effects of its transits happen immediately, usually but not always before the transit even becomes exact, and lasts long enough to be more apparently obvious than Mercury or the Moon. Being aware of the unpleasant effects of Mars opposite my natal Mercury, after finally recognizing after several years there was a correlation with the real experiences I was having, then was determined one year to resist any and all opportunities for conflict during this transit as it approached. Sure enough the day of this transit there was a highly unpleasant interaction which presented itself on social media, but I entirely resisted responding, which was incredibly difficult and required an enormous exercise in self control. Strangely, after leaving that opportunity *another* later that day presented itself, which also required great pains to ignore, and this pattern continued the entire rest of the evening and ended only once I went to bed. While of course social media is filled with opportunities for chaos and conflict I have been on social media almost every day while writing many parts of this book and not gotten into seriously conflicts or felt as enraged as what occurs under the influence of Mars, which occurred before I was aware of the influence, which I would find *after* the event and not before.

Mars square natal *Pluto* is also usually (mostly but not always) a disturbing or profound experience because Pluto represents power and control and usually in this transit we experience confrontation in relationships concerning power dynamics such as with business partnerships, family members, or romance. Because all humans experience Mars influences, a double-blind study with participants who have no idea about astrology will demonstrate the validity of these correlations if the study can correctly collect the requisite data and accurately establish when such commonly difficult transits to the natal chart by Mars in opposition, square, or conjunction with natal Mercury, Venus, the Sun, Ascendant, etc., knowing, by the way, that the effects of Mars transits often occur immediately when the transit begins to take effect, and not only at the exact aspect (where other planets might also do this or come during or at the end of the transit, depending on the nature of the planet).

Because Moon transits are also rapid they are easy to measure empirically, but the Moon in traditional astrology does not represent any power or influence on its own and is instead a mirror or reflector and as such reflects the energy of anything it aspects. As such, the Moon reflects all the astrological planets in a person's chart as it travels through it, but one of the Moon transits in my personal experience which is most obvious is the Moon transiting conjunct to the natal Sun, which is described by my preferred astrology source (*Astro Dienst*) as causing a "burst of energy, a time when the mind and body feel recharged," and as long as there is not some overly stressful drama occurring in my life this transit *always* causes a renewed sense of wellbeing and contentedness, usually even an impromptu underwear dance party around my apartment, even when I have no idea it is occurring (in spite of writing this book I do not constantly monitor my astrology, for reasons that will become more apparent later).

Similarly, whenever the Moon conjuncts my natal Pluto, which also represents transformation (because of the effects of power and control), I always have a highly emotional experience, not necessarily depressing but always emotionally the most remarkable compared to other times of the month.

Moon transits are also highly correlated to cyclical interpersonal relationship patterns, since we are typically in close proximity to those people, and if you've ever noticed that you or a partner have arguments with repeating periods on a regular (monthly) cycle it has nothing to do with the content of those conflicts but is because of these kinds of Moon transits which are as regular as, well, the cycles of the Moon, and there is an inextricable and undeniable cycle in every person's life occurs for the length of lunar cycles that could also reflect in study data. Because women have an estrus cycle and experience menstruation also on a near monthly (lunar orbit) schedule their emotional nature is often blamed on that estrus cycle, but in fact everyone has these monthly cycles and can be just as emotionally sensitive and volatile on a regular and cyclical patterns even without menstruation, but since there is an absence of outright physical evidence it is far easier to ignore or deny (being a man who loves men I should know).

Any time the Moon aspects with *Uranus* or Mars I also get a sudden urge to do something fun, unexpected, or exciting (though I don't always act on it, which is an important caveat in collecting data), though my natal Uranus and Mars are also conjunct so this likely exaggerates that influence for my personal experience than is typical of others.

When the Moon conjuncts *Neptune*, the planet of spirituality, is also when I usually feel most at peace, and there are many other such combinations with other planets and influences. But most aspects of the Moon are not so obvious, though, as its aspects to natal planets are so regular and so brief (at most half a day) they can just fade into the background noise of life, because all life at every moment is a sum of all the combinations and influences which are currently active, which is never zero, and can be as much nearly all of them.

Transits of the Moon through the signs themselves last longer than aspects to natal planets, for about 1-2 days, and might also be easier to quantify. Whenever the Moon is in the *twelfth* zodiac house for Libras, for instance, I always feel a sense of disconnect from the world around me, sometimes depressing, as if I cannot find fulfillment in relationships or goals, because the twelfth house is associated with spirituality, rest, and endings (and delusions). Though they are often a prominent focus of astrological practice I have never really been able to observe the correlations of the Moon's phases (full, new, etc.) on my life, nor eclipses, which does not mean there is nothing there, I just have not seen very obvious evidence of it. One interesting full moon correlation fully established by studies, however, is the increased rate of violent crime and suicide which occurs during a full moon, regardless of whether it is actually visible, which is yet another empirical dataset that proves the existence of metaphysical laws of reality of which we are willfully naive.

One undeniably uncanny proof of the validity of astrology as a concept (not necessarily all astrologers, remember) is also the fact that *both* my parents have their natal Moon in their fifth house which is described by the astrologer, Robert Pelletier (through *Astro Dienst*) as likely 'having a child who will become famous or notorious.' I first saw this description years ago before starting my website that would lead to publishing my first book and thought perhaps since I was the person in my family endeavoring to work in the film industry could possibly be the child it referred to, though some of my siblings also had dreams that could potentially take them into the public eye, we now know for sure it was me, though as to whether I am famous or notorious is, I suppose, a matter of opinion (lol), but certainly is true that I am one or the other as it says, generated by an automated algorithm yet also exactly correct (but also worth pointing out this did not become correct until their child was in his mid-forties, so for instance trying to prove its validity in those with the influence would be difficult since publicity could occur at any time in a person's life (and would only be relevant to anyone with children anyway, as the Moon in the firth house has several meanings referencing creation).

One of the most Universal, consistent, obvious, and measurable astrological influences is the *Saturn Return*, when the planet Saturn returns to the place in the zodiac it was when a person was born. This is actually the first *life crisis* and occurs around the age of thirty (about ages 29-30) during which we suddenly realize we have been an actual adult for years and are thus forced to evaluate whether or not we are on the path we wanted for ourselves. If the answer to this dilemma is 'yes' we recommit and work all the harder to achieve our goals but if the answer is 'no' we often make new decisions and new commitments in order to get there, but the later of which (such as in my own case) is often ineffective because the reason we find 'no' is due to distractions or problems which have derailed us from our desired goals in spite of our best efforts, often because of lacking skills required to otherwise achieve them which are not necessarily resolved simply because we try harder. When this transit occurred for me, long before I knew anything about astrology, I realized I was running out of time to find a partner, settle down, and start building a life and unfortunately because I was still very traumatized by my youth and lacked effective interpersonal skills required for evaluating good romantic partners ended up impatiently settling for someone who was volatile, unreliable, and did not have my best interests in mind which then later blew up in my face and caused even further deviation from my life goals. But the Saturn return is a consistent period of evaluation and first life crisis that *every* person goes through around the age of 29-30, the severity of the crisis depending on whether we have the tools and skills required to achieve our goals, or not.

But Saturn's transit around the zodiac also has a very specific influence on the general pattern of life in thirty-year cycles, where Saturn's position across the *Nadir* (the "bottom" of the chart, as discussed in upcoming chapters) also correlates with muted public success and often obligatory retreat from worldly progress. When my parents both had Saturn in this position their business became chaotic and fell apart and they even absconded again to Hawaii to start over. The years in which I began and finished these books coincided with this long and often challenging transit of Saturn, while most of my siblings had Saturn transiting more or less at the top of their charts and were busy with their careers and families while viewing me as a failure for not having any money. Everyone, regardless of life path, will be obligated to experience the Saturn cycle's influences where its transit across the bottom is a period of challenge which compels inner growth and introspection while simultaneously limiting outer, material success relative to that person's life path. Most interestingly, *both* of the two recent Presidents of the United States *lost* the Presidency while Saturn transited the bottom of their natal chart and were thusly forced into a period of obligatory introspection. Opposite-ly, when Saturn is at the "top" of the chart is a time when people usually reach a pinnacle of outward, material progress relative to their previous experience.

This pattern is also especially obvious with schoolchildren, especially because they have not lived yet very long and do not have complex responsibilities it is easy to see how those whom have Saturn transiting the top of their charts are more popular, extroverted, and active with school and peers while those who are bullied, shy, or introverted instead have Saturn transiting the bottom such as was my experience as a young teenager. A survey presented to children to rank the most popular other children would very likely return the most popular as all having Saturn from their Descendant up to the top of their charts, while the others will all have Saturn moving down and around the bottom. Likewise, when Saturn transits the bottom of the chart celebrities often fall out of the public eye so things like film appearances or plays of their music on streaming sites could be a great correl-ative study. As a specific example a recent, young, highly popular musician was a worldwide recognized musical talent but has, for the last several years, retreated from public view and been open about his internal conflict and search for purpose which, at the time of this writing, Saturn was going through the bottom of his chart.

During my parent's transit of Saturn through the bottom of their chart they *both* experienced loss and challenge in their professional lives and involuntary retreat from public life, rather than one having success and the other not. All but one of my siblings also started and grew their families before, during, and after Saturn's transit through the top of their chart, and the one with Saturn in the bottom found the experience of parenting far more distressing and disturbing than the others. During Saturn's transit of the Nadir a person will encounter increas-ingly difficult and frustrating failures, stagnation, and roadblocks compared to other life stages that seem to prevent us from realizing our material goals, and though I worked very hard during this transit to rebuild my life and even wrote and published two books during this time (and am working on this one as Saturn rises toward my Descendant) I found myself unable to cultivate a large following in spite of even having accomplishments as finding the cure for cancer or have much impact on the world and lived in small, humble apartments with very little personal possessions, romantic opportunities, and even friends, even when I tried to find and create opportunities to resolve those deficiencies.

As discussed in more detail in the upcoming entries on Saturn, the purpose of its transit across the Nadir is to focus our inner development, which is why I was able to write so prolifically, because with plenty of time to think, reflect, study, research, and ponder there was then an abundance of material. Not everything goes wrong during this transit, in fact this was probably the most important period of my life in which this lull in progress helped me accomplish the things I have, refining my knowledge and experience, got to know myself better, and resolved a great many past traumas and inconsistencies in my life, while the public recognition from this work which will surely come later will not be the actual accomplishment, but a reward (I suppose?). Indeed, if my first book, *Fuck Portion Control* had been met with wild success at the very outset most people would have consumed it years before it actually contained some of the most important discoveries and thus not benefited as fully as now when my works are more complete.

Relationships are also excellent opportunities for establishing the validity of astrology, and in fact the first time I realized astrology held at least some truth was when learning about Venus signs and the placement of Venus in the natal chart, which had been *disturbingly* accurate in their description of every one of my past love interests. Where most romantic astrological advice concerns the Sun signs, which can indeed affect conflict or compatibility but is not the primary influence, our Venus position is how we relate to love and partnership and is far more influential in the dynamics between both lovers and partners which are not romantic but still very close, such as in business (and as such Venus also rules many aspects of business). My Venus is in Leo, and one long term partnership was with someone whose was in Pisces which, while being described as also an incurable romantic by one of my other favorite astrologers have fully opposite love nature in which we do not understand the motivations of the other and they are unable to make me feel "special and unique" in love as my sign typically desires, and I resent a Venus in Pisces inability to be direct and plain, which is exactly what they were like. Many bad astrologers focus on Sun signs in romantic pairings and do not actually discuss Venus signs, which is the most influential, and even the most casual astrologer who understands Venus signs usually does a pretty good job at describing or characterizing the romantic *dynamic* between partners based on their Venus placements, not the success or failure.

Because Venus governs relationships there will also never be a relationship that does not fatefully begin with some involvement of transiting or natal Venus. When first meeting a past partner they had been serendipitously invited to join some friends at dinner, but Neptune was exactly opposite my natal Venus, which indicated delusion and deception in love, which was exactly how that relationship played out. But my future partner not only had Jupiter the planet of luck and opportunity conjunct his natal Venus it also retrograded over it during our first few weeks and months knowing each other during which I initially evaded his advances. During our meeting both transiting Venus and Mars were also nearing conjunct, which is often a powerful impetus for romantic meetings, which then reached conjunction almost exactly at the time I finally decided to give him a chance. Uranus, the planet of surprise and unpredictability, was also extremely close to his Venus and also in retrograde at the same time as Jupiter, so it may have also had influence (Uranus concerns both surprise and unpredictability).

In terms of astrology, people are most commonly familiar with the concept of *Mercury retrograde*, where retrograde is when planets appear to move backwards in the sky due to our position on Earth relative to their orbit of the Sun and different speeds at which the planets each move. Classically during Mercury retrograde there are often disruptions in communications, plans, and systems, and while there are often failures of these things outside of Mercury retrograde, since the influence of Mercury and other planets do not only occur during its retrograde, problems are often amplified or more numerous and obvious during its retrograde, encompassing even major events such as the recent Crowdstrike/Microsoft server failure that took down an insane number of government and private sector networks and computers. While people often focus directly on the exact retrograde period there is also an important *shadow period* before and after, from the point where Mercury will backtrack until it then fully exits past where it first turned stationary, and the retrograde itself always concerns things which occur in those shadow periods, with the resolution occurring during its post-retrograde shadow. The Crowdstrike fiasco of 2024 occurred three days into the Mercury retrograde shadow period in July and then resolved as the retrograde progressed. Similarly, the 2010 McAfee DAT 5958 update that also crashed millions of computers all over the world also occurred during Mercury retrograde, as did the famous 2021, six-hour outage of Facebook, Instagram, and WhatsApp social media apps.

Mercury retrograde also has themes depending on what house it occurs for people personally or the world at large, and during one Mercury retrograde I received a dreaded letter from the IRS indicating I actually owed more money than for what I had filed, which occurred in my eighth house of power and control which is commonly associated with shared resources as taxes (they are shared because it is both government and citizen money). Mercury retrograde is thus a rich field for possible and interesting study inquiries if designed and executed correctly. In fact, of 8 major outage events I found from 2024, 5 of them occurred within Mercury retrograde or its shadow period. It is a small sample but given Mercury was *not* in retrograde or its shadow for 70% of the days that year that kind of result would more than legitimize any other study if was done on less taboo subjects as astrology, and similar results from quality data would likely be similar for any analysis, where being large enough to make the news is probably a requisite criteria for consideration in the broadest, world-wide analysis.

Mercury is, of course, the planet of communication and knowledge, and because it transits the zodiac so quickly its effects are highly noticeable, and there are not only disruptions or developments during its retrograde but also any time it forms aspects with other planets during its transit. For instance, in the last few weeks editing this book there was a period of days in which lots of miscommunication occurred that felt much like Mercury retrograde but which was in fact semi-square (a mildly disruptive influence) with both Uranus and Mars at the same time. The influence of Mercury over communication and knowledge can seem so regular because as humans we are so conditioned to speech and communication dynamics as part of our innate existence, but specifically looking at Mercury transits and how they correlate with specific events in our lives reveals very obvious patterns of influence.

One of the most uncanny proofs for the validity of astrology is that the far majority of all couples present with close alignment of the great angles, the Ascendant, Midheaven (also the called the Medium Coeli), Descendant, and Nadir (also called the Imum Coeli) because the angles represent the life path, our life journey, and the events, occurrences, and people we meet along the way, so a person's life path would need to be at least somewhat similar to others to even meet, let alone start a relationship. My parents's angles are similarly very close, as are also my own with my former partner, and other couples whose data I have seen mostly present with their Midheaven alignment no further than one sign away from their partner's, and this pattern also appears so frequently in available astrological data that if this frequency rate occurred in any other study it would prove its investigation fully beyond a doubt.

But this association is stronger in relationships which are organic and spontaneous rather than those which serve ambitions and professional goals—for instance a very famous couple in politics do not have the alignment but the husband has it with a woman he made famous through having an affair while he was President. Some of those which are not so aligned instead have opposing placements, which is an interesting and legitimate deviation and likely reflects relationships of utility rather than organic pairing. But even more important when dealing with the angles is that the quality of the data must be *absolute* because they are determined by the exact *time* of birth, not just the day and location, and because this data can be easily incorrect it can return a completely invalid result, for instance there are two conflicting claims of birth time for Princess Diana, though interestingly neither is aligned with King Charles with whom she was famously recruited for a relationship of convenience, where instead Charles shares the same placement of Midheaven in Aries with Camilla.

As such too one quite regular aspect that occurs both often (twice a year) for each person and is very obvious is when the Sun squares the Midheaven. As the Midheaven represents our material, public, and outer life (as opposed to the interior, family, personal), when the Sun squares the Midheaven it is nearly always a day when we put in very long hours or have otherwise significant activity in our work and professional life than is typical of other days, the "square" aspect being one that is often inharmonious and associated with stress. There have been a few days on this transit where it wasn't quite so busy but the majority of them are and would return significant correlative data when tested across a wide group of people as to how many hours they put in each day during the year and how intense or busy they would rate each day which would reveal such correlations.

Because they are so close to the Earth, the retrogrades of both Mars and Venus are also the least often of all the planets, and whenever Venus goes retrograde people become spontaneously more sexual and less interested in the flowery, romantic visions of sexual intimacy and instead are just more interested in fucking, which of course is fine because sometimes too much lovey-dovey romance can actually interfere with things like procreation, and deep down many of us become highly aroused by the sexual attention of someone who just wants to ravage us for no other reason.

But this action of Venus retrograde also results in less kindness and compassion in common interactions—it's not enormous nor alarming, as Venus is not an especially aggressive influence, but incidents where people might otherwise be nice tend to otherwise be charged with self-preservation, selfishness, and short tempers, which may also help us to see where we may be excessively pleasing or self-defeating and offer us a chance to find more self-determination and set boundaries, but the pattern is there nonetheless.

The Venus cycle, however, is also a larger, longer, and more complex and influential cycle than most people are aware. Lasting 16 years, the Venus cycle is one of the major cycles which correlate with cycles here on Earth. Though Venus takes 8 years before completing its wandering path throughout the sky, the apex at eight years also coincides with a switch between the predominant nature of influential forces, events, and ideology throughout the world before embarking on another 8 year cycle of its ideological opposite, thus a total of 16 years, not 8. Interestingly, the Presidential election cycle in the United States is aligned almost exactly with this Venus cycle, so every election will occur when Venus is either in Libra or Sagittarius. In the former she appears as the *Morning Star*, although because of light pollution most humans now have no idea of this otherwise prominent sight in the previous entirety of human history, during which the status quo tends to be upheld and usually coincides with the retention of power, or if power does technically change hands it does not go to someone ideologically much different than those in power (i.e. it remains in the establishment). When in Sagittarius Venus is oppositely the *Evening Star*, and power usually does change hands and to those who are far different than the previous from whom they seized power.

As I write this book the elections in Canada and Australia have just occurred under a Morning Star with power being retained by those which already had it, but the election in the United States last November was an Evening Star during which power went to the opposition. This correlation with power transfer and the Venus cycle has been observed for thousands of years by astrologers, conquerers, rulers, and usurpers whom were often advised of this phenomenon in their political pursuits, but in reality the astrological cycles of Venus are not concerned specifically with who wields power but in the nature of sociopolitical balance. During the eight-year Venus cycle in which Venus occurs in Libra during U.S. elections she helps conserve and preserve, so ideologies of conservatives, authoritarians, fearmongers and warmongers, xenophobics, racists, and all those who resist change tend to be favored. But during the following cycle when Venus occurs in Sagittarius during U.S. elections she instead becomes more of a bold revolutionary whom favors change to prevent stagnation and thus those whom subscribe to hope, progress, and egalitarianism, with each of these two disparate themes reflected not only in those who rise to power but also the support they receive.

The purpose of the Venus cycle thus is not to win power but in the maintenance of *Karma*, which instead as is the true nature of Venus to achieve balance in the Universe, to maintain the breadth required of life in that there must exist good and evil together, not one or the other in perpetuity, to understand how power and control is determined by the Universe because such dynamics are in fact those of *partnerships*, not true power, which is instead the domain of Pluto, and how deluded we are when given power or expecting it since none of us in fact has power over the laws of cause and consequence to decide such things. As such, people whom deign to design their campaigns around the Venus cycle are deluded to the effect and nature of astrology, thinking they have some advantage or cheat code when the victors and losers in previous cycles had no idea of this influence and still won or lost, and knowing of the cycle can only be correlated with winning or losing, not causing it, which would instead put the power of reality within our own hands, which is never the case.

If we include even the superficial subversions of this pattern and only consider strictly if power changed hands during a Morning Star election in the U.S. Presidential elections or did not during an Evening Star election (which are opposite of expected), of the 61 which have occurred from the founding of the United States only 18 have subverted this pattern. That is a 71% rate of correlation which is absolutely astounding and effectively proves beyond a doubt the association of the cycle of Venus with power. If we did consider subjective qualifications of whether the *establishment* effectively retained or lost power it would be an even greater percentage probably approaching 85% or 90%, because in reality the influence is 100% there but it is *also* affected by other influences as nothing in reality exists in isolation and instead built upon *all* the laws which create it and not limited to only one at a time, but they exert considerable *influence* which cannot be denied and is also readily apparent even in empirical data.

Technical exceptions to this pattern such as occurred in the 2020 election may thus also be described as effectively maintaining power to the establishment, the candidate coming to power a man whom had been in Federal government since before I was even born (whom then also proceeded to run a campaign of genocide), but the pattern also technically held true anyway in both 2016 and 2024 to the surprise of a great many people (though 2024 really wasn't surprising if you were at all paying attention to the ongoing sociopolitical and geopolitical conflicts). Another apparent deviation of this cycle was the Presidency of Franklin Roosevelt which began during a Morning Star election amidst the pit of the Great Depression, which was obviously a major extraneous and more powerful influence which could easily override such patterns, and indeed the times in which this pattern has deviated from the expected pattern are during significant sociopolitical turmoil such as the elections of Lincoln and Grant at the Civil War, or the recent political chaos. But influences are never subverted, it is only that we do not fully understand them and wish instead for stability and predictability in order to control our own fears of the unknown, and although described as a political moderate Roosevelt was initially very conservative in his politics and was only nominated to run for the Presidency in the first place because of his demonstrable past as an establishmentarian who courted support from all political spectrums, including outright racists, and his first term and attempts at reform to save the country were entirely in support of the very institutions and policies which caused the depression in the first place, supporting in fact then the Venus pattern of *upholding* power during the Morning Star, giving banks shelter from their depositors and empowering the Federal Reserve with even more political power, and many of the policies and actions taken by Roosevelt were in fact designed and already put in place by the previous Hoover administration which greatly pressured and influenced Roosevelt, effectively making him simply a continuation of that power as the cycle describes. He also slaughtered excess livestock and paid farmers not to grow food while many Americans literally died of hunger, ended antitrust laws in an attempt to manipulate consumer prices and cut benefits for veterans by a whopping 40% which saw 500,000 veterans and widows removed from pension rolls at a time when these people were literally homeless, jobless, and starving.

It wasn't until 1936 when Roosevelt won one of the most lopsided victories in American history, a subsequent Evening Star election, that he truly reformed the government and implemented many needed changes, including the repeal of prohibition. By this time Roosevelt had lost the support of businesses because his other reforms began impairing their exploitation of consumers and workers, and the real, karmic shift in power described by the Venus cycle finally occurred, which is not concerned with technicalities but instead metaphysical laws of karma and balance. During this time other reformations occurred such as outlawing child labor, establishing minimum wage, overtime, and the 40-hour work week which most people today think as something that just fucking happens and was not in fact won through the literal blood and sacrifice of our great grandparents. The economy had again begun to decline into recession because of the failure of policies in his first administration in support of the establishment and Roosevelt began a public campaign against monopoly power and big business, and finally provided direct stimulus relief to workers rather than institutions and created more than 3 million jobs at a time when the U.S. population was under 130 million. He also expanded the National Parks and Forests systems, the Conservation Corps saw great success, and unemployment finally began to decline persistently, and the entire reason for the constitutional amendment restricting candidates to two terms was a reaction by the establishment to Roosevelt's populist reformations which directly restored the middle class, in order to prevent the middle class from ever again retaining power for long periods of time.

Of course, World War II also began during Roosevelt's second term, and in his third term, a Morning Star election that favors retention of power, the U.S. formally entered the war. For his fourth term Roosevelt again appeared to subvert the pattern of an Evening Star defeat but in reality he was rapidly declining in health, which he had been keeping secret from the public, and had been convinced to replace his vice president with Harry S. Truman, the man who would end up murdering innocent Japanese civilians with the first use of the atomic bomb, and the war created a massive paradigm shift in power both within the United States and without, culminating in the creation of the United Nations and the literal death of Roosevelt just one year into his reelection, demonstrating how such cycles as reflected in Venus are absolutely *not* a function of our own wiles and schemes but instead the will of the Universe in its organization as reflected in the cycles of Venus.

Because money and wealth come from others, and our relationship with others is ruled by Venus, some income and business cycles are also reflected in

Venus dynamics such as where it is located in our birth chart or what transits are actively engaged. During Venus retrogrades which occur every 18 months always also coincides with business dynamics such as the Hollywood writers and actors strike of 2023 which began exactly during the pre-retrograde shadow and extended entirely through the retrograde to resolve in the post-retrograde shadow. Again, this *does not* mean that Venus transits can be used to predict wealth, 'wind-falls,' or tragedy as astrologers so often use it, which is in reality nothing more than exploitation of people's common desire for and fears of material security, but instead to understand the nature of relationships to others which can also sometimes include themes such as income or wealth (and if any astrologers desire to be good and effective it is required to become a master at inventory practice in order to shed biases, fear, and insecurity which otherwise corrupts understanding and wisdom).

Other factors which can and do also affect wealth are any influences to the 2nd and 11th houses, for instance I was also born with Neptune in my 2nd house (not Zodiac house, but the houses derived from the angles) which indicates vagary and delusion around finances and fortunes during my lifetime, and indeed found the ability to even find opportunity to earn wealth confusing and unproductive for most of my life, and anyone with this placement tends to frequently demonstrate poor ability to manage finances and investment schemes unless it is transitioned toward a spiritual function which transcends the material to instead value that which is *immaterial,* especially in the service of others as described in the upcoming chapter on spirituality and Neptune. But my natal Venus is also located in my 10th house, which represents society, and appears to indicate I will eventually find wealth through public recognition, likely through this work and research I have done over the last decade such as is presented in my books, as opposed to say someone with Venus in their 7th house which would instead find wealth through a partner, or Venus in their 4th house for which wealth would come from home and family, etc., which could also be established in empirical data.

Mars being the planet of action and initiative its retrograde motion is similarly most commonly characterized by *stalling*, not necessarily of things failing as what Mercury retrograde often brings but just stuff not happening or not 'getting off the ground,' so to speak. It can be a period of boredom, frustration, or frenetic and undisciplined energy as we desperately try to get plans in motion but find many roadblocks or obstacles. The purpose of retrogrades is not to arbitrarily impede our progress, however, but to force us into introspection so that we might better understand what we are doing and recognize the flaws or inconsistencies in our work and learn new information, and contrary to how retrogrades are often characterized they are in fact meant for our benefit, that we may be more effective in the future. In nearly all Mercury retrogrades, for instance, I often finally solved problems and research that was giving me a lot of problems otherwise, so I actually began to look forward to them as times when I would find solutions.

When we fail to take time to plan or rest we often miss important life experiences, details, and truths, so Mars retrograde, which only happens once every two years (the least frequent retrograde of any planet as might be expected of the planet of *action*), also helps us slow down, take a break, and appreciate life if we are not obsessively ambitious. For those which are it can instead be a frustrating time, and at the recent Mars retrograde (which also was very near the Venus and Mercury retrogrades too), I found a great deal of frustration in getting my work progressed, but then as it began to resolve realized exactly how to finish this very book which, as they all three began forward motion again, wrote sometimes as much as 10,000 words a day. This is obviously not something which can be easily quantified but in personal experience it is very easy to observe the effect of Mars retrograde, especially because the sign in which it retrogrades relative to the placement of our Sun sign will then affect that area of our lives. For instance the recent Mars retrograde occurred the 10th zodiac house for Libras, which concerns public image and career, and I noticed specifically obvious slowdown in traffic to my website, reduced inquiries for assistance, and otherwise lacking in general interaction from the public towards me and my work, and when the retrograde reversed and got going to normal speed and through the post-retrograde shadow several new people signed up for coaching and traffic to my website and social media once again picked up.

Mars being the planet of action and initiative also triggers many events as it aspects other planets in its rapid transit through the zodiac, especially those in the world around us. For instance in the last weeks of writing this book Mars went inconjunct with both Saturn and Neptune and squared with Uranus the moment of the start of the Israel and USA attack on Iran. As discussed in the upcoming chapter on world events and great cycles the Saturn and Neptune conjunction is a time of endings for dysfunctional, oppressive governments and institutions and the beginning of new cycles which replace them, being officially activated by Mars transiting the conjunction while Uranus is the planet of surprise which further characterized the nature of the conflict's beginning, and will likely continue well into 2026 since that is how long the Saturn conjunction with Neptune lasts, with Mars being the factor which activated these major transits.

The Sun in astrology also serves to activate other aspects and transits which are occurring, both in our individual lives in relation to our natal chart but also other transiting planets in the general experience of everyone at the same time. An especially obvious, recent example of this was the Sun transiting opposite to Uranus, currently in Taurus in November 2024, activating its influence there more prominently. For Libras this is the eighth house of power and control which also affects things like shared resources, and a delivery of flour I received that day had exploded in the box during transit and went everywhere inside it and out when I opened it, causing me to lose some in the process. Surprisingly (although not so, since Uranus is the very planet of surprise and unpredictability) my request for a refund at the site of purchase was immediately refunded within seconds of requesting it, and I was able to salvage most of the flour, and being as expensive as the bag was amounted to about $60 of free flour even though it was technically a disruptive (opposition) event, fully characteristic of the planet of surprise. A Sagittarius I knew also asked that day for advice on some surprise, acute stomach discomfort which had started the day before and wasn't going away, as Uranus was Sagittarius's sixth house which governs health as part of our requirement to daily care for our needs and obligations, which I was then able to help them.

While Sun transits to every aspect of the natal chart occur every year they do not occur on the same day, due to the precession of calendar days which occur every year (the reason we have leap-year), so when we also have repeated patterns of behavior at different times of year this is also related to Sun transits and is thusly also quantifiably observable. During a transit of Sun square natal Mercury, for instance, I *always* have some dilemma concerning themes of communication— recently this was an a software bug in one of the 3D design programs I use and had to consult forums which weren't able to help, and then had to file a bug report with the company that makes the software. I also always have a transit of Sun square natal Saturn every year just before Christmas, and throughout the entirety of my life these days leading up to the holiday are always fraught with tension, which melts as soon as the transit passes just a day or two before the holiday, and the holidays are then afterward usually enjoyable. Some might say that, well, yeah the Christmas holiday would be stressful for you given your experience growing up and, YES, THAT IS EXACTLY THE POINT—that we can actually quantify metaphysical experiences through observational corollary with the world around us the same as what is observed in other natural phenomena, that it is not arbitrary and specious but in fact built into the very laws of reality as what construct the physical world, which can then be studied and understood to better help us understand the laws of the Universe.

As with something that is so complex and interpretive as astrological practice, doing so through scientific rigor in turn requires discipline and mastery and not specious and cynical studies infected with the bias of intent, which in turn do nothing to disprove astrology but instead the supposed work of the insincere weirdos who cannot even be bothered to properly understand the subject they are supposedly studying. The reason I write this book is because just as there were realities I observed which helped me elucidate the cause and cure of diseases like addiction, depression, and cancer there exists real and apparent correlations between the observable Universe around us to the metaphysical events in our daily lives and the ebb and flow of all existence, and studying them to better understand metaphysical laws of reality should be the subject of a scientific discipline and inquiry, because everything in the Universe exists at its pleasure, not apart from it as so many would otherwise prejudicially prefer to believe.

15. How to Use Astrology

There are ten planets in contemporary astrology—Pluto is still considered a planet in astrology, as the word *planet* means a 'wanderer of the heavens,' so this also includes the *Sun* and *Moon, Mercury, Venus, Mars, Jupiter, Saturn, Uranus, Neptune,* and *Pluto,* and the word 'planet' was later coopted by science to specifically mean massive celestial bodies which orbit a star and are large enough to clear their local orbit (this is still an errant definition since there are many rogue planets which actually no longer orbit a star), but for purposes of astrology the 'planets' are those major heavenly objects which transit the *sky,* as opposed to the *Great Angles,* the *Houses,* several *asteroids,* and mathematical points of interest like the *Nodes* of the Moon which do not.

The *Zodiac Belt* which encircles the Earth is the primary foundational feature of astrology, and is divided into 12 equal divisions each identified by their prominent constellation *Aries, Taurus, Gemini, Cancer, Leo, Virgo, Libra, Scorpio, Sagittarius, Capricorn, Aquarius,* and *Pisces.*

Astrology has been around a very, very, very long time and there is an incredible glut of esoteric and pedant trivia concerning its practice, much of which is truly superfluous and some of which is just wrong or misguided, and the best use of astrology concerns simply the prominent aspects and elements in order to understand our general life path and personality, and not to get 'lost in the weeds' or to use astrology as a distraction or coping mechanism in place of actually living our lives and being present. In reality, all planets *are at all times in aspect* to all other planets and elements, but it is at at the major aspects that we see obvious correlations with the events of our lives or in the world around us as what can be observed when plotted on a geocentric (Earth-centric) graph commonly referred to as an *astrology chart* (as shown in *figure 2*).

figure 2.

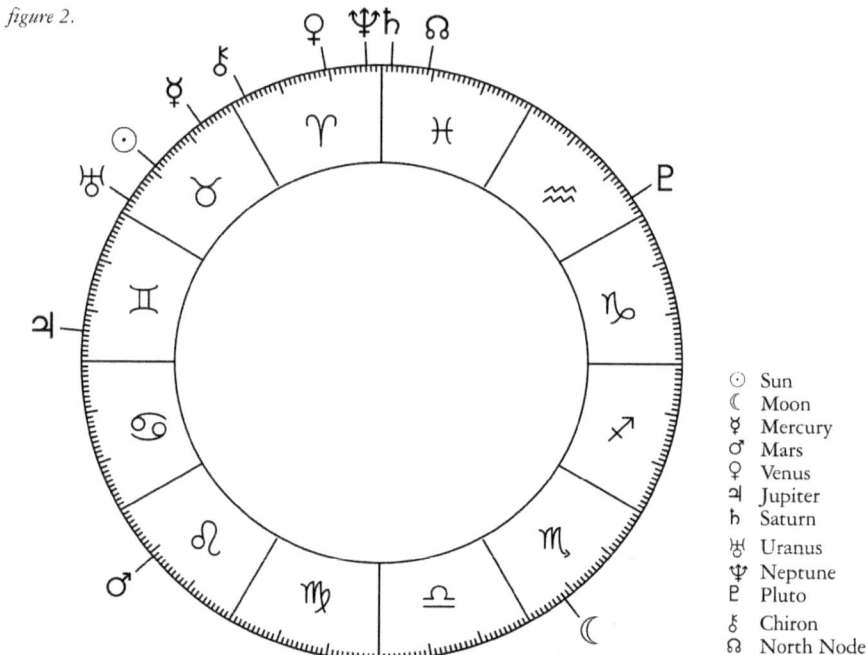

☉	Sun
☽	Moon
☿	Mercury
♂	Mars
♀	Venus
♃	Jupiter
♄	Saturn
♅	Uranus
♆	Neptune
♇	Pluto
⚷	Chiron
☊	North Node

*an astrology chart of the Zodiac Belt showing the 12 signs in divisions of 30°, divided into decans (10°) and single degrees

Each sign spans 30° (degrees) of the chart, and each of those are divided into *decans* (10°), with smaller tics for each degree, making it rather easy to identify the degree that any planet occupies within each sign (i.e. *Pluto is at 4° Aquarius*). Using this chart the degrees separating each planet can be calculated to then derive an interpretation of their influence. The prominent aspects of 0° (conjunct), 30° (semisextile), 60° (sextile), 90° (square), 120° (trine), 150° (inconjunct), and 180° (opposition) are of most importance and not only represent likely major periods of consequence but also the *quality* of their interaction according to commonly observed effects of each angle which are *most likely* to occur as established by past observation, correlation, and probability. The angles and aspects between influences are often plotted by astrologers using automated computational services and offered to consumers, such as at their websites (like *Astro Dienst*) where people can enter their birth information (be careful only to use reputable sources or you risk having your identity stolen!) and thus easily see what influences are occurring or in the past or future, without having to directly calculate them ourselves, but which of course is very easy simply by subtracting one angle from the other (an index of common glyphs which represent aspects and planets is at the end of this book, but they can also be found online).

Specifically, the conjunction (0°) is the beginning and end of all cycles and represents a harmonious marriage of the energies of each object of concern. Conjunctions are often described as auspicious events in astrology, and they often can be, but that is both dependent on the planets involved *and* their aspect to other planets that may also be factors at the time too. For instance right now at the writing of this book there are indeed quite significant events occurring all at once, from the square of Jupiter and Saturn to the conjunction of Saturn and Neptune is itself also *both* in sextile to Pluto *and* Uranus but also square Jupiter, indicating mostly favorable conflicts of power and revolution (which will resolve next year in 2026). As mentioned by the famous astrologer, Andre Barbault, the following conjunction of Saturn and Neptune in 2061 will instead be associated with a square to Pluto, indicating more direct and *unfavorable* challenges to power than this current transit, showing how it is not sufficient only to identify such aspects but all of which must be taken into consideration.

The semisextile (30°) is not often a focus to astrologers as it is not always extremely significant, but it does still occur and represents a time of minor opening or closing developments that are usually also in harmony like the conjunction. While this is not a focus it still should always be identified and mentioned because the events are not also trivial, but represent the opening and closing chapters of any cycle and thus are important to the theme and trajectory of the cycle as it progresses and can provide insight into what may lie ahead or greater insight about what lies behind.

Sextile (60°) is the next most important aspect in the cycle (or second to last in the close) and indicates quite prominent developments that are normally auspicious in nature, often the consequence of actions or events at the conjunction and semisextile. It is important to remember that astrology is *impartial* to our personal lives, goals, and experiences, and having an auspicious aspect as sextile does *not* also mean that what we are doing is necessarily good. Many who do evil will realize evil plans and evil behaviors at such aspects just as those whom do good will as well, and success from amoral behavior often deludes its actors into believing they can get away with it or that it will not have other consequences. But, just as bad things happen to good people, bad things also happen to bad people, and consequences in turn come at later, inevitable aspects which are at inharmonious angles.

The square (90°) is the first of these inharmonious influences, as the square energies between the two elements are at cross-purposes to each other and is the point where they begin to oppose one another, and represent the nature of conflict during which requires adjustment and adaptation to consequences of past actions accordingly. Square aspects almost never mean that we cannot make progress or grow but that such progress or growth often requires changing behavior or dealing with unpleasant or unfortunate circumstances. With highly auspicious planets as Jupiter or Venus, square aspects are often not very difficult but still require some discipline or resolve in order to realize its full potential.

Trine (120°) is the next auspicious aspect but where the sextile influence often happens spontaneously the trine aspect is often much more a result of past actions directly related to earlier influences, and can be thought of as the good consequences of past choices and behaviors, often of which also can require continued action and choices to best experience their effects. Because trine aspects are a result of past actions or what still continue to require actions they are not always guaranteed to be productive, or even very noticeable. For instance I recently had the transit of Uranus trine to my natal Moon which brings a surprising and unexpected desire to beautify, organize, and convert the home into an interesting environment and spontaneously bought organizers to make my closets and cupboards not only clean and organized but also aesthetically and functionally interesting, as well as more plants to grow and even a nice monitor riser for my computer which for the last three years had simply been sitting atop three copies of the first edition of *Fuck Portion Control*. None of this obviously would have occurred without action, and it would even be possible to let it go by without acting on those impulses, such is the nature sometimes of trine aspects.

The inconjunct (150°) occurs in the sign or house immediately flanking that which is in opposition, but is almost never described by astrologers because it is commonly considered less important. This is a grave mistake, however, which originates probably from a complete misunderstanding of the aspect, because it is in fact quite inauspicious and is specifically associated with impulsivity and often comes with significant consequences. For instance on the day I write this entry Jupiter is inconjunct Pluto during the war between Russia and Ukraine in which the latter yesterday launched a massive drone strike on the Russian air fleet, destroying many long-range bombers deep inside the Russian borders—while the planning of that attack occurred over an entire year, the response of the Russian leader is very likely to be very impulsive and poorly conceived, which is exactly the nature of this very significant transit. Because astrologers have deemphasized the inconjunct aspect not many people then know about this and so are not empowered to handle its influence when it occurs, which is similarly associated with impulsive actions in our own lives, where awareness can instead help us to make better choices or better handle events which occur during any inconjunct aspect.

Lastly, with the opposition (180°) any influences fully oppose each other, and while this is often presented as an inharmonious and inauspicious influence it is in reality a time of *evaluation* and *reevaluation* in which the consequences of the first part of the cycle become very clear, to either benefit or harm our interests, of which it can be either but most often is *at least* somewhat disruptive and inharmonious, and for those whom are very misaligned with their purpose or causality can be extremely disruptive and will always be a time of obligatory reflection during which we can change course and adapt if required or otherwise reinforce positive behaviors and choices validated by the test of this alignment. Because it is a test which helps evaluate the validity of choices and behaviors in the cycle it is nearly always a stressful transit, but whether we effectively emerge from that test or not is the relevant influence which can then be used to better understand our life path and choices and change course if required.

Also, while angles are important their influence is almost *never* the exact moment of alignment but will instead occur *before* or *after* the exact alignment, depending on the nature of the influences at issue. For instance most Mars and Uranus transits begin to have effects as soon as they get near the aspect, with Mars often concluding before the aspect is even exact, since Mars is the planet of initiative, but Uranus lasting most of the way through. Jupiter and Venus transits often don't even occur until after the exact alignment, and Neptune, Saturn, and

figure 3a.

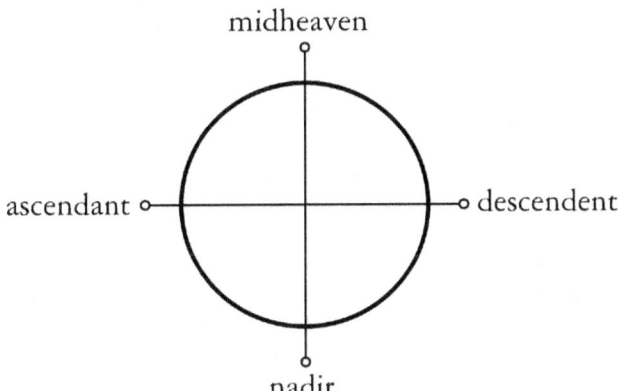

*the midheaven line is almost never exactly perpendicular to the ascendant line

Pluto tend to have influence through the entirety from beginning to end because their transits are so long, but are most influential across the center of the transit near exact.

In addition to the signs and planets, the entire sky's position in relation to the Earth's rotation is divided into the four *Great Angles* which comprise the ascendant system of the birth chart (*figure 3a.*). The *Ascendant,* which is the easternmost point on the horizon at the time of birth, the *Descendant,* which is the westernmost point on the horizon at time of birth, the *Midheaven* (originally called the *medium coeli*) which is directly above us, and *Nadir* (originally known as the *imum coeli*) which is below, where the Midheaven and Nadir are calculated from the intersection of the meridian and the ecliptic (planes that divide the Earth). Each are overlaid an astrological chart and displayed as if they were also planetary locations, often represented by the acronyms AC, DC, MC, and IC, respectively.

Where the placement of various planets at the time of birth determine the personality, the angles oppositely represent the general *life path* of an individual, where the Midheaven represents our material, public destiny, the Ascendant is the equivalent of the self and is thus *how* we get to our destiny. The Nadir opposite the Midheaven is our inner, spiritual, and home life while the Descendent opposite the Ascendent is our relationship to others, our partnerships, and our life path through them.

Although the angles can somewhat influence the personality this is not their primary function as often misrepresented by entertainment astrology, and it is more that the angles describe the actual journey we take through life wherein the personality comes more from the planets, each representing a specific aspect such as our love nature from Venus, emotional nature from the Moon, communication nature from Mercury, discipline from Saturn, etc. While mechanisms like the angles might at first seem superfluous or nonsensical in their justification or relevance, our position on the surface of the Earth in constant motion such as is achieved by its rotation is what we experience as *time* as discussed in earlier chapters, and thus is highly significant to our very experience of life and reality as reflected in the Great Angles, our position on the surface of our spinning planet compelling us forward through spacetime.

Which zodiac sign each angle is located indicates the nature of those placements and thus in turn the major course of our life paths, with the corresponding angle imbued with the properties, natures, and qualities of those zodiac signs as earlier described—for instance, because the Midheaven represents our public, material destiny, those with their Midheaven in Libra like Princess Diana (as stated by per personal astrologer) will have a life culminating in the concern for justice, fairness, and balance, while since the Ascendant represents how we get there those with a Libra Ascendant (also colloquially called the *rising* sign) instead reach their destiny *through* balance, relationships, justice, and fairness. My ascendent is in Scorpio and all those with this position at some point in their lives experience a significant life event which entirely alters the course and direction of their lives (which is the very reason this book exists for you to read, by the way), because Scorpio represents power and themes of death and rebirth.

The angles are also further divided into the 12 *Houses* (*figure 3b.*), and each also corresponds to the 12 Zodiac signs, with the Ascendant occurring at the very beginning of the first house, which is correlated with the sign of Aries, and so on around counter-clockwise, each house representing the same themes and concerns as their associated Zodiac sign, for instance having Saturn in the *eleventh House* portends a difficult time making and having friendships throughout life due to an overly serious estimation of them (since the eleventh house rules fortune and friendships), or Mercury in the first House represents a life of achievement through communication (since the first house concerns identity and purpose), or Mars in the first house a highly forceful pursuit of life goals, or Neptune a delusional or spiritual one, etc. Thus, the placement of the planets in the houses is more fateful

figure 3b.

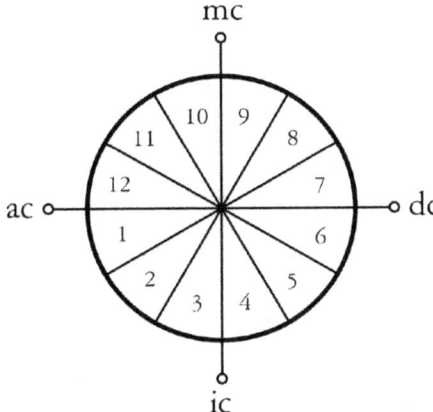

for our *life path,* where placement in the Zodiac affects *personality*.

Because the angles are determined at the exact minute of birth, even slightly incorrect records can entirely corrupt results and interpretation of the angles and houses, for instance Princess Diana's officially stated time of birth was different than what she reported to her personal astrologer (and considering the nature of her life it does seem the amended account is more likely). So it is important to remember when interpreting the angles and houses they are the data which is most vulnerable to error since they change so quickly. Slower influences like the planets will still have much greater accuracy even if the exact time of birth is unknown or incorrect, but if the birth time and location are correct then very significant information can be derived from it since the angles not only describe (not determine!) the course of the events of our lives but are also themselves affect-ed by the transiting planets, as the planets can aspect the the angles just as they do other natal planet points. For example, in my early forties and already engaged in my work and writing for a half-dozen years the first time any person with any notable social media following boosted my work was when transiting Jupiter was trine to my natal Ascendant. Jupiter is the planet of expansion, growth, and luck, and the attention boosted my book sales for several days from the usual one or two sales a day to more than one-hundred over a day or two and helped pay for a badly needed move. Similarly, at the outset of my work and first announcement in late 2017 of my book to readers who signed up to pre-order copies and helped fund its publication Jupiter was at that time also conjunct my Ascendant, demonstrating how transits to these aspects in turn then correlate with the activation and occur-rence of life experiences which progress us along our life path.

Many astrologers slightly misunderstand these cycles and might describe transits to the Ascendant as a windfall or gift, but in reality they mark *milestones* and *developments* along our life path and do not simply occur out of the blue or for no reason, often in consequence to previous life experiences or choices to later then experience the consequences along the way, repeating cycles over and over with similar themes and concerns throughout the entirety of our lives. Twelve-years earlier in 2005 when I was twenty-four and Jupiter was also again conjunct my Ascendent, before I knew anything at all of astrology, was my move to Los Angeles to begin that chapter of my life, another auspicious beginning of a new Jupiter cycle and end to the previous which began in 1993 when I was twelve or thirteen years old (and as such have no idea what that might have entailed since it was so long ago and I was so young).

Jupiter being the planet of opportunity and growth is more relevant to that type of life progress, however, and not all planets are as auspicious as Jupiter (even Jupiter can be inauspicious, however, as discussed in upcoming chapters), and many aspects to the angles can also likewise bring stressful events as much as those which are benefic, but they do very much reflect life events which are still critical to our life path.

If it happens that the time of birth is not known it can possibly be *"rectified,"* which is the process of reverse-engineering the location of the angles and time of birth such as through the kinds of events I experienced, for instance if an auspicious event occurs which is determinant of the life path such as when our goals or hard work begin or bear fruit it may be the interaction of Jupiter or another planet to the Ascendant, and if other planets also aspect that position in your natal chart it can be surmised through the nature of those events where these angles are located and thus the time of birth. Even if there is a vague idea of the time of birth that can be a useful starting point for estimating them, but should be taken with that knowledge it may not be correct, and work to remedy the exact time through analysis (not all transits have effects at the exact aspect, remember, but may occur before or after depending on the nature of the planet). The most useful influences to rectify the birth time are the more regular, constant transiting planets like Mars, Moon, and Sun, as there are specific types of life events which correlate exactly with them.

Karmically and fatefully the Ascendant is equal to the first house and our relationship with the self *(figure 3c.)*, being correlated with the sign of Aries, while the Descendant is equivalent to the seventh house and our relationship to others (close partnerships such as in romance or business), correlated to the sign of Libra which is always opposite the self since other people have their own interests, thoughts, goals, etc., and why we often find friction in dealing with others (though as discussed in my book on psychology not all conflict is bad!). Similarly, the Nadir is equal to the fourth house and the inner life as correlated with the sign of Cancer, including the spirit, mind, home, and family, while the Midheaven is equal to the tenth house and the sign of Capricorn which concerns the public, outer life, and material and professional success.

While the Nadir is associated with home and family not many people (astrologers) realize the Midheaven can *also* concern the house and family when they are a function of status and material success rather than our inner nature, and the Nadir specifically tends to be more concerned with our soul, spirit, and the inner, personal world which can definitely include family if it is insular from our public and material life, and ironically many people will still find partners and build their domiciles through Midheaven transits but not during Nadir transits because of the

figure 3c.

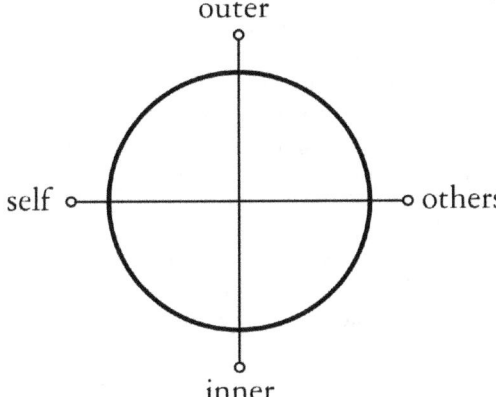

inharmony inherent to our choices and pursuit of those things as functions either of materialism and success or inner self and spirituality. But because the Midheaven is opposite the Nadir most of us experience stress and conflict with our family and home when we are experiencing outward and material success, being overly consumed by work and career, and later the opposite when transits concern the Nadir and oppose the Midheaven and find it difficult to achieve outer success and being obligated to attend to our family and inner, spiritual needs. Generally the aspects of any planet or other influence to the angles thus also affects those life paths as development of the self, our relationship to others, our spiritual self, or material success.

There is, by the way, nothing inherently mystical or special about the astrological birth chart that makes it different than any other point in time—birth just happens to be the only universal experience from which astrological study can be *anchored*, since while other life events like marriage or the loss of parents occurs to most everyone those might occur at different points in life, and the only other universal experience, death, is not a useful point at which to begin. Although the Ascendant is also always exactly opposite the Descendant, and the Midheaven is always exactly opposite the Nadir, these pairs are *not* usually exactly perpendicular as misrepresented in figures *3a, 3b,* and *3c,* and the midheaven line will be more or less oblique or acute to the ascendent line depending on the position of our birth along the lines of latitude and the time of year, with those being born farther to the poles having more extreme angles of the midheaven axis which then also changes the size of the houses within those angles accordingly since each quadrant is divided into three houses. This can mean that some themes and life events for some people are more or less evenly spaced across the time of our existence while for others they may be comparably far greater or lesser spans of time.

Overt astrological influences like Sun signs are also more obvious when we are young, as all lives are a journey from one place to another and along the way we develop and grow to then complicate and enrich our personality and character. Because of their youth and relative inexperience, young people are some of the most obvious incarnations of Sun signs, but in those who are much older the influence of the Sun sign nature gives way to reflect more of our Moon sign, which is the inner and emotional nature. While the outward, Sun sign is every bit as valid an aspect of our personality it could be said for this reason that the Moon sign is our true sign, not the Sun, because it is through our inner nature that *we* experience our life, and our outer, Sun sign is how the world oppositely experiences us, or in other words how we interact with the world around us, which is separate from how we experience it.

If done by a skilled and experienced astrologer who actually knows what they are doing the birth chart thus can be quite instructive and useful for understanding ourselves and our life path (again, be very careful to avoid scams and predatory services). My chart as assembled automatically by *Astro Dienst* and interpreted in prewritten, automated text from astrologer Robert Pelletier described me as being likely to create and impart knowledge, specifically to teach and help people, that my early life would be marked by much change in domicile (I've moved over thirty times since being born), and my fortunes in life would only come *after* middle age—all of which is expressly true and very specific. It also contains advice for me about getting an education to better compete with my peers but obviously that did not happen due to my early life circumstances and I view that kind of advice as Pelletier recognizing aspects in the birth chart which indicate the absence of formal education and difficulty competing with peers but then offering advice, where in reality that is only useful if we have opportunity to receive it beforehand, but also which is not likely to change the outcome since we cannot in fact bend fate to our will, instead the most useful part is recognizing the reality which is that such a path was literally not in my stars.

As has been mentioned many times already in this book, in order to effectively use and understand astrology and, most importantly, minimize delusion and other risks which inherently accompany practices as astrology it is absolutely *required* to practice inventory therapy. If not, astrology and astrologers can easily mislead us (intentionally or not) into reinforcing unhelpful and unproductive control and coping mechanisms, especially those which are based in fear, selfishness, self-doubt, self-pity, insecurity, unreasonable expectations, and anxiety of the future. Such control and coping mechanisms can otherwise motivate us to make unhelpful or even self-destructive decisions in response to what we see or understand about astrology which, because of the nature of control mechanisms, may seem rational or in our best interests but instead are in fact nothing more than efforts to control reality and fate, which we cannot do and so become ineffective (sometimes even disastrously). No doubt I will even cause that to some who read this book, just as my birth chart interpretation caused me to feel sorry that I had not gotten an education instead of realizing that it was in fact part of my life path and in fact is far more relevant to what I eventually achieved than if I had done so, which I eventually realized as a product of having done much inventory on my insecurity about lacking formal education.

Worst of all, one of the most productive and useful reasons to use astrology is to understand that the purpose of life is simply the experience of it, and for those of us with unresolved trauma these simple truths are not sufficient an explanation to satisfy our dissatisfaction and control behaviors, so we then continue searching for other answers that will satisfy our desire for excitement, entertainment, control, or to fuel personal delusions for life when in fact we have already come across the answers and disregard them for being too simple. Practicing inventory while also studying astrology can thus be extremely productive, both learning new and effective life skills and resolving past trauma which interferes with our understanding combined with knowledge about our life path and real, lived experiences as not something arbitrary but in fact highly relevant to our personal journey, which then imparts unmatched wisdom and insight and improves our personal effectiveness.

The fateful relationship I had which began as both Venus and Mars conjuncted by natal Sun while the transiting Sun was also conjunct my natal Venus might have been an auspicious combination if not for transiting Neptune, the planet of spirituality but also *illusion* and *delusion*, which was opposite my natal Venus portending the illusory nature of my conception of this person. At the time I did not know so much about astrology, but had I the occasion to know Neptune was opposite my natal Venus I would probably not have necessarily avoided that relationship but instead been paranoid and even more suspicious of my partner because of unresolved trauma and control behaviors because, again, *it is impossible to be consciously aware of that which is unconscious* such as the trauma and control and coping mechanisms which affect our behavior, so without practicing inventory it is not even possible to be aware of what it is we are actually doing, nor how to use astrology effectively, and the effect of much practice results instead in the increased ability to be more self-honest, learn how to distinguish between reality and delusion, and therefore to be more effective in our choices and decisions.

Part of the delusion of behaviors such as trying to avoid the negative effects of a Neptune transit is believing that we can in fact avoid such experiences such as through choices like avoiding a relationship. But the challenge of Neptune transits is not to better control our lives, which is usually the entire context of entertainment astrology, and the fundamental purpose of most choices, but instead to better *respond* to the fated experiences that we absolutely will have regardless of what choice we make. It was certainly possible to have avoided that relationship altogether, but I might then have just had another take its place and make the same mistakes anyway, and the third option between avoiding it or participating which anyone ignorant to the effects of inventory therapy might not have recognized would be to instead have *compassion* for the person who despised themselves so much as to feel the need to deceive me to their real nature and needs and desires, and to be able to perceive that true nature through the trappings of deception and have compassion for them, *not* contempt. Truthfully, I willingly went along with the charade, so desirous was I for a loving relationship, like many of us often do, believing wrongly my intentions justified my choices.

Because part of the delusion was also not having compassion for my own insecurity, I also had none for him, dissociation being one of the only tools I possessed to handle life (as is the case for most people), which in turn made me ineffective in all my choices, *regardless* of what they were—to leave or to stay—as such do most of us similarly fail to be effective regardless of what choice we make because we do not in fact truly understand our *reasons* for making it and thus remain entirely unaware of other available choices, until we do the work to resolve the cause, which are our unresolved experiences of trauma and resultant control and coping mechanisms, *not* the study of astrology (or any religion or other philosophical discipline). So if you have not yet done much inventory therapy, you may as well stop reading right here, as the rest of this book will do you no good if you refuse to *practice* and use inventory therapy as a self-care tool.

Likewise, Neptune is also the planet of spirituality and its transits do *not* doom us to delusion, but only so much as we are intransigently delusional and unable to be self-honest, which is a skill that is practiced by such behaviors as sitting down to do inventory, and refusing to do that is instead the practice of delusion, so such a transit could mean having deeply spiritual experiences through conflict in the recognizing of humanity in a person trying their best but possessed of very poor estimation of their self-worth, rather than resenting them and trying to control them in turn. What I was not able to do, which I should have done if possessed of effective life skills, was to instead focus on my own life and behavior and help others through example (although I was still to young and inexperienced to have much wisdom on this), but our trauma and coping mechanisms turn our attention away from our own behavior and experiences and instead toward that of others, due to feelings of inadequacy and insecurity, fixation on others or other outside factors (such as disease or misfortune), rather than ourselves, being the heart of delusional behavior.

Being aware of the quality of transits while also having a tool such as inventory then instead might have helped me find compassion for both him and myself rather than discouragement, which would also have empowered me to either effectively separate myself from that relationship or instead to help him in turn find his own confidence by teaching compassion by example, not disdain and resentment, because it is only when we recognize the humanity in ourselves that we can then find it in others, which is achieved through actionable self-compassion like inventory.

A desire to subvert reality and transcend cause and consequence by engaging with astrology, science, or religion is a natural human motivation because life is hard and often contains some unspeakably horrific realities which shock the mind and wound the soul. But atheism submits that there is nothing in control of our lives so that we may pretend ourselves to have that control instead, which we do not, and in so doing is no different than religion by a different name. *This is not the purpose of astrology*, nor the correlation of celestial events with our own existence and, quite the opposite, astrology is a force that is greater than anything we can control or divine and the purpose of astrology *is not to placate our fears*, to subvert responsibility or consequence, or to help us ignore or excuse our weaknesses, shortcomings, and trauma. Being presented as such is part of the reason why astrology is often dismissed and denigrated because its attempts to control reality is often so transparent as to be ludicrous, which many people rightly recognize, and it is silly to think that we can somehow subvert fate even if we were able to predict events in the future. Astrology is instead (or should be) quantifiable, provable, and empirical, and used to describe reality as it is and not as we want it to be. Recognizing patterns in nature, the cosmos, and our lives is statistics and probability. Those are quantifiable. Delusionally using astrology to subvert fate is not. Words and descriptors notwithstanding, the planets and stars do not determine anything—they merely correlate with the laws which govern them as well as ourselves, and correlations can be used and are used to make statements of probability, based on those statistics, not absolutes and certainties to entertain our delusional desires for an uncomplex life.

Of course, bad astrologers can not only be unhelpful but also cause outright harm. Some astrologers see certain transits as opportunities for violence or severe misfortune, and in turn advise patrons as if we have power to avoid it. Delusion does not only accompany delusions of grandeur and prosperity, but also fear, and if transits occur regularly which are described as unpleasant experiences can saddle patrons with debilitating fear or inflame phobias and trauma by implying the power to prevent or subvert such experiences is within our control, rather than wisdom on how to handle such unfortunate occurrences. It is true that auspicious life events can occur during certain transits, such as those involving Jupiter, but these only occur after years of hard living and trying our best to grow, develop, and succeed, and are very often the fruit of past labor, learning, and reformation and do not occur like a lottery prize happening out of the blue for no real substantive reason, and presenting otherwise is simply preying upon the fears and dreams of people rather than trying to help them.

But it is not only the fault of astrologers that we are susceptible to such propaganda, because they too are only humans trying to survive in this world, just as ourselves, and it is our responsibility to be informed and practice wisdom and eschew delusion (which is done through practiced behavior like inventory therapy and not willfulness, which is itself a delusion!). Early in my recovery from alcoholism and cancer my grandmother died, and I was terrified to attend her funeral because nearly all of my aunts and uncles had over the years demonstrated themselves to be virulently homophobic and either callously indifferent or outright hostile. I looked at my transits which would occur during the weekend set for her funeral and one described a likely unpleasant encounter, especially with men. Already nervous about going I made the decision not to attend, naively thinking I had bested fate and felt much relieved by it. A few days later on the weekend of her funeral while in the car with my sponsor I tried to share my insight into the cause and remedy for his acne, the skin being an excretory route for iron, because he had been complaining constantly about it. Unexpectedly he shouted at me to shut the fuck up, and our relationship soured from that point forward, and I had the very experience portended by my astrology in spite of my efforts to avoid it, as the quote by Jean de la Fontaine at the front of this book (and Master Oogway) explains, it is *impossible* to avoid fate, and we often meet it in the path we take to avoid it, and trying to use astrology for that end only ever leads to greater disappointment and even more ineffective behavior since doing is so is motivated and informed in the first place by fear.

While astrology can never be used to subvert fate or avoid consequences, it *can* be used to make wiser decisions, be a better person, find peace and purpose, and connect with things of a spiritual nature which can then in turn make us more effective in our lives. But incorrect interpretations of astrological phenomena can also mischaracterize or harm a person into believing things about ourselves which may be untrue which are instead a product of interpretations by people whom do not fully understand what astrology really is.

This is not entirely consequential because nobody, not even astrologers, can bend fate, but in doing so might be unhelpful in helping others understand and appreciate the tool which is astrology and thus through their behavior prevent others from becoming more empowered. So when practicing astrology much care should be put into every characterization of every aspect as well as in our consumption of it, recognizing at all times that all astrology is man's attempt to describe what we observe, and there is much room for error in translation that is also influenced by personal bias, past experiences, and even poor communication.

For what it's worth I have never experienced violence or other harm during any transit which has been described as such, and the occasional advice to "stay out of bad neighborhoods" which sometimes appears in writing is in fact pretty xenophobic, possibly racist, which only serves to enflame anxiety and prejudice rather than helping people become more effective in their lives. At several times in my life I have even lived in "bad neighborhoods" in which I would go on long walks to get sunshine, even though camps of heroin addicted homeless, whom are just people trying to get by, and while an occasional person would direct inappropriate comments my way never once did a 'potential violence' transit materialize, and nearly all the acute harm which has befallen me so far has come almost entirely from those I've known and had relationships, or even as a result of my own actions.

There is also often a manufactured controversy that regularly appears concerning astrology when people make the mistake of assuming astrology is based on the *sidereal* zodiac, which means based upon the constellations, where in fact astrology as it is practiced is based on the *tropical* zodiac, which is based on the seasons. The name, configuration, and identity of each zodiac sign is merely named after the constellations associated with each sign when it was made, but is not dependent on them in the least.

While the use of astrologers and their content can be helpful it is highly profitable for everyone to understand these concepts separate from actual astrologers or sources of astrological information, because there are both basic principles in these dynamics and actively occurring influences which no astrologer can adequately communicate to every individual which uses their content, instead making broader, general content which appeals to the most number of persons (it's a lot of work!). Just like my book *Fuck Portion Control* which emphasizes that every individual needs to understand how the human body works in order to overcome health problems since health professionals are incapable of doing that for every person, we cannot possibly understand our life or things of a spiritual nature if we do not individually understand these principles on our own, separate from professional astrologers, whose information should instead help support and promote our own, personal knowledge, not replace the responsibility for study and learning (and certainly not relying on entertainment as our primary source).

So if there is a fundamental underlying reality to astrology, how do we use it and what relevance can it have for our lives? As I mentioned previously, astrology is often employed as an entertainment, and most often the daily horoscopes provided in social media, media outlets, even from serious astrologers is often nothing more than content and merely intended to attract engagement, which is done most effectively by triggering people's emotional response, not telling the truth or being useful. While daily astrology sometimes can be used to understand themes occurring in our lives there is almost no influence which occurs on a daily basis which is so consequential that we should need to consume astrology daily. The Moon is the most rapid transiting planet and all of the aspects it forms with our birth chart last at most a half day and typically concern our emotional response to the world around us, and repeat every lunar cycle so they occur quite frequently throughout the year (about 12.4 times, to be exact). Instead, it's most useful to understand the long-term influences such as major transits to the natal chart or those which are occurring in the zodiac which affect things like society, politics, government, economy, etc., and the most important astrological transits are those which happen over weeks, months, and years, and even decades, although daily entertainment can be fun if not taken too seriously.

Although the Zodiac traditionally starts with the sign of Aries and moves counter-clockwise to the last sign of Pisces, in practice and reality there is not any such a thing as a beginning or end to the Zodiac. Much like existence itself, it is instead a continuous cycle which builds on the events before it, and as such also do all features of the zodiac from the planets to the angles and houses, and though many aspects are described to begin and end they are more accurately described as building upon the past and adding to our already lived experiences, never really being over but instead new aspects which are added to our life.

Again, this is the kind of thing I find quite compelling about astrology as a practice and science because while religions often elevate mythological figures into positions of worship, astrology is instead based on the reality we see around us such as the stars in the sky, the cycles of life which are apparent in all things, and the various patterns we see in nature rather than any abstract and insubstantial mythology, figures, or institutions. Reality, after all, is actually based in the laws of physics and nature which construct existence, so why shouldn't our personal philosophy, study, and sense of purpose also arise from it?

Because we are mortal humans, however, it is logical to have a beginning as a place to start, as such with the sign of Aires which represents *the self*, such as in terms of birth and beginning but also purpose and identity, since we need a physical body in which to inhabit and some identity or purpose from which to operate. Of course, the body is also the temple for soul, which is governed by the previous sign of Pisces and the house of spirit, so you can see how instead of beginning with Aries the Zodiac could instead begin with Pisces, since we need a soul (energy) to inhabit a body, without which the body will be lifeless. But there is no observable point we can use to anchor the Zodiac from the spirit, so instead it is anchored at the birth of the body, which is a universal experience for all humans and so an anchor point from which everything else can be understood beginning in the sign of Aries.

The structure of the Zodiac itself also reflects immutable laws of reality and progression of existence as constrained by those laws. For instance, laws of causality and time are reliable constants which move in a constant direction which no other aspect of reality may subvert, and this is reflected in the progression of the Zodiac which, like our perception of reality, moves only in one direction and is always preceded by necessary events. An example of this is that first the Earth must have been created before any life could exist upon it. Or that a couple must first meet before they can create a relationship, fall in love, and produce children. Children also must come first before the responsibility of caring for them, and knowledge must first be gotten before it can be applied. While this may seem like quite a basic and obvious conception of reality this is the fundamental construct of the Universe which allows it to exist in the first place, without which it would not. You can imagine some other kind of reality with no causality in which a random child appears out of nowhere and is yours but you don't know that because it comes from a future relationship you do not yet know about, nor even perhaps do you necessarily know what a child is because it came before the step of acquiring knowledge. So because order underlies our reality the Zodiac wheel progresses counterclockwise. It could have been constructed the other direction but instead reflects the observational direction of celestial objects relative to our location on the Earth—a geocentric model—as observed by the ancient peoples who started it, and it is also geocentric not because of any philosophical reason but instead because of *Relativity*, in which our experience is entirely relative to that of which occurs around us such as from our point of observation here on Earth, rather than a heliocentric model.

Colloquially the *Sun sign* is the place in the Zodiac where the Sun was located at the time of birth, and is the most prominent aspect used when classifying persons since, like the Sun, our most obvious and outward aspects are the brightest and most easy to identify, but any of the planets or aspects will take on the qualities of that sign in which it occurs. For instance, having the Sun in the sign of Libra means our Sun sign is Libra, but having Venus in the sign of Aries means our Venus sign is Aries, and since Venus represents matters of love and partnership a person with this placement will be direct and ambitious in matters of love and partnership such as is characteristic of Aries, while a person with an Aries Sun will instead have these qualities in their general outward personality as experienced by others. This explanation of astrology makes it quite simple for anyone to understand, which is that any aspect takes on the nature of the place it is located, and so long as anyone has a birth chart they can in turn infer the nature of any feature simply by understanding that principle and knowing the properties of each of the 12 signs. One important variable to note is that having any planet or element on the *cusp* (the very last degrees of any sign) combines some of the qualities of the abutting sign, so a person will instead have a mix of traits from both rather than the one in which they are born, but will be primarily the sign in which it is located.

Each of the signs is also categorized into the four *elements* of *fire, water, earth*, and *air*, which represents their qualities and generally describes their natures, as well as three different *modalities Cardinal* (instigators), *Fixed* (consistent), *Mutable* (adaptive). While the element of each sign is real and relevant to our nature it is not so important as is often implied in entertainment astrology. The modalities are quite illuminating, however, because they generally also apply to how we handle life, being fixed and stalwart, adaptable and malleable, or catalysts for action and change.

Being the first sign, Aries is the first *House of Self*, which concerns things like our purpose, identity, action and initiative, and sometimes the nature of the body itself. As with all signs, having any other planets or aspects within the sign of Aries imbues that influence with these qualities, so for instance if the Sun is located in Aries then the Sun sign is Aries, as is the case with the Moon, Venus, Jupiter, etc, and will be imbued with the qualities of the house in which it is located.

The second sign is *Taurus* and the second *House of Means*. Though tradition-ally associated with money there are many things we find of value besides money such as health, energy, or even the opportunity itself to earn money which are not effectively described by traditional attribution. This sign is also separate from the concept of *wealth*, which is instead governed by the eleventh house of fortune, so in my efforts to resolve discrepancies I found the term *'means'* a more useful attribution for this sign and house since 'means' can and does also encompass not only any resources but also literally the means by which we obtain them. Thus all placements and transits in Taurus and the second house concern means, money, and other resources.

The third sign is *Gemini* and the third *House of Knowledge*. Like other houses I have slightly retitled this house attribution, traditionally described as the house of communication, which it does encompass, but more accurately deals with all aspects and dynamics concerning *knowledge* just as the second house does not only concern money but also the means of its acquisition, so this sign concerns the acquisition of information, its application and use, and then also its disbursement, which is technically the definition of knowledge, not communication (which technically leaves out the *application* of information that is very much a feature of this house). This house also rules siblings too, however, probably because they are people whom learn at the same time as we do, and I've also noticed it rules child-hood friends, people we grew up with, whom are like spiritual siblings. All transits and placements to Gemini and the third house thusly concern knowledge and all that is relevant to getting, using, and sharing it.

Cancer is the fourth sign of the zodiac and the fourth *House of Home*. Because home is where the family resides this sign literally represents both the place of domicile such as a house or apartment as well as the family from which we come from, and all placements and transits to Cancer and the fourth house thusly concern family, the inner life, and even the literal place of residence.

The fifth sign is *Leo* and the fifth *House of Creation* traditionally attributed to a seemingly broad and disparate range of themes such as pleasure, children, sex, and art. When studying astrology I realized this seemingly disparate set of themes all in fact shared one, singular concept—*the act of creation*, such as the creation of children, art and creativity, and casual sex which is the creation of relationships, etc. It can also probably include projects of any kind or any attempt to create an organization or entity, a business, restaurant, etc. Thus all placements and transits in Leo and the house of creation concern the act of creation and creativity such as in art, projects, new sexual escapades, children, etc.

Virgo is the sixth sign and the sixth *House of Labors*. Like the fifth sign, Virgo has also been sorely misattributed by tradition, beforehand thought to be the house of work, health, and even pets it is in reality all those things we are obligated to do on a daily (or highly regular) cycle to care for our needs and responsibilities, thus aptly described by *labors* required of our regular, daily lives.

This is why the house traditionally encompassed such disparate themes as pets and work, which seem to have nothing to do with each other, but do in the sense that they are obligations to which we must attend on a daily or near daily basis. Because we get sick if we don't care for our health many people with transits through the sixth house will neglect either work or health as they feel pressure to address both and often find consequences such as getting sick. As thus all placements and transits to Virgo and the sixth house concern labors to thus affect things like work, health, pets, chores, responsibilities, etc.

Libra is the seventh and middle sign, the seventh *House of Partnership*, and being the middle is also themed with balance although, as mentioned previously, people commonly misunderstand balance as being exceptionally disciplined or well-rounded when that is not at all what balance means, which is instead equal divisions of a whole. In fact, being exceptionally disciplined is not balance, because real balance would have more or less equal amounts of discipline and undisciplined, of work and play, love and hate, pain and joy. This sign could very well be called the house of balance, but Libra opposes Aries, the sign of the self, and thus represents the *Other* which is reflected in our relationships and partnerships, the very foundation of which requires balance otherwise they do not function, or even exist, when either party takes or gives too much or too little. This sign also rules business because business is effectively also a partnership which similarly requires balance, and does not only concern romantic relationships (although that is its primary concern). Relationships like parent to child are not covered by this house because that relationship is inherently imbalanced in nature, with parents having more responsibility to care for children which are thus under the purview of the sixth house. Any placements or transits to Libra or the seventh house thus concern all types of partnerships from romance to business.

The eighth Zodiac sign is *Scorpio* and the eighth *House of Power.* Often called the house of sex and death and associated with transformation, rebirth, and things which are taboo Scorpio in truth represents only *power* and *control*, where all things like sex, death, and the taboo are simply battlegrounds for control and power dynamics. This is especially obvious because many transits to the eighth house will not have anything to do with sex or death but instead our ego, egomaniacal control mechanisms, and general power dynamics with others or with life around us. This sign also represents death but only because the Universe is in charge of death and consequences which cause it which we do not have power to transcend, and death often occurs when we try to control things over which we have no control such as health and biology, or the behavior of other people. But death is also *metaphorical,* and after experiencing consequences of choices or life behaviors we are often reborn anew with greater wisdom and experience which reshapes and reforms our identity and purpose, experiencing a rebirth through trial and struggle. Because of this power dynamic, the eighth house is *also* associated with shared resources like taxes or wealth shared with a partner, because those are also other occasions for control and power. Thus any placements and transits to the eighth house concerns control and power and all the various dynamics and occasions for it.

Sagittarius is the ninth *House the Mind* which encompasses philosophy, though, expansion, travel, adventure, and growth. Where Gemini is the house of knowledge, that is rote knowledge and its application and Sagittarius is instead the more existential knowledge as wisdom, philosophy, ideology, and experience. This also includes the law and higher learning since higher learning is often about the creation of knowledge or experiencing knowledge, not just the learning and utility of it. Thus do all placements and transits to Sagittarius and the ninth house concern matters of the mind as philosophy, wisdom, travel, adventure, etc.

The tenth sign is *Capricorn* and the *House of Society*, traditionally considered the sign of institution, the public, and career (not the same as vocation), as Capricorn in turn embodies the opposite of Cancer, the home and inner life, to instead encompass everything outside the home such as what is composed of the general society of man and its associated institutions, rules, authority, governments, religions, etc. Thus are all placements and transits to Capricorn and the tenth house concerned with all aspects of society.

Aquarius is the eleventh *House of Fortune* and associated with wealth, luck, and friendship since friendship is also an aspect of fortune which is not only a function of money. Indeed while money is fake, real friends are not a construct of society but other actual persons whom can in fact provide an increase in security, companionship, interaction, and community entirely separate from institution and order. Aquarius is also the sign of rebellion, progress, and technology, because the fraternity of man is the most powerful force within humanity, and rebellions require friends, and when governments and rulers are overthrown it always involves some aspect to Aquarius or associated eleventh house. In fact, we have entered the Age of Aquarius, a common new-age phrase but which has real meaning as discussed in the earlier chapter on evidence for astrology, which is a more than 2,000 year long period that affects and guides the general path of all of mankind and likely began with the advent of democratic representation as the primary form of worldwide government and overthrow of most established monarchies in the previous centuries, and now also results in the breathtaking advance of technology that we naively enjoy such as the internet, computers, cars, planes, satellites, etc. Thus do all transits and placements to Aquarius or the eleventh house concern wealth, fortune, friends, surprise, technology, rebellion, unpredictability, etc.

Traditionally the last sign of the Zodiac, although there is no such thing because it is an endless cycle, is *Pisces* and the twelfth *House of Spirit* associated with spirituality, endings, illusions, delusion, secrets, and all that is unknown. Because Pisces is opposite Virgo and the house of Labors it also associated with opposing themes of rest, surrender, illusions, and the intangible. Whenever the Moon transits the twelfth house every month people enter a couple days when nothing seems to happen yet we feel a heightened sense of sadness or despondency, which is the time during which we should pause and take reflection on our life and connect to the spiritual which, when we do not, instead feel empty or lost. As such, all placements and transits to Pisces or the twelfth house concern either spirituality or delusion, and illusion, dreams, endings, rest, introspection, secrets, and all that is unknown or unknowable.

Because the Sun is primarily our outward personality and how others experience us, our true personality instead is more reflective of the Moon sign placement which, depending on what sign it is located, will take on the qualities of that sign as described above and indicative of our true personality, our emotional self, and how we experience the world around us (more info on each sign is in the upcoming chapter on birth, signs, and destiny). Similarly, the placement of all the other planets will also take on these qualities according to the sign of their location. Mercury is our communication nature, Venus the love nature, Mars is our initiative, Jupiter growth, Saturn structure and responsibility, Uranus our spontaneity and friendship nature, Neptune is our spiritual nature (or delusional nature), and Pluto is our control nature. Referencing a personal birth chart, it is thus possible to see all the placements of the planets—the Sun, Moon, Mercury, Venus, Mars, Jupiter, Saturn, Uranus (which used to be spelled Ouranos and so is actually pronounced Our-ran-os, but you can be immature if you want), Neptune, Pluto, the North Node of the Moon, Chiron, etc., which can then be inferred by lay persons as to the nature of the placement of these planets and their corresponding part of the self to which it relates simply by which sign it is located.

Because planets are also often *retrograde* this means that they can be retrograde at the moment of birth which alters their meaning than from when they are not retrograde (this is usually indicated in birth charts with a subscript 'r' next to the planet), and reflect inherent contradictions in our personality that everyone faces as conditions of being alive. Although untalented astrologers often try to present these retrogrades as curiosities or even talents they are in fact some of the very reasons for our individual, personal weaknesses and insecurities which are part of the challenging nature of life which, if we fail to properly understand or delusionally mischaracterize in order to make ourselves or others feel better (in other words, lying), handicaps our ability to be effective. For an example of this phenomenon those born under Venus retrograde, which only occurs about every eighteen months, are often more suspicious of relationships and intimacy and frequently self-sabotage or avoid romantic partnerships in favor of casual affairs—or oppositely exert greater control behavior in them (including control behaviors like being very passive or giving too much, which are still control behaviors), which then has highly destructive consequences than when we otherwise relent control. While this retrograde might sound ominous it simply reflects traumas we have experienced which prejudice us to be more skeptical, cynical, or fearful of romantic dependency and the relenting of control that partnership demands, which are no different a requirement than for any other of the billions of people on the planet, and practicing inventory on our fears, failures, and resentment of dependence and lack of control over life and people can help resolve that kind of dilemma to oppositely find satisfaction in our dependency, fallibility, and vulnerability and learn to trust others. After all, the problem wish such influences is always our *perception* of such things, informed by our past trauma, and not how they really are.

Mars being the planet of action is the least often retrograde, and those born under its retrograde are similarly conflicted by fears of asserting themselves or demonstrating aggression and initiative. This is not shyness but more typically those who don't erect boundaries or push back when required, refraining from asserting themselves in conflict even when it is required, which likely resulted from early childhood conditioning when these self-care boundaries were torn down by abusive, neglectful, or opportunistic adults. Very often, attempts to assert boundaries when possessed of a Mars retrograde is likewise done through ineffective means such as by overcorrecting and taking it too far, and similarly the practice of inventory surrounding these fears and trauma can help empower more effective skills which lie in the middle of extremes and the acceptance of our vulnerabilities and limitations. Like Venus retrograde, the conditions of life are no different for this placement than the rest of humanity, and it is only our perception which is the problem, when corrected through such introspection can then be resolved.

Mercury and the other planets are retrograde more often, so most people have at least one or more retrograde planets with those besides Venus and Mars being most common. Mercury retrograde in the birth chart indicates problems and unwillingness to ingest information, since Mercury is the planet of communication and knowledge, and every person I personally know with retrograde natal Mercury are insecure about their own powers of intelligence and get angry and resentful when they do not understand words or have a hard time comprehending concepts. The thing about intelligence is that information can be learned by *anyone*, and this is similarly a perception of our circumstances which originate from our experiences and insecurities, not reality, and like other fears and insecurities can also be overcome through trauma therapy and recognizing that learning is a part of everyone's life, that there is nothing wrong with being stupid (some of my most favorite people are dumb as rocks!), and coming to accept our limitations as a mortal human rather than resenting them brings freedom, peace, joy, *knowledge,* and togetherness.

When Jupiter, the planet of growth, opportunity, and expansion is retrograde it is indicative of those whom least believe in themselves and thus find progress and achieving of goals in life to be a fraught experience. In reality, progress is often difficult for most people, and every person on this Earth will at some point experience heartbreaking loss and humiliating failure, not only you, and coming to terms with that reality through inventory and recognizing that you are not, in fact, special (which is ironically part of the problem) can be immensely helpful.

Saturn being the planet of rules, structure, and discipline its retrograde in the birth chart can manifest as resentment or fear concerning authority, structure, rules, and institution, and many with Saturn retrograde are often always finding ways to be relieved of the responsibilities inherent to systems, institutions, and society or expect others to take responsibility for things which in truth are our own. Of course there is nothing wrong with daydreaming and fantasizing but it is often used as an escape from our fears and resentments, and when instead these are inventoried we find it easier to accept these parameters of life and even find the kind of fulfillment and adventure in them we though instead came from their absence. Everyone in fact chooses to participate in systems, it is often not obligatory as may seem, and just as we choose to check out of expectations and responsibilities we can also check in and get them done.

Uranus retrograde can manifest as a frenetic, erratic, and unreliable nature. While this is probably the least consequential retrograde we can still find it difficult to establish solid and peaceable relationships if others cannot rely on us, and is likewise borne from a fear of closeness, dependency, and loss, since Uranus is the planet of humanity, the group, surprise, and revolution. We may also find ourselves with many unreliable friendships but the problem is not those people but instead our own insecurity and unreliability which motivates the choosing of shallow and insubstantial friends, and inventorying fears of intimacy, vulnerability, rejection, loneliness, humiliation, etc., can greatly help with this.

Retrograde Neptune in the birth chart can indicate secretiveness or delusional estimations of our behavior and its consequences. Secretiveness is often a control mechanism meant to prevent rejection or consequences, but we cannot really prevent any of that anyway and being this way instead removes us from the closeness and bonds of deep and meaningful relationships, so might as well practice inventory about it. The opposite of secretiveness, which is ultimately a fear of other people, is close spiritual intimacy which is instead achieved through vulnerability and relenting of control. Of course, many people are secretive because our behavior really is harmful (although sometimes things we think are harmful are not actually and simply a result of childhood trauma), in which case this is also a delusion to think we can do harmful things to others without incurring consequences, and inventorying our insecurities and fears can illuminate what parts of life retrograde Neptune has made us delusional which can then be amended and remedied through spiritual closeness.

Because Pluto is the planet of control and power, its retrograde in the natal chart correlates with distinct fears of not having control which then very often manifests as controlling behaviors seeking to keep and maintain control, but which can cost us dearly in harmed relationships and lost support and opportunities, and inventory should be immediately practiced around fears and resentments of letting go of control, not having control, and even others having control over us.

Chiron is an asteroid which represents healing and, since its discovery in 1977 has been rightly included commonly in modern astrology, with astrologers astutely recognizing that it specifically correlated with experiences and processes of healing, both mind, body, and soul, and experiences of Chiron transits thusly encompass experiences like learning to accept oneself, recovering from betrayal, or literally finding the cure for an illness or receiving of aid which does the same. Thusly, natal Chiron retrograde is probably common in those with significant life health challenges (such as myself). Often described also as hypochondriacs, I do not believe there is such a thing because many conditions such as my cystic fibrosis are often inobvious and hidden even from those most skilled at medicine, indeed went until the age of 40 before it was ever discovered, even when I went to the emergency room for a year-long cough which had not improved, and the person who suffers from seemingly mysterious or nonexistent illness can have conditions not widely or effectively diagnosed. Even if hypochondria truly is a psychological illness it is still itself an illness, and thus indicative and ironic that hypochondriacs while perhaps misguided in their fixation are not wrong in the fact that they *are,* in fact, ill. Many have alcoholism and addiction which, as I discovered and discussed in my book *Fuck Portion Control*, is a literal neurological disorder. Many gestational periods for disease also involves literal colonization by opportunistic microbes which derange the commensal gut microbiome, ruin digestion, and cause general excess of stress even though no specific disease has yet developed, and Chiron retrograde likely indicates such problems in life. As mentioned before, no humans are subjected or liberated from the conditions of life and existence which govern every other, and most of us at some point do become ill, even severely, and being sick or struggling with chronic conditions is no more or less painful and inconvenient if it occurs at the first of life, the middle, or the very end, and it is our *perception* of illness and mortality which is the greater burden than the actual state of illness, a lesson I know finally at middle age after a lifetime of disease, which I attribute to the practice of inventory therapy and learning how to have compassion for myself and my mortal condition, which is also the antidote to the fears and insecurities which accompany retrograde natal Chiron.

In addition to the placement of planets, signs, houses, and transits it's possible to also look at the lives of people whom are slightly older than ourselves for a preview of most transits we will encounter—not those dependent on signs but instead on natal planets, especially those which are slowest, because our natal placements and thus the active transits will not be much different between us and those whom are slightly older than ourselves. For instance, although I am fifteen months older than two of my siblings we have nearly the same placement of Neptune, Pluto, and Uranus, because those planets are not only very slow moving but also retrograde for about half the year. This same can also be true of transits to the signs and Sun placement in that we will experience transits to our sign as those with earlier signs in the Zodiac belt experience, Taurus after Aries, Aries after Pisces, Pisces after Aquarius, etc., and at the time of this writing many Aquarius and Pisces are now experiencing the same Saturn transits of the first and second zodiac house as Libras like myself experienced about twelve to fifteen years ago which were so fundamental to my now identity and purpose.

There is one problem or discontinuity in astrology which has me puzzled, however, which are the planetary rulers and their assignations. Before the discovery of Uranus, Neptune, and Pluto the signs Aquarius, Pisces, and Scorpio were instead attributed also to Saturn, Jupiter, and Mars, which instead are the rulers of Capricorn, Sagittarius, and Aries, and this was remedied (except for astrologers stuck in the past which continue to use both) once they were found. While these signs and their incorrect rulers shared seemingly related characteristics they obviously were *not* correct in the first place since they in fact had other rulers that we simply did not know about. For instance Saturn transits always concerns discipline, structure, rules, and authority, so its attribution to the sign of Aquarius, which is in fact none of those things, is entirely and utterly incorrect. The new rulers thus made far more sense, *BUT* there remain four signs which still share ruling planets, and that does not make logical sense considering how every other sign has only one ruler. Taurus and Libra share Venus, and Gemini and Virgo sharing Mercury. Similarly, while it can be said that Taurus or Virgo share similar qualities of their supposed rulers the other signs Libra and Gemini seem to be far more justifiable. For instance, if Virgo is about labors and our daily responsibilities, how does the planet of knowledge and communication fit that? Mercury is not constant, often in retrograde, unlike the consistent nature of Virgo. Venus rules balance and relationships which gives Libra an almost annoyingly diplomatic, indecisive nature but Taurus are strong-willed and uniquely singular, and part of the shared rulership is justified by Venus's rule of beauty.

All that is required to resolve this glaring contradiction are *two* additional rulers to fully rationalize that system, with each sign having only one ruler and thus a consistent law, and one of them is staring us right in the face—or, rather, is right underfoot—*Earth* itself, which is more the natural ruler of means and resources such as what Taurus and the second house in fact represents, being literally the primary means upon which we depend. Taurus is literally always compared to by astrology as having qualities of the literal Earth (separate from its quality as an earth sign, but it has that too), and would thus be more relevant to the steady and singular nature of Taurus, and is also associated with beauty because the Earth is Venus's twin and also possessed of immense beauty. Taurus also loves to take care of others just as the Earth cares for us, and the true nature of a Taurus could easily have been confused for that of Venus since Earth is so similar. I'm not sure how this would fit into the astrology system, as Earth does not orbit the Zodiac at all, but that could also easily account for the fixed and steady nature of Taurus anyway (except when they get mad and can explode like a volcano).

The only sign without a unique ruler would then be Virgo but that could very well be the as yet undiscovered *planet* X currently suspected by astronomers to orbit out past Pluto in the Kuiper belt, and then every sign has a singular ruler! The undiscovered planet will by far have the longest orbital period of any, which could easily be relevant to Virgo and the house of Labors, labors being the most consistent and unchanging aspect of life—the constant obligation to participate in it—because of the likely orbit of planet X would be the longest and most consistent of any in the Zodiac. Like this hypothetical planet, Virgos are unchanging and often quite distant, though they make steady friends and relationships are also not always reachable, and are more bothered when there are kinks or disruptions than other signs, which is not a character of Mercury which is the most changeable of the planets, Geminis being naturally quite unbothered by disruption, so Virgo, I declare, is ruled by Planet X (please God chose a better name for the planet—preferably one that is female so Venus isn't the only one up in the heavens—*Hera* would be a good one). At the end of this book is a handy index of all the important elements of astrology, as well as their glyphs and brief summary of their function, and I have taken the liberty of making this presumptive change to state the Earth is ruler of Taurus and Planet X (Hera) as the ruler of Virgo, but anyone can use the traditional ruler if they prefer.

There are many, many, many other aspects to astrological practice, tradition, and history, since it has existed for literally thousands of years. While some of the traditions not stated here might be valid I have only taken the time to cover those which are actually *useful*—much of the remaining esoterica could be interesting to astrologers studying astrology but most of it is pedantic, trivial, often trite, and very often completely wrong, imagined, or even harmful or misleading, and using astrology or any other related metaphysical practice should only ever be used to serve our best interests and help us grow and understand ourselves, not distract us from reality, cause fixation on minutiae and trivia, or serve to engender control and coping behaviors, the very opposite of what astrology is actually good for which thus can invalidate it as a practice when contaminated with superfluous or myopic fixations. Remember that astrology is supposed to be the characterization and interpretation of real phenomena, cycles, and laws which govern reality, and we are not the ones capable of defining reality, only the Universe can do that—Astrology is a question asked of the Universe, not a statement.

The Helpful Rock

A young man was one day out for a walk in the bright sunshine, but not paying attention to where he stepped struck his toe upon a rock, which sent him tumbling into the dirt and bushes. Upset at being so careless he sat down near the rock to assess the scrapes and scratches.

"Hey thanks a lot!" he shouted angrily at the rock. "My life is hard enough and you had to just jump out and trip me!"

But the rock did not reply, for it was a rock. Still, the young man was filled with indignation and didn't care if the rock could listen to him or not.

"Things are rough with my love," he said. "We aren't getting along. Work is demanding, I don't make as much money as I need, and I don't know what to do with my life!"

The rock listened patiently to his problems, and the young man buried his head into his hands and sighed deeply. "Why is life so difficult? Why do we have pain? Why am I so afraid of rejection? Why am I paid a pittance while my boss takes so much for himself?"

The young man sat near the rock for some time, letting the Sun warm his skin, the light breeze caressing him gently. Clouds danced at the edges of the sky, and birds occasionally whirled overhead. The music of crickets in the bushes was rhythmic and constant, and the grasses and wildflowers seemed to sway in place with uncommon patience. As he watched the sky pass overhead, the vast potential for his future suddenly opened up before him, and his face softened, "You're right!" he said to the rock. "I *am* doing my best. I try my hardest, I show up when I'm needed and take opportunities as they come. Life is just hard—so what if I'm not good at it? I'm just one person and I can only do my best, and as long as it's my best I don't need to worry if I succeed or fail because I can always keep trying!"

The young man got up and brushed the dust from his clothes. "I'm going to get a better job. Or start my own business, or do something better for myself. I deserve to be treated with kindness, and even being on my own would be fine because I can treat myself the way I want instead of expecting others to do that for me."

He smiled at the rock, "Hey thanks for the talk," he said before heading home. But the rock did not reply. For it was a rock.

16. Birth, Signs, and Destiny

Astrology is often presented in ways which make life seem metaphorically much like a winding road marked with blockages and obstacles, and when we navigate those obstacles they remain behind us as we progress down the road of life. But astrology, and life, is instead more like traveling with a large suitcase into which we add life experience after life experience, and we do not remove something from the suitcase and leave it by the wayside but it must remain in our suitcase until we reach our destination in order for us to have it when we get there. Meaning specifically that a transit is not merely active or influential in our lives while it occurs but instead influences our life from the point of initiation and forever thereafter, because life experiences are added once they occur and we carry them with us for the rest of our lives, building and building up our possession of experiences until the suitcase is enormous and our journey literally slows from the weight of it, but we indeed have much accumulated life experience to show for our journey.

Some life events may be hard to put in our suitcase, but we cannot just take them out because we don't like them, nor can we put things in the suitcase which we don't have. Transits are also not esoteric concepts—they are tangible, lived experiences and though many may hurt us or alter our life course they are always part of our life experience, and even though new experiences may also teach us how to heal or to cope with pain or frustration the experiences themselves can never be wiped from our history, and are valuable to our life story and are part of the complex metaphysical journey which is life and destiny. Transits are thus not momentary encounters like passing through a town, but are achievements, the building of a house, the collection of collectables, each one adding upon the last and forever being part of our life from that moment on from which it occurs, each transit one more step in the staircase or one more wrung on the ladder of life.

This truth is reflected in my journey which even led to these books in the first place, for if things had gone well for me none would exist. Indeed even after finishing the first version of *Fuck Portion Control* and having several years of recovery in Los Angeles I still found it difficult to put together any meaningful conception of a life, and if instead had made good friends and found my way I would never have moved back near my family and had the experiences which led to completing my second book, the literal journey compelling me to write them, with no plan or intention on my part as I would rather have been doing so many other things, a result of life directing me where to go as reflected in the transits of the planets and this period of my life culminating now in this third book.

In early astrology astronomers only knew about some of the planets, so attributions were made to signs and events that were not sufficiently accurate. Some astrologers believe that astrology before this time was legitimate and that the discovery of new planets simply reflect astrological destiny. But this is just a pitiful coping mechanism meant to defend against criticism and weaknesses of astrology, and the planets like Uranus, Neptune, and Pluto we did not know were still in existence, as were the laws the Universe which form existence itself, and the planets are merely a *representation* of the forces we observe and do not themselves have any magic power, them as affected by the laws of nature and reality as we are and therefore reflect correlations between them and us.

For instance the same forces which determine when, how, and where we meet a new lover are also those which influenced the position and behavior of Venus— not meaning there are separate forces but the complexity and nuance of physics and metaphysics having differing effects on different aspects of reality resulting in both the formation and position of Venus as well as the meeting of a lover or the course of that relationship. Indeed there is strong evidence too of a yet undiscovered planet (planet X), which in my opinion is probably the true ruler of Virgo as previously discussed, and its discovery might further clarify parts of astrology which still remain insufficient. Early weaknesses of science did not invalidate it, and criticisms and inquiry improved science so that we now have such accomplishments as we do today, as is and will be the study of the metaphysics which underlie astrology.

The first Sun sign of the zodiac, *Aires* being the house of the self is the most ambitious, direct, and headstrong of all the signs. Ruled by Mars, the planet of action, those born between the dates of April 19 and March 21 will thus be an *Aries* Sun sign and are thus full of energy, directness, and ambition, although these can be somewhat tempered or complicated by other natal aspects especially related to trauma we experience as children. Aires are often entrepreneurs, initiators, planners, and highly determined and capable, but that description also implies that Aires could be shrewd which they are absolutely not—being the first sign, Aries are the most uncomplicated of persons, and their ambition, plans, and directness is usually pure and without cunning or duplicity. Indeed, are willful participants in fun and games as well as hard work and productivity, and their energy is often so abundant and infectious they can easily share it with others who may be lacking. If it sounds like I favor Aires, I do, as the one boy I dated whom was actually fun was an Aires. Libra and Aires are exactly opposite on the zodiac, and for us this produces a balanced and thrilling relationship complimenting each other's strengths and weaknesses (which can often be the case for any opposing signs), and because they are a cardinal sign (instigators), if not frustrated by excessive trauma Aries are one of the most fun signs of the Zodiac. Indeed there are few others which can match the sheer energy, drive, and independence of Aries (though which of course can be altered by other astrological influences) as seen in Harry Houdini, Vincent van Gogh, Leonardo DaVinci, Stephen Sondheim, Aretha Franklin, Lil Nas X, Dorothy Height, Elton John, and Lady Gaga.

If excessively burdened by trauma Aries can become markedly intransigent, which imperils the ability to form good relationships, which require giving and not only taking, because as much as we desire to be independent we are still human beings fundamentally dependent on others for many of our needs and wellbeing and cannot, in fact, do everything on our own, and taking into consideration the needs of others can thus be most helpful. Because Aires are typically driven by ambitions and high in energy but also being the least likely sign to engage in manipulation of others, lacking an outlet for their energy either through real barriers or psychological trauma and poor self esteem can lead to intense restlessness and even self-destructive behavior such as substance abuse and interpersonal volatility. Coming to terms with the fact that we cannot always make the world the way we desire, accepting the limitations of mortality, learning to be patient and content with the progress which can be made, learning new psychological skills, and directing excess energy into physical activity or challenging creative outlets can help prevent that frustration and direct Aires energy into healthy, productive endeavors. Leadership comes naturally to Aires but lacking the shrewdness of other signs they may not realize their own abilities to lead.

Anyone born between April 20 and May 20 will be a *Taurus* Sun sign, Taurus being the second sign and house of means associated with money, material resources, and things we value. When possessed of strong convictions Taurus have no reservations fighting for what they believe in, but this requires conviction, which not all persons have, but if they do there is no one better to have in your corner than an actual bull (the taurus constellation is a bull). Being also the very sign of materialism Taurus often has plenty to spare and thinks nothing of sharing their resources with others, giving their home, food, and even money if needed to help those they love and care for. Because they are generally so good-natured no one should ever resent their Taurus for being stubborn, as they more than make up for it. Yet one of the more intelligent signs of the zodiac, it is hard to pull one over on a Taurus—they know what they want and are not afraid to get it, and this also makes them great sex partners, probably the best of the entire zodiac, and their charm will completely disarm you for good or ill, but you will enjoy every moment of it.

As discussed earlier, Venus was traditionally assigned to Taurus, but that means sharing a planet and I am pretty convinced that Earth itself is actually the ruler of Taurus, which is also a fixed sign which means they are not given to changeableness the way others are (Venus also has no behaviors that imply it to be fixed), which can be mistaken for reliability or alternatively attributed as stubbornness (one reason the sign is represented by the bull), and because this sign does rule materialism Taurus can sometimes also be shallow when burdened by unresolved trauma. Other means embodied by the sign of Taurus are independence, freedom, autonomy, stability, etc., and those born in the sign of Taurus are usually agreeable, predictable, grounded, independent, and prioritize materialism, handsome, clean, and appreciate beauty and refinement but are themselves too strong-willed to be truly precious. As an example are Taurus Sun signs John Cena, Lizzo, Adele, Channing Tatum, Cher, Barbara Streisand, Karl Marx, and the legendary Keith Haring. While materialism can sound like a shallow aspect Taurus can and does usually also value people and relationships just as much and thus are as likely to freely share wealth and means as acquire them, and there is in fact no more generous sign than Taurus if they are free from excessive trauma and myopic, selfish control behaviors, otherwise they can also treat others as useful rather than inherently dignified and can inadvertently misunderstand and take for granted relationships or the feelings of others which can then be a significant liability, and resolution of trauma will create the more considerate version of Taurus.

The third sign is *Gemini,* the house of knowledge and communication, so people born from May 21 to June 21 will have their Sun in Geminis and are bright, energetic, full of talent, and with plenty to say and do. Ruled by *Mercury,* Geminis can be the most lucky of signs, as all other humans respond to such mercurial talents and many of the greatest music artists, writers, and performers of stage and screen are Geminis because of their innate talent for communication as in the case of Marilyn Monroe, Stevie Nicks, Morgan Freeman, Che Guevara, Prince, Anne Frank, Judy Garland, Josephine Baker, and Sir Ian McKellen. Most astrologers errantly consider Gemini to be a sign of *inherent* intelligence, however, which is very strange because many of them are not at all, and the acquisition of knowledge also requires its absence in the beginning, otherwise there would be no need to acquire it, so those born in Gemini are not necessarily intelligent but do present as bright, adroit communicators and conversationalists whom have heightened potential for great intelligence but do not, as a rule, possess it without also applying themselves to that end just like everybody else. Unfortunately many Gemini are distracted by too many interests to devote the study and learning required for the intelligence they think they should have, and in a world full of insecurity where being stupid is considered a moral failing a Gemini can lose sight of more useful and important talents like brevity, joy, friendship, fun, togetherness, and art, and the only reason this becomes a liability is when a Gemini does not accept who they really are.

Gemini is a mutable sign and considered one which possesses a dual nature, and has the distinct ability to adapt their outward, effacing nature to reflect their environment separate from the person they are on the inside more than others signs are capable. While some might describe Gemini duality as dishonesty the Gemini intends instead to be politic, to better relate with and connect to others through relating and understanding, although when overly emphasized due to excessive, unresolved trauma this can prevent Geminis from forming deep relationships with others since relating requires sharing ourselves too in spite of the risk of injury, or can even lead to opportunistic corruption and scheming to the loss of a person's humanity, being too preoccupied and talented with talking they never stop to just listen. Because of this talent a great number of traumatized Geminis are politicians, as they instinctually understand how to wield language and communicate to people, even though doing so is often opportunistic or even duplicitous, and not many politicians are admirable so I will not list Gemini examples besides Jeremy Corbyn, but anyone can independently see what I mean just by searching for others. As discussed in the chapter on evidence for astrology, Geminis more than any sign have potential to succeed in fields of science or communication because of their natural talent for learning, communicating, and creating knowledge.

Cancer is the fourth sign of the zodiac associated with the fourth house of home and family, and those born from June 22 to July 22 will have their Sun sign in Cancer. Cancer is ruled by the *Moon* which is the ruler of emotions and our inner self, and so Cancers are thus naturally caring and sensitive and love to care for their loved ones through functions which create a home, such as through cooking meals for those they love or constructing a nice home, so Cancers often become skilled chefs once they are exposed to culinary practices (but not necessarily professional chefs because the act of cooking is one of love and care, not profession). But because Cancer is ruled by the Moon, and unless more prominent astrological configurations interfere, they are the most moody and emotional of all the signs because in astrology the Moon has no significant influence itself but is instead a mirror of other energies, so Cancers take on the qualities of the world around them and, if they are very traumatized (which is extremely easy to happen to a Cancer), literally feel the events which happen in their lives as if they are actually those events and not a separate person from them (this is called *pseudoempathy* as I discuss in my book on psychology). The problem for Cancer is we can and do also affect our surroundings very much, and if Cancers are unaware of this fact of life their experiences of disappointment and misery as adults are often a result of their own behavior and choices. Rather than intellectual knowledge, Cancers are also more oriented toward emotional intelligence and, for better or worse, intuitively understand the feelings of others better than most signs (my Moon is in Cancer so nearly every romantic partner I've ever had, including even Sun sign Cancers, accused me of being overly sensitive).

Because Cancers are both sensitive and a cardinal sign (instigators), when burdened by unresolved trauma Cancers are the most critical of others, which is a preemptive defense mechanism meant to shield their own sensitivity and insecurity, the problem being that doing this can easily destroy relationships and drive others away, if not literally at least emotionally, because putting up emotional defenses prevents people from drawing close to us and feeling safe in our presence. The key to satisfaction for Cancer is learning that it's not only okay to be imperfect, but desirable, and the very things that about others which seem to annoy or bother us are the very things we will miss about them when they are gone. Life would be boring if everyone was just like us, and others have their strengths just as they have weaknesses, and having self-compassion as discussed in my book on psychology can also help us have compassion for others. But self-worth is often elusive for Cancers because a mirror cannot easily see itself, and while Cancer can be quietly romantic and passionate, fear of letting people know us truly can prevent the fulfillment of passion and intimacy. Spending time resolving past experiences of trauma with inventory therapy can help Cancer transcend the defensive behavior which prevents them from finding the satisfaction they so greatly desire, and coming to terms with our vulnerability, having compassion for ourselves, and learning how to set healthy boundaries (not just boundaries!) can help resolve those destructive behaviors to create the version of Cancer which is most caring, compassionate, and sensitive.

Grand *Leo* is the star of the zodiac (and will even revel in reading that), ruled by the Sun and associated with the house of creation, born between July 23 to August 22. Like their ruler, when free of trauma Leos embody light, joy, creativity, and thrive on attention, love, and highly value appearance since that is a mechanism by which to gain attention. Leos are often described as natural leaders but their leadership ambitions are rooted in fulfillment by attention and admiration, so while they can be charismatic, charming, beautiful, and even kind and loving their leadership skills are not really the point, and function exceptionally as communicators, spokespersons, actors, models, creators, and compatriots. While this can seem to be superficial it is Leo's inherent desire for love and connection which motivates attention seeking, so it is usually innocent and unoffensive and people are usually more than willing to give you exactly what you want. Leo always takes care of their appearance, is the best dressed of the entire Zodiac, and very often striking in their appearance because Leo is ruled by the shining Sun.

My natal Venus is located in Leo, so my love nature is to be the center of attention and to make my relationships beautiful and prioritize love. Likewise, Leo is naturally outgoing, affable, adventurous, sexual, and confident, although when young they may be intimidated by their own power and appeal until they get more comfortable with adulthood and their identity. Sometimes the desire to be at the center of everything can lead to professions like drug dealing, however, where people are forced to come to Leo for what they need. There is nothing wrong with wanting to be loved, validated, and seen, but when that becomes the only reason for living it can rapidly devolve into manipulation, chaos, and destructive behaviors. Luckily, Leos usually value love above all, so this is relatively rare, but still a potential liability if a Leo is frustrated by a sense of inadequacy. Because attention is gained through beauty but lost through its absence Leos can also sometimes be some of the most superficial of all the signs if burdened by a significant amount of unresolved trauma, but because they are so beautiful and full of energy they are often forgiven for such behavior even when it really should be corrected, and because many parts of life can be ugly (including, at times, ourselves) it is imperative for Leos to learn compassion for themselves, especially when beauty is fleeting. Otherwise Leos are a fixed sign, and so can always be depended on for pure fun and flirtation, and are bright, sexy people who enliven life and help others feel welcomed, invited, and understood, for example Leos like Madonna, Viola Davis, Maya Rudolph, Helen Mirren, Martha Stewart, Shawn Mendes, or Barack Obama, and Leo can easily be a true star since the spotlight is already so natural for them to attain. But simply being the joy of life can be enough for Leo, if they also learn to accept their own insecurities and shortcomings and that people will love you not in spite of your flaws, but because of them.

The sixth sign being associated with the house of labors we can see how the Sun sign of *Virgo*, those born August 23 to September 22, are people which exhibit consistency (without qualification of being good or bad), responsibility, helpfulness, pragmatism, perfectionism, and are often helpful and oriented to service. Because of a natural instinct for responsibility, Virgos like Idris Elba, Zendaya, Keanu Reeves, Kobe Bryant, Michael Jackson, Beyonce, Billy Ray Cyrus, and even Richard the Lionheart are often seen as hard workers capable of staying with tasks long after others would give up or retire, and naturally skilled at jobs which require both keen attention and dedication such as teaching, science, or anything that requires consistent practice and dedication. Because of their perfectionist or pragmatist nature (because of the nature of labors which are everyday and never-ending), Virgo's weakness is that they are also the whiners of the zodiac whom, no matter the medium, will find some weakness or flaw which aggravates their rational and perfectionist nature, although this is often done in such a way as to even come off as endearing or humorous.

As also earlier discussed I suspect Virgo's true ruler to be the very likely but undiscovered Planet X (whatever it may be called…such as *Hera*) which makes sense because of its exceptionally long orbital period it will have that would easily reflect Virgo's nature of tireless labor, perfectionism, and distance. My natal Jupiter is located in Virgo so my growth goals and behaviors are likewise consistent and meticulous, for instance I have never had need to set New Year's resolution because I always make changes immediately when they are needed. Virgos are also skilled communicators and often possess a plethora of friends, acquaintances, and business relationships, though unfortunately the quantity of relationships they have does not necessarily reflect quality, as while finding friends and associations comes easily to Virgo due to their innate charm, agreeableness, and gregariousness, trauma and insecurity can prevent cultivation of deeper, more meaningful experiences and create a more shallow and superficial, distant personality which, like for all signs, can instead liberate our greatest potential when addressed.

Having the Sun in *Libra* for those born from September 23 to October 23 is the middle sign associated with the house of partnership, relationships, and balance, and Libras are often engaged in fun, partnerships, and value beauty, love, refinement, and justice above all else since justice is the authoritative arbiter of balance. It is because we embody the actual definition of balance that Libras are incredibly indecisive because we can see in every choice an equal abundance of justification. At the heart of conflict in a Libra's life is romantic partnership, because a person alone is by definition unbalanced, and Libras are quite frequently and usually oriented toward pairing, whether romantic or not. Libras with excessive, unresolved trauma who do not know how to show themselves love will frequently make excessive demands on their partners and use relationships as salves for our fear of isolation and loneliness, and we can and should be better partners by learning skills and behaviors through which we can give as much or more to others than we take for ourselves, not to serve our own egos but because we have usually taken too much already. Being the center of the zodiac Libras are thus also the most annoying, having enough shared traits of the rest of the zodiac to not be simple but not so much as to be very complex. We are lukewarm, flighty, and oftentimes very shallow. Being in the middle is the entire purpose of a Libra, to be diplomatic but non committal, to satisfy everyone and no one, to siphon attention so long as it does not siphon too much and draw attention to our desire for attention unlike Leos who are unabashedly upfront about it. Thanos, the Marvel villain who eliminated half of Universe with a snap of his fingers, was probably a Libra. Libras will often value relationships for the sake of them and not because they are actually good or helpful, and can sometimes mistakenly see good in things which really should be discarded or abandoned entirely.

Interestingly, because Libra embodies balance very nearly every male who is Libra is usually softer and more effeminate than is stereotypical of males while Libra women are very nearly always more masculine and aggressive than is stereotypical of women. This often imparts devastating beauty since a balance of gender tempers their extremes, and tend to be refined in tastes and manners but cannot deny themselves gratification in sex and consumption since austerity is inherently an unbalanced principle. This is also because Libra is ruled by *Venus*, the planet of love and beauty, and so not only have her weaknesses but every strength as well, if only we can learn how to marshal them, and when liberated from trauma can possess an unmatched capacity for altruism, empathy, and to engender peace such as have Libras Mahatma Ghandi, Oscar Wilde, Sting, John Lennon, Snoop Dog, Serena Williams, and Mark Hamill. When Libras withhold love, either from others but also themselves, we otherwise become petulant, myopic, and the most annoying and insecure of signs as we become the self-appointed arbiters of justice or balance according to our biased, personal interpretation. While we can use our talents to help others, it is also not our job to corral people or make everything alright (other people must take responsibility for their behavior and wellbeing too). Ironically, balance also means not having too much of any one thing, so while Libras are the very embodiment of relationships we can also become rapidly resentful when they are not in balance and others expect too much or too little of us. It is important to remember that everyone is trying their best, and when people are needy or distant in relationships it is borne from fear, trauma, and insecurity, which is best resolved with compassion and kindness, not judgement and resentment which is in reality just an insecurity of our own, and resolving trauma can result in a truly magnanimous personality.

Scorpio is the eighth sign associated with the house of power and control, ruled by *Pluto*, and those born from October 24 to November 21 will have their Sun in Scorpio. Those born under this sign are themselves the embodiment of power, and are often but not always sexual or romantic conquerers who thrive on passion or dynamic interpersonal relationships, and are often naturally intense and probing, often dominating, demanding, highly discerning, and sometimes destructive, for if Scorpio does not channel their desires into productive endeavors like intimacy, creativity, and productivity it can very rapidly translate into domination and power games, especially since Scorpio is a fixed sign which can be very inflexible. Sex is a vital part of the human experience, but the last several thousand years of religious desecration of sex, intimacy, and relationships has perverted reproduction and sex to things which are shameful and embarrassing, when in fact our deepest connections to other people are achieved through real intimacy which comes when we abandon power and control of others such as what occurs in healthy sexual encounters or other closely intimate (not necessarily sexual) relationships. Scorpio has every right to be as sexual and passionate as they want to be, it does not need to be hidden, excused, nor leveraged for control of others, and being honest about noncommittal preferences or engaging in ethical polyamory instead of misleading others or conforming to societal expectations (when you don't really want to) just to get sex can ironically get *more* sex without also hurting anyone (famous Pele was a Scorpio and apparently a great guy apart from all his duplicitous philandering). Someone has to make babies, after all. Or love. If others are embarrassed about sex do not let it influence your own passions because Scorpio is blessed with the very tools for creation and the willpower to get it done, and conquering another not through force or coercion but the very allure of your being and the promised depths of bliss is something to be proud of.

Being so passionate and powerful, Scorpio is also a skillful radical, friend, and compatriot if true self-esteem can be found amongst the rubble of trauma as demonstrated by the likes of Georgia O'Keeffe, Joni Mitchell, Elizabeth Cady Stanton (a famous women's rights advocate), Marie Curie, Ryan Reynolds, Weird Al Yankovic, Whoopi Goldberg, and Trace Ellis Ross. Unfortunately as the natural sign of power it is often difficult for Scorpios to achieve subversion of selfishness, and frequently result in destructive personalities if burdened by unresolved trauma. Learning how to responsibly and effectively direct Scorpio energy is their greatest strength, since all humans possess sexuality and the power to create, and the responsible use of power can be highly productive and advantageous. Because of these themes Scorpio also embodies the transcendental heights that are achieved when we subvert baser instincts, and can thus also be associated with abstinence and other aspects of reality which eschew the mortal in favor of the spiritual. When this is channeled positively it results in an alluringly passionate and seductively dynamic person full of charisma, charm, and wit, but if impeded by unresolved trauma can instead be controlling and myopic egomania, especially since Scorpio is a fixed sign (whichever type you are is entirely up to you).

The ninth sign is *Sagittarius*, associated with the house of the mind and ruled by the planet *Jupiter*, for those born November 22 to December 21 will have their Sun in Sagittarius which encompasses philosophy, though, expansion, travel, adventure, and if you know someone who is always traveling and loves seeing new places, especially to a fault, they are definitely a Sagittarius (or have a Sagittarius Moon or Ascendant). Because they are naturally inclined to experience life and other people Sagittarius becomes restless if they are not constantly experiencing new and novel things, or can become tiresome if they are prevented from freedom and adventure. This has its obvious drawbacks since there is a fine line between adventurousness and plain instability, and many husbands, wives, friends, and children get left behind by Sagittarius who cannot find expansion and purpose in static institutions such as family, relationships, and community, and travel and adventure can also occur in the mind or where we live and does not always require literal travel. Sagittarius are also never ideologues, since that requires some degree of naivety, which they are not, and because experience is more valuable than the material they often don't prioritize salary and status but instead jobs which help them better experience the world. For this reason too Sagittarius can be hard to pin down, and are best paired with people who share similar life goals and priorities. My natal Neptune is located in Sagittarius, so one of my most favorite, healing, spiritual experiences is getting in the car and going on a nice, long road trip anywhere, the journey itself being part of the experience and not just the destination (which occurs too quickly by plane). Because the ninth house embodies growth and new experiences, its natives, love traveling, meeting new people, cultures, having new experiences, and think nothing of hopping on a plane to go on an adventure.

While naturally intelligent, Sagittarius is also a mutable sign and their otherwise amiable and freewheeling nature can instead transmute to erratic, unreliable, and irresponsible nature if there is unresolved trauma and insecurity, but otherwise Sagittarius are great fun, always down for friends, and the most adventurous sign in the Zodiac. Like Gemini, knowledge must be earned, so if Sagittarius does not pursue learning they can be just as dumb and insensitive as anyone else. But when applying themselves to thought and knowledge can be the greatest of philosophers, thinkers, and humanitarians the likes of Mark Twain, Jane Austen, Frank Zappa, Pope Francis, Tina Turner, and Sinead O'Connor. Sagittarius are one of the braver signs, like Crazy Horse who was a leader of the Oglala Lakota tribe who fought to resist colonialist theft of Native American land. Indeed it takes a degree of bravery to travel the world and seek adventure so freely, but travel is just the easiest way to experience novelty, in fact it's the laziest way to stay occupied—just jump on a plane or schedule lots of business trips, but finding how to expand your mind and experiences without shirking responsibility is quite the challenge for Sagittarius. Be creative. Jupiter being the ruler of Sagittarius they are often empowered with the blessings of Jupiter as they transit the great adventure that is life, just make sure in your haste to see and experience things you don't leave people behind.

Dear *Capricorn,* the tenth sign and the *House of Society*, one of the few peaceable signs of the zodiac, with members like Jim Carrey, Dolly Parton, Betty White, Elvis Presley, Denzel Washington, and Stephen Hawking embodying the dedication, work ethic, and brilliance of which those in this sign are capable. Ruled by *Saturn* for those born between December 22 and January 19, Capricorns are often very rules-based and find comfort in structure, discipline, tradition, authority, and consistency, while also being unmoored by spontaneity and instability, although they are also a cardinal sign their instigations are most often authoritative in manner rather than fun and lighthearted, and often obtain positions of significance won through determined participation in organizations, government, corporations, etc. Capricorn are often reserved, even shy, feeling like their power only comes when protected by the group or institution, and rarely abandon institutional structures in which they are born or strongly affiliated. Capricorn are extremely reliable and stable and make for great romantic partners for those seeking consistency and reliability, when they do decide to settle down.

As all signs have their strengths so do they have weakness, and much of Capricorn's apparent steadiness is actually paralysis from a deep fear of life and others when burdened by trauma, and excessively seeks shelter in institution, tradition, and rules which may in fact hinder growth, resolution, and intimacy, being too serious and fearful should instead also find value in personal experiences, desires, and needs, and realize that any one person is often more capable, even on our own, than we often give ourselves credit, but that it is also not our job to control life and make people and events conform to rules and expectations or avoid heartache or misfortune. When unburdened by trauma Capricorns this fear, over time, can mature into dependability and consistency and produce the most lovely of persons and the very rock on which the foundation of humanity is built. Above all a Capricorn must learn it is okay to be afraid, to be in pain, to carry trauma, and that life is never consistent or reliable and we should not expect it to be, that forces both of order and chaos exist together and not only one or the other. Life can often be unsettling and scary, but if Capricorn can transcend fear and trauma and harness the strength inherent in Saturn they more than any other sign can conquer life and its challenges.

The penultimate sign, *Aquarius*, associated with the house of fortune and friends is ruled by *Uranus,* the planet of revolution, technology, and progress, and those born with the Sun in Aquarius from January 20 to February 18 are likewise nonconformists, even from an early age, and often have eccentric personalities whom do very well in theater, science, writing, music, etc. They are also rebellious, unique, smart, optimistic, and always up for a good time, especially if it has a purpose (like sport or acting), or involves sex, and are always a source of surprises both good and bad. Aquarian boyfriends and girlfriends often show up with flowers or cards, or will take you by surprise in a sauna, but they will also casually announce an affair like it was something they picked up at the grocery store (the ones which are disloyal, which has nothing to do with any one sign). Just like Uranus, Aquarians are unpredictable (the wonderful Tallulah Bankhead was a great example), and for those who prefer the safety of predictability this can be either exciting or disturbing, and unpredictability is more destructive from Aquarians with an abundance of unresolved trauma and dearth of self-esteem, where those without them will simply be fun, spontaneous, and enlivening. All Aquarians are intelligent, but their unwillingness to adapt and change often prevents the development of wisdom (Uranus is the planet of doing rebellion, not receiving it, where the quality of rebellion is inherently inflexible and forces change in the world around it, but does not itself change), and so their intelligence is often ineffective in making significant improvements to their problems. Strangely, while Aquarians are often beautiful, romantic, and gifted in sexual endowment they are usually a bit graceless in bed which, considering the erratic nature of Uranus, makes a lot of sense. While this can be off-putting for more refined or prudish signs, sometimes you want to just get your brains fucked out, something an Aquarian is not only good for but very willing.

A fixed sign, Aquarius's nature is very consistent in its inconsistency, but it is also the sign of humanity, so Aquarians can be compelling trendsetters, influencers, and general catalysts who keep humanity from stagnating. Vanessa Redgrave is an Aquarius known for her electric acting ability, but whose early advocacy for the Palestinian people during their genocide by zionists at a time when doing so was extremely dangerous for a person's career was notably brave. One strange and consistent dichotomy in Aquarius is a marked ability to see the goodness in humanity but the worst in those closest to them, such as a partner, since the sign encompasses humanity as a whole which by definition does not celebrate the individual. This can lead to Aquarians having many, many friends and acquaintances but strained romantic relationships as they hold those whom they love to unreasonable standards not required of the rest of humanity. This is probably a specific result of poor self worth and unresolved trauma in Aquarius which motivates them, ironically, to seek conformity and unmoored by nonconformity. Tallulah Bankhead for instance turned down a role because of its cockney accent, afraid her British friends would think ridiculous, and the role instead went to Betty Davis which won her immediate critical acclaim at at time when Tallulah was trying to conquer Hollywood. Similarly in relationships, if unable to resolve the dearth of self esteem in themselves cannot then understand why someone should love them, and for inflexible Aquarians the strengths of spontaneity, assertion, and self-will easily become weaknesses in a sign driven by the energy of a planet like Uranus. Aquarius is also at risk of seeking popularity as an antidote for insecurity, which can be especially destructive because popularity is insubstantial and illusory. But when this trauma is resolved Aquarians can't help but entertain, disrupt the status quo, and inspire others with their fearless commitment to being interesting, talented, and unique. Especially in this age of Aquarius marked by technology, progress, and humanity, Aquarians thrive and find themselves especially effective if they direct their Uranian energy into productive, compassionate, and honest endeavors and can ascend to considerable heights the likes of Galileo Galilei, Franklin D. Roosevelt, Frederick Douglass, Virginia Woolf, Charles Darwin, Bob Marley, LeVar Burton, Oprah Winfrey, Harry Styles, and Abraham Lincoln.

Sweet, miserable *Pisces*, does anyone really know you? Do you really know yourself? Associated with the house of spirit and ruled by *Neptune* for those born February 19 to March 20 are highly intuitive, often meek (sometimes even to the point of shrinking), and more connected to the spiritual undercurrent of life. All signs in the zodiac build on the one before it, so Pisces is an amalgamation of all and this complexity can be overwhelming for humble Pisces, who are humble only when they also know compassion for themselves else a Pisces personality becomes peevish and shrinking under all that weight. The intuitive nature of Pisces can sometimes confuse them into feeling unfulfilled by religion because, like most religious members, mistake religion for spirituality, where religion is actually an institution and philosophical, not spiritual, where spirituality is instead found through inward introspection and personal spiritual experiences such as discussed throughout this book. The outside, real world can often be too harsh for Pisces, so taking time to care for personal needs is very important to feeling safe and fulfilled. But because of their singular connection to the undercurrent of reality, Pisces are uncommonly contemplative, insightful, or overwhelmed. Pisces are extremely agreeable and make excellent friends for those who desire real connections and friendship, although when burdened by unresolved trauma Pisces can be excessively timid, secretive, delusional, unscrupulous, or attracted to more powerful, charismatic types that help them feel safe but actually imperil their wellbeing. Being assertive is almost impossible for young, timid Pisces but once they overcome their trauma and realize not only is there nothing to fear but that their spiritual power is actually the greatest in all the Zodiac they can do truly incredible things.

While it's certainly possible for Pisces to be hurtful or harmful just like any other sign, true egomania is exceptionally rare, so natural is their connection to the spirit and undercurrent of reality. Every human is powerless in life, and when Pisces come to terms with this indefatigable reality, which is a core tenet of spirituality, they become the greatest of all great people like Albert Einstein, Linus Pauling, Kurt Cobain, Harriet Tubman, Elizabeth Taylor, Liza Minnelli, Copernicus, Jack Kerouac, and Nina Simone. Naturally drawing strength from the intangible, Pisces do not have the drive, power, or ambition of many other signs such as are ruled by Mars, Jupiter, or Saturn, so Pisces become particularly powerless when trying to be like the other signs. Don't be them. Be you. Be wise, kind, perceptive, and imaginative and there is nothing in life you cannot do.

In addition to the Sun sign, the placement of the *Ascendant* is the primary astrological indicator of our life destiny, as correlative to the *self* and the first house and the first sign of Aries the Ascendant thusly represents the destined life path of the self and our life purpose. As discussed earlier, most astrologers often include the Ascendant (also known as the rising sign) in the personality, but in fact the Ascendant is more applicable to our life path which, while having some effect on the personality, is not as influential as the Sun and Moon sign placement, and very specifically the Ascendant concerns the nature of how we try to get what we want from life, which is in turn informed by our overall sense of purpose as concerns the First House, the Sun, etc., and the *Midheaven* is our ultimate destiny, the Ascendent describing how we get there. I have mentioned how having the placement of the Ascendant in Scorpio such as I have indicates a dichotomic life wherein the first part before a time and event of forced evaluation will be entirely different than than the one which occurs after, wherein the life goals and purpose are changed by some trauma or significant event, but all the while a shrewd and passionate pursuit of life goals, whichever they are.

Of all the common astrological traditions, the Midheaven (Medium Coeli) and the Nadir (*Imum Coeli*) are some of the most widely misunderstood, usually described as the career and public image and home and family life, respectively, are in reality the outward life and inward life and far more than simple career or family themes. While the outward life can include career and vocation it is instead rather the entirety of everything which is the outer life and even includes how we are perceived by others as an effective and material person, and even houses and family can be part of the Midheaven if they are a function of the outer life rather than our inner selves. Likewise, the inner life often includes home and family but our inner, immaterial life which the Nadir truly represents is especially concerned with our spiritual growth as a person, our humanity, humility, wisdom, skills, trauma, ideology, etc. This mistaken confusion occurs because in terms of the Solar zodiac the sign of Cancer, which rules the home and family, sits in the same representative position as the Imum Coeli relative to the Ascendant as Cancer does to Aries, the beginning of the Zodiac. But the Sun from whose position the Zodiac fourth house is calculated represents the *self*, and the fourth zodiac house of home in truth then houses the self, not the family, the body being the home of the soul, and so is not necessarily about home and family but our entire inner world of the spirit, mostly that which is unconscious and often hidden from our understanding and conception, not least of all too from other people.

The Nadir can thus include home and family but only so much as it is a reflection of our inner life and spiritual development, and represents the family *from which we came* but not necessarily our current family. Very often at the same time our partner also will have a similar journey since the angles for partners are very often near each other, thusly often to correlate our experience with theirs but which is not necessarily harmonious nor inherent to house, home, and family members which in truth are very often also subverted during those influences because partners and our place of domicile are so often functions of material. We also only consider the Midheaven the "peak" of life in the natal chart because most of us fear material insecurity and rejection by others and so desire wealth and success as control mechanisms for avoiding those things. In fact, the Nadir could be considered the true peak of the chart if inner, spiritual development and wisdom was revered as much as material success, especially considering that immaterial skills, insight, and growth liberate us from poor decision making and dependency on money and status to live more effective and fulfilled lives, not because of what we have but in spite of it. In fact, while the loneliness of Saturn's path across the Nadir as discussed in the upcoming chapter on Saturn can be unsettling and frustrating I grew to greatly enjoy and cherish the years of rest from the rigors and responsibilities of more active and demanding life, to recharge and renew along with making great progress in the work and research undertaken during this time, and it was quite easy to meet my needs, spending all of my spare time doing only what I wanted, having much peace and quiet, especially relative to the tumultuous years of past relationships and associations associated with the other three quadrants. In fact, wasting any transits to the Nadir, especially that of Saturn, spending it in resentment for the limited material success and opportunities is an extremely wasted opportunity to grow spiritually and deepen wisdom and character that will not come around again for thirty years (since that is how long it takes Saturn to transit the Zodiac).

The primary driving force for what we do with ourselves and our time and resources is *purpose*, and those without purpose often dawdle or wander, searching for meaning, fulfillment, and direction such as predominated my early adulthood. The problem is that most of our sense of purpose is different than those of others, so one sign's source of fulfillment will not be a useful guide for all the other signs, and is one reason why self-help guides and influencers do not often in fact help people find their own purpose since this will be different for each person such as what is described by the locations of the Sun, Ascendant, and Midheaven. Part of this discrepancy is that stress and challenges in life are not barriers to fulfillment but the very *mechanism* which makes fulfillment possible, since without insecurities and weaknesses we would have no need for fulfillment in the first place, and repeated life experiences which cause us stress function in truth to point us in the direction of purpose and fulfillment such as what occurred during my twenties and thirties (especially my thirties). As I have before mentioned, life only exists because of the conflict between creation and destruction, and although many of us would rather it not exist the joys and highs we experience in life only occur at all because of their opposite experiences of pain and loss, and blaming other humans for the reality of life, even that of our being here, is entirely founded on the mistaken idea that any human has any control over reality whatsoever, which is instead a product of the Universe which intended us to be here and to experience the growth which occurs from unfortunate experiences as much as those which are fortunate, and if your parents did not birth you some others would have, and we are here *against* our current will as a human being for reasons we do not fully understand but which likely have to do with simply experiencing life as we do. As such, having a sense of purpose is facilitated first by *not* having purpose, which is another way to say that there is nothing wrong with not having a sense of purpose since this is first required to get one, which will be different for each person according to their unique birth conditions as described in the birth chart, especially in the location of the great angles.

The Midheaven placement thusly represents our outward, material destiny and the nature of our ultimate successes in life, while the Nadir placement indicates where we feel the most comfortable and safe. When I was in my twenties my primary goal for myself was to make film and video games as part of the machine of these creative industries, to tell fun stories that helped alleviate my own and others' fears and pain, but for the purpose of creating and not my own fame or fortune. But try as I might there seemed no effective route for me to realize these goals and aspirations and in part this was because my of the placement of my Midheaven in Leo, meaning a destined life path of individuality, public prominence, and creation. Because I was insecure and felt more comfortable in the safety of a group with the Nadir in Aquarius, which is the sign of the group, I often sought to fulfill my goals and dreams with and through others but wherein I was constantly rebuffed and ruined, such as when after working on art and animation for some major gameshows had an idea of my own that a friend stole and sold its development to a major network (it didn't go anywhere though because he stole it and didn't understand how to execute it). It was not until I forged my own path entirely independent of others (through force of circumstance and failure) that I finally found this success and direction which also will likely hurl me into the public spotlight in spite of my reservations to such a life.

What's most interesting about the great angles is the often a shared alignment of the axes between couples, since fate and destiny as what the angles represent would need to be at least be somewhat aligned for people to even meet (at least, for organic pairing and not relationships of utility or convenience), and I find it most common for couples to have alignment of their Midheaven and Ascendent axes not more than one sign away from that of their partner as discussed in the chapter on evidence for astrology. This placement could be misunderstood to mean that our relationships are predestined and thereby lead people to place more emphasis on them than is merited, to be sure they are special and may be predestined but all relationships, no matter how special, are predestined in the same way that what we had for breakfast this morning is also predestined, and thus we come into contact with them as opposed to other people with differing life paths which prevent us from even encountering them instead. Like all aspects of astrology, this information is a tool most useful to understand life, not control it. All relationships also pass away and end just like every other aspect of life, and being delusional to that reality causes greater suffering, not less, because it is inevitable that relationships end. This also makes them more special, and if we rationally understand and accept the limitations of mortality we are not only *not* taken by complete surprise when such cycles end but spend the time we have with others more present and thus more fulfilled since we are not constantly living in the future or past in anxious expectation of loss or obsession with failures and mistakes. All life is a cycle, and as I write this book while having been single for about the last decade of my life know that Saturn is currently on its way to my Descendant which is a more favorable time for meeting romantic partners, especially since the people I would meet are also experiencing a similar influence required to facilitate our meeting that will, in the future, eventually bring us together as the cycle continues onward. Understanding thus the placement of the great angles to our life path can be used to understand our lives, the Midheaven describing our destination and the Ascendant describing how we get there, moving from the comfort zone of the inner self as described by the Nadir to the destiny of the outer self in the Midheaven, the Descendant opposite the Ascendant representing our relationships with others.

The meaning of these placements is thus when the Ascendant is located in Aries the sense of purpose is singular, uncompromisingly determined, and constantly in motion in tireless pursuit our goals and dreams, the sense of self and what we want to achieve clear and unwavering even from the beginning of our life, although being so self-assured this placement can often unintentionally disregard others in the quest to achieve or refuse the cooperation of others, since Aries is naturally opposite Libra, the sign of partnership, where the Descendant is located, and acting without regard to others can often get this placement into trouble until they learn integrity. There is nothing wrong with pursuing goals, even uncompromisingly, but committing offenses in the process will only impair success as the laws of causality come for their due. Once the person with this placement achieves success the establishment of a relationship will be more likely and could be a long-lived and important personal relationship, since the Descendant which rules the Other is also in the sign of Libra to which it metaphorically correlates. Sometimes the goals for this placement can even seem naive, but there is nothing wrong with that as naivety is often a vehicle for trying things we would not otherwise, and powered by the relentless determination of this Aries placement is more likely to succeed in part simply from sheer determination.

When the Midheaven is in Aries this person is destined to eventually be recognized, for better or worse, as a singular, inexhaustible individual of uncompromising force, determination, individuality, and creates a commanding individual as inherent to the nature of such life goals as exemplified in Celine Dion, Julia Roberts, and Angelina Jolie. But this placement also means the Nadir is in Libra, the sign of partnership and balance, so those with this position can resent being subordinate to others, not because they are incapable of subordination or collaboration but because other people can rarely assist with the exceptional energy reserves of an Aries Midheaven in pursuit of their goals and ambitions. In other words, few people can keep up with you, but you also resent having to slow down for them, which can ironically leave you all alone out in front with nobody with whom you can share your successes once they are achieved, and this position misleads its natives to mistake romantic partnership as being a goal or task of utility, which can and does leave the romantic or family partners feeling managed, utilitarian, unfulfilled, marginalized, and even neglected, and care must be taken to recognize that relationships and family do not make us weak or vulnerable, that vulnerability is instead a strength, and only after that is realized can this position create true greatness.

In Taurus the Ascendant brings a purpose achieved through materialism and the creation of resources, and because Taurus is also determined and steadfast this purpose is formed from the beginning and never wavers. Taurus is not a naturally selfish sign, however, and if trauma can be resolved and the desire for superficial wealth subverted in favor of more enlightened conceptions of material existence this placement can result in the uplifting of the material conditions of many, although of course they can be shallow just as anyone else if burdened by excessive trauma. The Descendant here, however, is located in Scorpio, the sign of power and control, and this placement may indicate exceedingly prominent control behaviors towards others, the behavior of possessing others, which will likely cause much conflict until maturity and wisdom changes valuation of others from a possession to a priceless gift.

Those with the Midheaven in Taurus are likewise destined to achieve material success, and all their lives will feel competitively driven to surround themselves with money, luxury, and aesthetic riches. There is nothing inherently wrong with this placement, but the Nadir being located in Scorpio causes this placement to feel more comfortable with control and power, rather than letting go of it, so many with this placement whom have abundant unresolved trauma excessively empha- size both power and materialism, even to the point of being interested in nothing but money and devaluing other important aspects of human experience such as intimacy, relationships, spirituality, reputation, integrity, friends, family, stability, experience, wisdom, health, energy, intimacy, and joy. The dysfunctional aspect of this placement is fundamentally driven by fear and unresolved trauma surrounding material wealth, personal worth, and power over life which the facade of financial security can mask, and become so pathological as to entirely cause abandonment all human virtues. So it is important for those with this placement to remember that money is not the only valuable resource and that many other things are also valuable resources to cultivate, without which we in fact jeopardize our wellbe- ing in the singular and myopic pursuit of something as shallow and transient as money, and that the higher nature of Scorpio is to subvert human desire for control and power in service of a higher purpose. Recognizing this reality and cultivating all the important sources of value in human life can result in the creation of much beauty, integrity, and success as in the case of Margot Robbie, Marylin Monroe, and Tina Turner.

As Gemini is the sign of knowledge and communication the placement of the Ascendant here describes a life of indefatigable curiosity and learning, lots of talking, and probably a life in media, communications, or poring though books and academia and likely a very intellectual and learned life. But since intellectual- ity is the antithesis of feelings and emotions this placement may take for granted those more personal and emotionally vulnerable experiences. This placement means the Descendant is in Sagittarius, the sign of the mind, travel, and adven- ture, and it is likely that this placement finds many relationships and lots of travel and exploration, or a lover in a land far distant. Since neither of these signs concerns emotions, the danger (when trauma is not resolved) is an inability to establish deep, emotionally fulfilling relationships and to intellectualize experiences which instead require feeling. There is of course nothing wrong with being single, and part of the problem of this placement may be in thinking we are obligated to find a relationship, when that instead can simply be whatever we want to do and prioritize. Just make sure to secure a family, whether or not a relationship is part of that, since we are all inherently dependent on others and cannot truly handle life alone.

Having the Midheaven in Gemini similarly means a life destined for success through the act of communication, such as oration, writing, politics, acting, music, even to the point of unmatched greatness when this placement is supported by other aspects, which can sometimes endow its native with the ability to achieve great feats of talent and ingenuity like J.R.R. Tolkien, Charles Dickens, Ernest Hemingway, Jack Kerouac, Kurt Cobain, Keanu Reeves, Madonna, Kathy Bates, Melissa Etheridge, Franklin D. Roosevelt, John Carpenter, Dolly Parton, and many, many other famous people. That is quite the pressure to live up to, though, and in fact no person should strive for greatness but instead simply show up for opportunity and do our best, because greatness does not come from attempting to be great, but instead when people simply do their best at what they are great at. This means the Nadir in Sagittarius opposes Gemini's bookish dedication with freewheeling diversions and portends potentially shallow avoidance of deeper philosophical exploration, so those with this placement can often take refuge from their fear of the intellectual in superfluous distractions and meaningless ideology, presented even as a fear of being perceived as stupid. The thing to remember with this placement is that there is in fact nothing wrong with being stupid—stupid people can be some of the funniest and most endearing of individuals whom liven life and help others take themselves less seriously, and it's okay to find balance between ignorance and knowledge, mundanity and greatness, and the sharpest of minds recognize that our flaws are not meant to make us weak and incompetent but the very things which make us unique and interesting. As I state in my book on psychology, the only true marker of intelligence is *curiosity*, and flaws can give us strength of character which then can in fact be wielded for success, while those whom attempt to hide their flaws and insecurities and are incurious about things they don't know or understand become tiresome, ineffective, neurotic, controlling, and never grow as individuals. Having compassion for yourself such as is achieved through inventory can really help this position reach its highest potential through acceptance of our weaknesses and vulnerabilities rather than denying them.

In Cancer, the sign of home and family, the Ascendant indicates a life oriented toward the home, marriage, and growing and securing a family and all that requires and entails, with great satisfaction being found in the family and cultivation of a home. But this placement also means that the Descendant is in Capricorn, the sign of institution, rules, and authority, and because Cancer is a sensitive sign this may also make for excessive control and anxiety around family and the home and the forceful, authoritarian implementation of rules and order to turn the home into an institution which can instead upend peace and tranquility which other members cannot wait to flee, and you risk ruining the good in pursuit of the perfect. Both of my parents have this placement, actually, which is the reason our home was so (ironically) lacking in warmth and kindness, because when burdened with unresolved trauma the negative iterations of each sign becomes more prominent than the positive, and instead of a stable, loving home ours descended into an institutional, authoritarian facsimile. Just as our personal transits of the fourth house can likewise be stressful because we are both dependent on family members but they are also their own persons capable of rebuffing our desires (or leaving altogether) it can be highly vulnerable to invest in a family, but like all things in life the effort to control or prevent misfortune actually increases its likelihood, because it is not within our power to control life in that way, and surrendering those things to the powers which do in fact control them (fate) and being willingly vulnerable (if you have not done inventory you absolutely do not know what that really means) is the proper way to experience this placement, which will secure family and bonds through vulernability, where control instead destroys such things.

Since the Midheaven is naturally associated with the tenth house and sign of Capricorn, the Midheaven in Cancer can often cause much stress on the journey to our life destiny of home and family. This placement does not always mean the literal home and family, however, but like other transits to this house it can also be the home and family of all humankind or other larger structures which encompass any kind of family, and rather than the meaning of homemaker, although it can be that, instead embodies the metaphorical home and associated nurture, care, and shelter that home provides. Because the Nadir here is placed in Capricorn this can cause much mental anguish, stress, and volatile interpersonal conflict if care is not taken to find compassion and healing for ourselves and our past trauma on the way to our destiny, but those whom have achieved this feat have become the very bulwarks of humanity such as Mahatma Ghandi and Coretta Scott King. The strength of this placement is that the power of the outward self is found ironically within the inner self which, because of the opposite placement of the Nadir in Capricorn we are naturally suspicious and afraid, feeling more comfortable in rules and order than intimacy and humility and thus often forced by life circumstances and experiences to peer into that unknown that we otherwise resist greatly, but can find ourselves within the family of man if we but look. Cancer is represented by the crab, which is an apt metaphor, as Cancers of any type retreat into themselves for protection when offended or stress, but for the Midheaven in Cancer the inner self in this case (the Nadir) is inextricable with the public such as in the case of Mrs. King, in which because of circumstances we cannot actually detach ourselves and retreat from the influence of the public on our personal sense of self. For this Cancer placement, being a vulnerable individual is quite frightening, but while relying on institutions can be comforting it is also a liability, so it is important to remember that in fact all people are individuals and at many times in our lives we become starkly aware of the separateness between us and others, sometimes painfully so, but that there is also strength within the individual not dependent on others. If we but trust in our value as a person we can find power in vulnerability and thus the strength to meet life head on.

Since Leo is the Sign of the Sun, an Ascendant in Leo inspires purpose through fame, individuality, and creativity, and this placement likely motivates a shining personality inspired by attention, art, projects, creation, and other children of the mind and heart like music (as a form of self-expression rather than technical mastery). Because Leo uses tools like beauty and attraction to achieve attention, however, there may be excessive preoccupation with looks and style, and it is important to remember to be kind to yourself, that you do not always need to look beautiful, as inner beauty is just as alluring to other humans as outer. This placement also means the Descendant is in the sign of Aquarius which rules fortune, friends, technology, and revolution, and romantic relationships may be more friendly than is typical of romance, which is certainly a positive placement, but the negative side of Aquarius is to see the flaws of humanity in the individual, and could also create an excessively critical orientation toward others if we are burdened with unresolved trauma and control behaviors.

Having the Midheaven in Leo like Ethel Cain, Harry Styles, Prince, Joanna Lumley, Charlie Chaplin, Linus Pauling, or myself have undeniable flamboyancy and almost obligatory individuality. But because Leo is the sign of creation, those with the Midheaven here are also often preoccupied with building or creating something unique that has not perhaps existed previously, and commanding attention *through* individuality (as opposed to individuality being a result), and because of the complimentary position of the Nadir in Aquarius, the sign of rebellion, progress, and technology which informs the inner self, often revolution-ize or at least revitalize whatever they concern themselves. Because this position is naturally individualistic it also causes restlessness and resentment for authority such as answering to an employer, not because Leos are incapable of collabora-tion (quite the opposite they can be very good compatriots, having the Nadir in Aquarius), but because authority and institution so often impairs the creativity and production required for the successful realization of our life purpose. This position does, however, also cause frenetic energy which can result in ineffective or chaotic pursuit of goals before we eventually find something that actually works, but even after having achieved recognition can be notoriously inconsistent, restless, undisci-plined, and scattered, but the purpose of which is to be an agent of creative chaos. This can be frustrating for others who value stability and consistency but the act of creation is inherently undisciplined, and frenetic energy, while chaotic, can and does set many things in motion. Unfortunately if the native with this placement is given to evil their charisma imparted by this position is a facilitator of their destructive goals in which their personality becomes the focus rather than their work, and they instead create nothing useful or in many cases cause destruction as the embodiment of the worst of the Aquarian traits. But usually this placement is typical of bright, intelligent, and eclectic creative revolutionaries which can best achieve their goals by first resolving trauma and insecurity.

Virgo is the sign of labors and thusly the Ascendant here is of those which are very hard working and tireless, often eschewing style and flair for responsibility and consistency. As however the sign of Virgo is unmoored by inconsistency, when burdened by unresolved trauma this can sometimes result in peevishness if care is not taken to relax a little and not be so serious, and the Virgo rising can steam-roll over others in their relentless pursuit of goals, even if those goals are not well defined. Remember that there is plenty of time in life to achieve, and that fate will eventually meet us regardless. Because Virgo also embodies healing this can also mean themes of healing which occur during the lifetime, for ourselves or others, when pain and frustration are redirected from superficial pursuits to those which are more holistic. The Descendant here is also located in Pisces, the sign of spiritu-ality but also delusion, and when we are not clear about our goals and purpose we can unfortunately deal with our relationships and others in a delusional manner, and all these conflicts are resolved through the practice of inventory which can elevate delusion to spirituality and peevishness to determination.

A Virgo Midheaven placement similarly indicates a destiny that will be achieved through tireless work ethic and of a person who is perceived as dedicated and reliable. While this placement may sound mundane, the Nadir placement in Pisces, the sign of spirituality, the unknown, and endings, informs the work of those with this placement such as famous, beloved filmmaker David Lynch, whom can derive much insight and perception into the depths of human existence. When empowered by compassion and spiritual insight those with this sign can be incredibly effective in structuring their lives in ways which not only benefit themselves but everyone around them such as demonstrated also in Chrissy Teigen and Elvis Presley. Insecurity in this placement otherwise stimulates delusional obsession with productivity simply for the sake of it, and without regard for whether it is actually helpful and productive as in the case of some truly abhorrent past world leaders. Like all Midheaven placements, this one can cause significant internal strife for those who feel a conflict between what they feel obliged to do or be and what they feel capable, in which case the practice of inventory can help relieve us of onerous and inappropriate stress, shame, and expectations and help narrow, define, and enlighten the sense of purpose. As the Nadir in Pisces is a deeply spiritual influence those with this placement can mistake their need for and connection to spirituality as something that is satisfied in rules and order such as is brought by religion, but religion is inherently opposed to spirituality and is instead a control mechanism, and coming to terms with our real spiritual nature can empower us to have real spiritual experiences in daily life rather than relying on others to do it for us.

In Libra the Ascendant indicates a purpose of balance and harmony, and is often especially concerned with justice. But remember that balance does *not* mean what most people think it means, and instead is the equal apportionment of the whole, so this position makes for people who are not especially ambitious or obsessed with any one life goal, but instead tend to spread themselves out through several different areas of interest and focus, and very often concern themselves with social justice and advocacy as in the case also of Linus Pauling who spent much of his life advocating against war and military intervention (but he's not known for that, since it was his Ascendant, and is instead known for his quixotic personality and bold reformation of science due to his Midheaven in Leo and Descendent in Aquarius). Oppositely, becoming unbalanced will send them into a crisis, and it is important to remember that not everything can be okay at all times, and that we can get through surprises, challenges, and extremes just like any other sign. This placement also means the Descendant is in Aries, which means that we may have a tendency to treat other people with excessive aggression or impropriety, to steamroll over their emotions and needs, and probably a tendency to go about life on our own. Again, there is nothing wrong with being single, but we all need other people and instead of being simply an arrogant loner this position can instead be a commanding leader who helps others instead of bowling them over.

Because justice is concerned with balance, those with the Midheaven in Libra have a destiny most often committed to social justice such as in the case of civil rights icons Bayard Rustin and Malcom X, and compassionate reformers like Alexandria Ocasio-Cortez, Adlai Stevenson, and Eleanor Roosevelt. Princess Diana also had this placement, and was single handedly responsible for helping to alleviate public fear surrounding HIV and AIDS when she publicly and compassionately embraced suffering patients, as did the queen of allyship who also had this placement, Elizabeth Taylor. By the way, if you're wondering why these examples are often celebrities it's because that is the data most available to use as examples, but regular people who are not famous these same themes are also easily observed, such as a man I know with a Midheaven in Libra placement whom also has many friendships and associations that effortlessly transcend social taboos and prejudices, which more greatly enrich the lives of those with this placement, because empathy and compassion comes to this native so readily and thus broadens their circle of friends, family, and associations. The Nadir in Aries, however, can cause some reservations for engaging in relationships, finding it more comfortable to think only of our own needs and independence or to oppositely control partners due to our fear of losing our independence, but this often resolves itself as a person ages and has life experiences which define or reinforce the sense of personal worth and recognition that being in a relationship does not and should not compromise our own individuality, which only occurs as a result of unresolved trauma, insecurity, and control and coping mechanisms.

As discussed earlier, the Ascendant in Scorpio destines a life which will transit the metaphorical themes of death and rebirth (not literal death), in which the frenetic desire for control and power over life will result in a traumatic life experience that will totally alter the course and direction of goals and purpose later, though both before and after there will be the same potency of spirit. It is said that this placement can be tragic, and it does not itself destine us to a necessarily benevolent life afterward which can in truth also be potentially malevolent, the only guarantee being that our purpose reflects the nature of Scorpio which embodies metaphorical death and rebirth, of both the higher and lower natures and transformative effects as a consequence of transgressing laws of power and causality. While this placement can be stressful it also portends profound life experiences not common to others, and can through the forge of life imbue an uncommon ability to heal through the sharing of experiences and wisdom that can result. This placement also means the Descendant is in Taurus, the sign of material things, by which relationships and other people are usually regarded as highly valued, even more than material resources, though in those with unresolved trauma this may also commute to possessiveness and regarding others as possessions.

As Scorpio is the sign of power and control those with the Midheaven in this sign are likewise destined for power and are commanding persons capable of much achievement in their lives. Notable figures with this uncompromising placement are Marie Curie, Hasan Piker, Jane Fonda, and Barak Obama. Mrs. Curie, famous for discovering radiation, also mentored and promoted many women into the profession of scientific research at a time when women were usually denied the opportunity for education and work in the sciences. That which is sexual, gender, and taboo is often attributed to the sign of Scorpio since sex is in reality a convenient vehicle for control (since nearly everyone has sexuality, gender, etc.), so Jane Fonda successfully leveraged her raw sexuality to challenge the slaughter of Vietnamese people by the United States government and military during the Vietnam War, and Hasan Piker is famous for being very comfortable in his sexuality and championing the rights of women, gays, transgender and nonbinary people, and subverting gender expectations as a mechanism to help others and promote change. Alexis Arquette was a transgender woman with this placement who refused to be silent about her identity and helped forge more awareness for those who are trans and the political and social persecution often directed at them. Interestingly, J. Edgar Hoover, the notorious head of the FBI under Nixon also with this placement who persecuted many Americans for Communist or Socialist affiliation (in complete opposition to the ideals of the Constitution and principles of free speech and association) was also apparently transgender, and his cowardice in hiding his true self (due both to social pressure as well as internal conflict) was likely redirected into the persecution of others using his position of authority to feel control in his life at the cost of his own humanity.

Carrie Fisher too had her Midheaven in Scorpio and was also famous for defying sexual and gender stereotypes and expectations, similarly using her sexuality and power to help liberate women further from oppressive social norms and expectations. Matt Damon, who also possesses this influence, once famously challenged a bigoted, right wing reporter about their lies concerning teachers and education, a profession largely filled by women, and without those with this placement there would likely not be as much progress and equality as what we currently enjoy and can easily lose without such advocates. The Nadir being located oppositely in Taurus this placement also causes conflict about comforts and material security, and more than several members of British royalty inter-estingly have this placement but instead of helping to liberate those whom are oppressed instead fixate, like Hoover, on power and control for the sake of it and hoard wealth and personal comfort even as many go hungry and struggle to make ends meet. This likely occurs, ironically, due to feelings of insecurity about how their material wealth was achieved, not from work and success but instead taken, hoarded, and inherited (not that earned wealth is any less culpable, as no rich person achieves obscene wealth without taking more than they give). It is import-ant to remember with this placement that there is nothing inherently wrong with material wealth, pleasure, and beauty, but that these only become problems when they are excessive, calloused, hoarded, wasteful, or immorally gotten, and that control for the sake of it will not bring fulfillment but can instead be its own cage in which we are unable to escape and spread our wings. Redirecting our fear to trust in life, be grateful for what we have, share our success with others, and fight for our rights, equality, and peace of mind can make us very powerful indeed.

The Ascendant in Sagittarius, the sign of the mind and adventure, is likely to motivate a life full of fearless travel, exploration, and the active seeking of adventure, constantly on the move and motivated by new experiences, meeting new people, and trying new things. These are probably types of people whom are photojournalists, field zoologists, airline pilots, surfers, etc. While this life can be exciting and thrilling it is also the antithesis of settling down and establishing security, so relationships and other institutions that require staying put might suffer if concessions are not made. The Descendant in opposite Gemini is similarly concerned with knowledge and communication, and relationships may be highly communicative and founded in conversation, dialogue, discussions, etc. Famous Oprah Winfrey has this placement, which made her a natural as an interviewer and presenter, and Frank Zappa and Nelson Mandela also. Remembering that while both both Gemini and Sagittarius concern knowledge and learning this requires beginning without knowledge and the seeking and acquisition of it, and when burdened by trauma this placement can cause an insubstantial life of instability, unreliability, and listlessness, but which can be corrected by practicing inventory therapy.

The Midheaven's placement in Sagittarius is probably the least complex, as the sign of Sagittarius strongly compliments personal growth and expansion, being ruled by Jupiter, since the purpose of the Midheaven is the outward self and pursuit of goals, dreams, ambitions, and vocation is very well served by the energy of Jupiter, and those with this placement find it very easy to act on their goals and dreams and realize them in their life journey, often quite early too, such as David Bowie and Harvey Milk. Having this placement with the Nadir in Gemini means it can be quite easy for this placement to neglect structured and disciplined learning, however, which would otherwise impart useful knowledge and protect against spurious or delusional thinking, such as one famous person with this placement who uses their platform to spread unhelpful and sometimes even harmful 'alternative' wellness information and products simply because they are alternative and not necessarily legitimate, safe, or helpful. Martin Sheen has this placement but is oppositely studious and tenacious, so his acting is singular in its ferocity of wit and intellect, and of course David Bowie was an unmatched musical talent because this placement promotes expansion of the mind and exploration of knowledge that is otherwise impaired by too rigid and structured of ideology as might occur with other placements or when this placement is burdened by unresolved trauma and insecurity.

Capricorn is the sign of rules, authority, and society, and an Ascendant here often finds purpose in rules, institutions, organizations, career, and doing what is expected of us. This can also ironically cause much stress if this is not possible, or we view it as difficult or impossible to do, and being serious or feeling that seriousness and conformity are required can make life far more stifling and insular than it really is since while rules can provide safety and order they can also limit that which is possible and confine us to only one dimension of reality. Chaos is as much a rule of reality as order, so the myopic obsession with rules usually ignores that which is inconvenient to our purpose, and when burdened by unresolved trauma here can result in a joyless, fearful purpose bent on making everything and everyone conform to our personal, arbitrary conception of rules. The Descendant here is in Cancer, however, the house of home and family, which is a perfect alignment for the building of relationships and a home if the better nature of Cancer is upheld and not subverted by excessive criticism and insecurity. Later liberation from expectations when seriousness is finally tempered by profundity (by practice of inventory) then produces instead an exceptionally enjoyable person who is grounded but also able to recognize all the laws of reality and not only those in service of control.

Likewise, having the Midheaven in Capricorn is one of the most structured and disciplined influences and very usually results in those whom are preoccupied and concerned with society, institutions, and government, so this naturally contains people like Martin Luther King Jr., John Lennon, Bette Midler, and Otto Warburg (the scientist who discovered the Warburg effect while surviving Nazi Germany as a gay, jewish man). Mostly this influence tends to orient people towards the useful function of government and society but often it can also simply cause preoccupation with rules and institutions (when burdened by unresolved trauma) such as in the case of Robert F. Kennedy whom also had this placement and tried to portray himself as a civil rights advocate but in reality helped promote the persecution and surveillance of Martin Luther King Jr., authorizing the FBI to wiretap King and the Southern Christian Leadership Conference. With this position, the Nadir is in Cancer, so I believe that obsession with institutions and society is motivated by insecurity in the self and home of origin, expecting institutions and society to protect the native from the insecurity of the self, which in reality occurs in everyone regardless of sign. Unlike the reverse of this placement, the Midheaven is oriented out toward the sign of institutions and public so this native is most naturally drawn toward a life of public service, membership, and institution, which when founded in compassion and equality can result in much effectiveness but when founded in insecurity and concern for the institution will results instead in frustration. It is important to remember with this position that the inner self is where strength is really found, and turning inward to nurture the self and resolve trauma and control behaviors can make this position quite productive indeed. One interesting person with this placement is George Lucas whose decades-old, billion-dollar, worldwide phenomenon, *Star Wars*, is actually an allegory for the consequences of empire, politics, governance, and even trade, and is a very creative outlet for this kind of placement in which he strove to help people understand how society and corruption can cause harm to many people.

In Aquarius the Ascendant is in the house of fortune which also rules friends and revolution, so those with this placement have a friendly but nonconformist sense of purpose and are often indomitable trendsetters or revolutionaries like Zendaya, Aaliyah, Nina Simone, and Karl Marx. This placement is opposite Leo, where the Descendant is located, and as such may not have immense regard for relationships or others as what is required to maintain long and deep relationships and may prefer the autonomy being single brings but will very much enjoy collaboration and creativity with others.

Aquarius is indeed the most quixotic of signs so those with the Midheaven placement here are destined to be misunderstood or unique in their aspirations and realization of their goals and dreams, such as Jules Verne, Orson Welles, Gene Wilder, Toni Morrison, Mick Jagger, Virginia Woolf, Pamela Anderson, and Charli XCX. Aquarius is the sign of humanity but Leo where the Nadir is located is the sign of the individual, so those with this placement also typically find their inner self to be more or less quite strong and unique but their outward lives intertwined with the rest of humanity (and not revolutionary as sometimes demonstrated with Uranian themes because the Midheaven is the destination whereas revolution is always a means or journey). While celebrities might seem to be the exception, those with this placement tend to enjoy or feel safe as workers, engaged alongside other people in pursuit of common goals, even if our dreams and desires are not to be working in a dentist office our entire lives. Tab Hunter, for instance, was a famous heartthrob of early film with this placement who was happy to stay a closeted gay man so he could integrate into the larger society through his success and benefit from proximity and inclusion with others. The problem with this position is also that we can make too many sacrifices in service of belonging and, like Tab, doing so can rob us of an entire lifetime of inner development and even intimate relationships with others, because the latter requires risking emotional vulnerability and greater potential heartache. Embracing our uniqueness, living openly and proudly, and risking loss such as do others with this placement like Cate Blanchett and Miley Cyrus can reap tremendous rewards because a more shameless embrace of our individuality is also more compelling than conformity. Kristen Stewart and Julia Child are other examples of this placement of those who unabashedly embraced themselves, even when pressure to do otherwise was significant, to find not only great freedom but great success in the process, since humanity as a whole is in fact mostly compassionate, loving, weird, and accepting, and far more interested in authenticity.

A Pisces Ascendant is all their lives in search of meaning and uncovering the mysteries of life, spirituality, and existence as seen in Gertrude Stein, Allen Ginsberg, Mary Tyler Moore, Laura Dern, and Vincent Price, or when frustrated and traumatized are haunted by knowing there is greater purpose but seemingly unable to find it because of deluded or dispassionate investigation, and regardless may still have some delusion in their pursuits in proportion to the degree of unresolved trauma. This placement also sits across from Virgo where the Descendant is located, in the sign of labors, so there may be much diligence and care put into the maintenance of relationships and other people—Gertrude Stein for instance hosted a salon frequently attended by her iconoclast contemporaries such as Ernest Hemingway and Henri Matisse, and a four-decade relationship with Alice Toklas that was no doubt a result of much commitment and consistency.

The Midheaven in Pisces similarly portends a destiny in constant conflict between spirituality and worldly obligations, having the Nadir also in Virgo, the sign of earthly labors such as work and health, because spirituality is the polar opposite of mortal obligations, so more than any Midheaven placement this influence results in people who feel fundamentally in conflict with their life purpose, obliged to spirituality but restrained by ego, fear, and the demands of material existence. Those whom overcome this like Lily Tomlin and Robert De Niro can understand the higher purpose of existence, but this internal conflict can also be so severe that many with this placement even develop disorders or other problems like Karen Carpenter, George Michael, and Judy Garland due to the magnitude of its responsibility, but which they are also loathe to share or ask for help because Pisces is also the sign of the unknown, the illusory, and delusions, and those with this placement may not even be aware that they have a problem in the first place nor understand that many people would help if they only asked for it. As discussed in my book on psychology this can be difficult because we intuitively know that others can and do exploit weaknesses or the asking of help for their control and power, but we are *not* responsible for the behaviors of others, and allowing that to inform our decision making only further imperils our wellbeing.

It is also important with this placement to remember that people are not responsible for the nature of spirituality, reality, or the laws of the Universe. Those are made by the powers which constructed the Universe, which also includes the humans in it, and we ourselves are likewise not responsible for those things either, and resentment of the limitations of life, relationships, and mortality because they do not meet our unreasonable expectations will merely cause us loss and heartache and limit understanding and enlightenment. Oppositely, appreciation for what we do have and resolution of trauma can help us find great insight into the human experience such as achieved by others with this influence like Steven Spielberg (actually, that's the only other example I could find because most all the others, even people I know personally, have all pushed others away in response to their insecurity and fear). Spirituality means surrender, subverting the ego and facing the depths of existence, which is understandably frightening, so refusing to relent control is particularly dangerous for those with their Midheaven in Pisces, and significant work resolving trauma and control mechanisms using inventory therapy can create the space to peer into eternity and embrace human mortality which will subvert the potential liabilities of this position to transmute them into strength of humility and insight no other position can match.

After the Great Angles come also the 12 *Houses*, into which the angles are divided, which can be confusing because we might refer to a planet's Zodiac position as being in a particular 'house' of something, when in fact that is not a House, but a house, (I sometimes call the Houses the *Ascendant houses*, since they are derived from the angles, beginning with the Ascendant). The reason this distinction is important is because, contrary to what is normally done in most astrology practice, is that transits to the Houses are more about the life path as opposed to outside forces which happen to us such as represented in the Zodiac placements and transits to our natal planets. For instance, a Pluto transit through our fourth zodiac house will cause us to experience control impulses from our home and family, such as loud and obnoxious neighbors or demanding landlords, or unaccommodating or antagonistic family members, but when Pluto transits the fourth House of the ascendancy this instead motivates us to design, build, and implement plans for home and family, to pursue and establish places of domicile and what they are and look like as well as the members which might also make up our family.

This is no less concerned with our destiny and purpose, however, since we do not control those influences any more in the Houses than in the signs of the Zodiac, and you will remember that even our emotions are biological hormones, and our thoughts and dreams are also products of our past and all its influences both good and evil. Yet because these influences originate from the inner self they are very often unconscious. For instance, from about 2009 to 2026 (a full 17 years!) Pluto has been transiting my Third House which correlates with the sign of Gemini and thus concerns knowledge—both the gathering, application, and distribution of it—and around the beginning of that transit was the first I ever began to truly investigate my health problems, and ever since have been engaged more or less with the accumulation, creation, application, and sharing of knowledge through my life's work. Though it was a Pluto transit, the Third House is not especially difficult (unless we make it so), and since these transits concern our motivation and purpose, which itself does not meet much obstacles other than our own willingness and the actions and choices we take, it feels as our natural thoughts, interests, and ambitions, which are often simply and naturally acted upon. This difference between the Houses and the Zodiac is often implied when good astrologers know what they are doing, but with the way that influences are sometimes described it can be easy to misunderstand that transits through our Houses entirely concern our inner motivation and the plans we make and actions we take separate from the effect of outside influences which influence our life path. This is one reason too why the placement of the Midheaven and Ascendant (and thus also the Descendant and Nadir) are so consequential is because they very directly concern our life path. That being said, following all those transits through the houses is not always a pressing concern because they are what we often intuitively understand we are trying to do anyway, such as when engaged in the writing of books, rehabilitation from illness, pursuing relationships and career goals, etc. But they can help explain many of our subconscious and existential experiences by identifying what it means for our past, present, and future.

As such, placement of planets in the Zodiac construct the personality and the self and transits to those placements are things from life which affect us, but their placement in the houses describe the life path that we take through it. For instance, Neptune, the planet of delusion, illusion, and spirituality, was in my Second House at the time of my birth, so means, money, and income have always been an illusory and confusing part of my life, including even what to do in the first place to earn money, and from the time I was an older teenager I knew what interested me but not how to translate that into an actual career plan. Indeed even what things were of value at all was initially confusing as during my childhood there was much emphasis that only money had inherent value, but money failed to solve any of my many problems. However, all the time I sensed the incongruity of what I was taught and what I observed reality to be and eventually life taught me that far more things such as health, love, energy, relationships, and most of all *time* have more value than money, eventually fulfilling Neptune's primary role in this placement which is to learn the spiritual meaning and spiritual application of whatever House it is placed, in my case means, values, and resources. But Neptune's placement in my birth chart is thus also in my third *zodiac* house of knowledge and communication, third from the sign where the Sun is located, which meant that my spiritual or delusional nature was inherently based in knowledge and communication, for instance that my spiritual nature demands investigation, application, and dissemination and cannot simply be blindly justified, unproven, or without usefulness. Since this house also rules siblings I also had delusional and spiritual experiences of them, believing they cared about me as much as I cared about them, naively trying to maintain those relationships even when they actively pushed me away or undermined the relationship, but also helping some of them with their life problems and imparting the tools and insight which assisted my own life. I literally even also spoke with a lisp the first eight years of my life (speaking being a literal part of communication), which arose out of not knowing how to place my tongue properly when forming that sound (Neptune also rules ignorance, the unknown, and secrets).

Because Mercury, Uranus, and Mars are all located in my ascendant First House, which governs purpose and identity and is also associated with the Ascendant, my life would be predominantly concerned with learning and communication (Mercury), with forceful determination and action (Mars), but also a great deal of unpredictably, instability, and revolution (Uranus) including even that of my own physical body as was the case in having autism, cystic fibrosis, cancer, and being surprisingly tall, etc., my body itself being literally unreliable and unpredictable over the course of my life as the house of the self also determines the literal physical body. But these planets are thusly also in my second zodiac house so my spontaneity nature (Uranus) *also* concerns how I relate to finances and other means, my learning and communication nature (Mercury) is of great personal value, and my drive and initiative (Mars) often concerns material things such a comfortable home (meaning blankets, not size) and nice, quality possessions. Currently Uranus is also transiting my Seventh House which in turn means that rather than love and relationships being unpredictable and unreliable that I myself am unpredictable and unreliable in relationships, while Saturn, the North Node, and Neptune are all in my Fifth House which rules creation so I have been very structured and disciplined in producing works, such as this book, but which also are oriented toward a spiritual, service function (and probably quite a bit of delusion), but because of the North Node is also related to Karma as described in the later chapter on Karma and Healing. So the placement of planets both within the Ascendancy Houses and the Zodiac 'houses,' as well as their active transit through those houses, affect us equally, the latter in our personality and the former in our life path.

Just as astrological transits are cyclical so too are patterns seen in the zodiac and houses themselves. It has long been recognized that different groupings of signs share related qualities, such as the Cardinal, Fixed, and Mutable signs. But those groupings also relate to the cyclical nature of our experiences as each house corresponds also with two others in those groupings which comprise shared themes of progress in the *beginning*, *middle*, and *conclusion* of life cycles, in four sets of *Triads* (called Triplicity in ancient astrology, but who wants to use that word when Triad is right there?) which also correlate with the elements Air, Water, Earth, and Fire which are the qualities of each of those signs. The first triad of the fire signs, Aries (1), Leo (5), and Sagittarius (9), are the *body*, *heart*, and *mind* in which we are *born* (self), *create* (creation), and derive *experience* (mind). The second triad of Earth is Taurus (2), Virgo (6), and Capricorn (10) and comprises what resources we value (*means*), how we get those things we value (*labor*), and what is borne from our values and labor (*society*). After that is the third triad of Air with Gemini (3), Libra (7), and Aquarius (11), which comprises the learning of *communication, partnership,* which requires communication, and the community (*friendship*) which arises from both communication and partnership. The fourth triad then being Water and signs Cancer (4), Scorpio (8), and Pisces (12), comprise the *home (*beginning), *life* and its journey of transformation (middle), which then culminate in things of the *spiritual* (endings). These patterns of transits from their beginnings to conclusions can bring much elucidation to the purpose of transits and placements in our life experiences and the literal stories we live from beginning, middle, and end, catalysts which cause us to make choices which then result in consequences which shape our lives and our identity to then later pay off in either good or awful consequences as cycles repeat over and over (or with the longer transiting planets, our broader story arcs lived only once). Which not only occurs in our individual lives but also in the cycles which surround us in family, society, economies, and history (as further described in the upcoming chapter on great cycles).

While the qualities of each sign (water, air, earth, fire) is relevant to our personalities and the types of experiences we have I do not find it of much importance to be concerned when using astrology in a practical way, except maybe in understanding (*not* controlling) relationship dynamics, and was more a convention of ancient astrology which looked for meaning and prognostication in every possible iota of correlation they could find. Especially because the practical use of Astrology should not result in excessive use or reliance on it, the qualities of each sign are more of a fun way to understand things like interpersonal dynamics, such as the mixing of Fire and Water signs which often create sizzling steam, or the general incompatibility of Water and Air signs such as discussed in the upcoming chapter on love and sex.

Every house also has a relationship with the house before it, for instance if we don't know what we value in second house experiences we have no direction for where and what to seek and accumulate knowledge in the third. This process is called *succession* and is practically relevant for our life paths upon which new experiences are built from ones which preceded, and clues to our life path and insight to our past and present and future can be derived by looking at the succession of these cycles and how they karmically feed one into the next. The sixth, seventh, and eighth houses are also what I call the Zodiac *Gauntlet*, which is a long, long stretch of the most potentially stressful life experiences which in some cases can last for half a lifetime during longer transits like Pluto, Neptune, and Uranus where in rapid succession we experience themes of health and work, then relationships, and power and control where there is a logical progression from problems with daily labor and meeting our obligations which affect our health and income which thereafter then affects our relationships which in turn affect our desires for power and control, and recognizing especially this consequential succession can help us become more effective during the many years these transits occupy (such as currently being experienced by the signs like Cancer, Leo, and Virgo, with Libras shortly to follow in the years ahead), which otherwise can be especially debilitating. The following ninth house is a period of reflection in which we (should) evaluate our previous experiences in the gauntlet to inform change and grow our personal philosophy and worldview, and failure still to do that results in profoundly delusional worldviews and control behaviors which then cause problems in the following tenth house of society and our public image and so on around in endless cycles, continuously building on the previous in the pattern of succession that is a logical process of advancing through life which, if we are aware, gives us the opportunity to be more effective in these experiences.

Likewise, more useful than anticipating the quality of our experience (i.e. predicting the future) is *understanding* how we are made and the purpose of reality, life, and experiences. Most of us stumble around life using our limited skills and resources as best we know how, thinking we have useful tools when in fact they are often simply reactionary control mechanisms and evolutionary instinct borne of unfortunate traumas and pain and are in fact acting quite *ineffectively*. Very often, the unique attributes of the condition of our birth as described by signs, planets, and aspects are in turn exploited by those of other signs which in turn serves to undermine our self confidence and effectiveness in life, being problems of incompatibility, misunderstanding, and difficulty relating to each other because of our uniqueness from each other ('*it takes all types*'). For instance, those with the Sun sign of Scorpio are the sexual conquerors of the zodiac, and those who are born as Scorpio are often targets of sex shame perpetuated by more prudish signs like Gemini whom struggle with their own conceptions of sexuality. Or the eager sociability and deft communication of Gemini may be shamed by judgmental Libras as shallow. Or the waffling search for balance exhibited by Libras be described as indecisive by more decisive signs like Aries, or the driven Aries be condemned as impersonal by sensitive Cancer, or Cancer's sensitivity criticized, well, also by other Cancers (lol).

This dynamic is most notable in parent, child relationships too because although a child is born of parents they have their own unique personality and life goals (their astrological composition) and may in fact be entirely of incompatible signs with the parents, but parents being selfish and traumatized will think it's the child's job to adapt to their needs and wants rather than maintaining balance, compassion, and empathy and appreciation for the child's individual otherness as discussed in my book on psychology. The factors which make up a personality are demonstrated in no less than twenty different astrological influences, so all humans share similar characteristics but are often expressed in different ways, with differing intensities, and from differing perspectives. Even twins whom are born within a few moments of each other have slightly different astrological profiles, and I have heard it said (I can't remember where or by whom…probably Carl Jung?) that in the moment we are born we thus embody the state of the Universe at that very moment in time, which brings with it all the potential variety that occurs from moment to moment. Understanding that our personality is derived from the state of the Universe at our birth can help parents especially understand why it might be difficult to relate to our children, and have instead compassion for ourself in not relating to them rather than resenting their otherness, that such variation is not only interesting but also good and necessary, without which we would all be cyphers without personality and life would be exceptionally boring.

Yet, those who think astrology is a way to control life rather than understand life might then use astrology to plan when their baby is born. But, like the quote from Jean de la Fontaine that man often meets his destiny on the road he takes to avoid it, whenever we as humans try to alter the course of fate it will still occur in ways we least expect, since laws of reality cannot be subverted, and since Astrology is also always only our *interpretation* of reality and natural phenomena, not those events themselves, the information used to make such ridiculous choices may in fact be entirely wrong anyway but we don't know that because we overestimate our own powers of reason in arrogant attempts at controlling reality itself. While I have said it many times, while Astrology can help us be more informed and make better choices from a place of understanding, and even to anticipate the general future, it absolutely can never be used to subvert fate or reality, and the information provided in this chapter is meant to help people understand our personal natures and how life works, not to manipulate people or reality, which always bring destructive consequences. As earlier discussed, there exists both fate and free will together, in harmony, in the construction of reality, and the choice is not to control life or to abandon it (which we can't do anyway), but to do our best and show up for opportunity, because life will take care of most of it as it always has and always will, and our responsibility is in fact a much narrower piece of life than most of us realize, which is never to make people love us, to give us jobs or money, or to prevent bad things from happening, but simply to experience life and to do our best, where our responsibility over fate stops the moment we make a choice, that we can never control outcomes or consequences, and the position of the Great Angles and the houses in the birth chart can help us derive insight into those things over which we do not have control, to understand life and better inform our decision making that we may in turn make more effective decisions, leaving the rest up to the powers that be.

The Nodes of the Moon also concern fate, but they are related to Karma and so are discussed in that upcoming chapter.

17. The Imposition of Saturn

Gathered in my tiny kitchen during the latter days of the coronavirus pandemic and serving them soup and sandwiches made from my homemade bread my cousin and their partner exclaimed how excited they were for their oncoming thirties, as it was going to be the decade everything finally came together for them. "Yeah, that's what I thought at your age too," I said. "Then I got cancer, lost my relationship, and spent the latter half of that decade just trying to put everything back together."

I now realize how dispirited that conversation probably was, but one of the things that was truly so stressful and demoralizing about my thirties was that, like my younger cousin, I had believed all my life that our thirties were the decade in which we finally reached adulthood, success, and figured out many of life's mysteries and problems which plagued our youths, and the disparity between what I expected and what actually occurred was probably as vast as could possibly have happened and was even more demoralizing because of the subversion of my completely naive expectations. They probably didn't want to hear that, and similarly traumatized in their youth by volatile and abusive parents they now have not spoken to me in the three years since passed.

Now that I am halfway through my forties looking back on the adults I knew in childhood, none of their lives lives actually materialized in their thirties either, and the fact we thought it was different was nothing more than the combination of childhood naivety and the tendency of adults (usually) not to share the details of their personal lives with children (which they should not be sharing anyway). As a very skilled chef my own father opened his dream restaurant in his thirties and while it went alright for a while he started too big, too fast in a very small town and it was an incredible stress which ended up failing and plunging them into years of debt which then motivated many of our later, frequent moves from house to house and town to town.

No, in fact, almost *none* of the adults I have ever met have found purpose or fulfillment in their thirties, still exploring what it means to be an adult we have not yet come upon the adult experiences which bring maturity and wisdom and in reality our thirties is the time of our lives in which all of our *misconceptions* of both adulthood, life, and ourselves and the absence of important life skills become severely challenged by life and the law of causality. Before 2010 when I was approaching my thirties I had previously found some success as a burgeoning motion graphics artist and animator, enough even to lease a fairly expensive car (which should instead have gone into savings for a home). But the 2008 financial crisis disrupted industries everywhere and the film and television networks used the excuse to cut budgets and wages even more than they had already (which is a reason why the industry sucks so much right now). My industry had already been gutted since the 80s and 90s and the average salary was fully *half* that made from comparative work by earlier generations, and suddenly found myself again having to struggle for income as I had in my early twenties.

During this time I also met the man who would become my future ex fiancé and thought my life was coming together anyway, throwing myself a grand bash for my thirtieth birthday which many friends and acquaintances attended. Surely my thirties were going to be as spectacular. Instead, I found not only my relationship a farce and plans for the future waylaid but my very physical health and wellbeing completely destroyed. Part of the delusion of growing up to adulthood is simply assuming that we will be *healthy* adults, when in reality there are a host of environmental, dietary, and pathogenic factors that can and do rob adults of our health, now many years into maturation with sex hormones which are specifically the target of parasites. While sexual activity also increases our exposure to pathogens they can also be got from anywhere (if plans to be an abstinent germaphobic popped up in your mind) from the food we eat, to surfaces we touch, even in the water supply such as has been occurring with *Blastocystis* since the 1980s, and being healthy is never, ever a function of avoiding exposure, because that is in fact entirely impossible, nor from disinfection which mostly only kills our healthy microbes, but instead having an intact body with working defense pathways and a healthy microbiome which protect us from pathogens. Thinking, however, we must diet and exercise we abuse our bodies which then causes consequences that we try to treat with yet more harmful choices and before long find ourselves dealing with the consequences such as cancer, fatigue, insomnia, depression, etc., exactly the opposite of what we expected to find.

Standing in our home out of which we had to imminently move, alone, at the age of thirty-four, having lost my love, my dogs, my home, my health, and my hope was rudely made aware that in fact our thirties are *not* when it all comes together. Instead, life comes together in our *forties*, or can, after we have picked up the pieces of our mistakes and grown in wisdom which itself cannot actually come until *after* we have made mistakes and had appreciable life experience. It's one of the reasons why this also does not occur until our forties (or later in many cases) because the young do not want to hear that danger lies ahead in their attempts to control everything and everyone as they were taught by their own parents which desire the same (since parenting/child psychology is an entirely *unconscious* mechanism as discussed in my book on psychology), and as was the case for myself and others many people must learn the hard way that the law of cause and consequence is as unyielding as gravity.

The extent of my exploration into astrology in my early thirties was knowing I was a Libra, but as I put my life back together and recovered from cancer, alcoholism, depression, and began my research into human biology and psychology I also came across astrology that made me think there was actually something real to it, as I was not too surprised to learn about the nature of Saturn and that whole time, beginning around 2009, Saturn had been transiting my first zodiac house which challenged the sense of self and life purpose, which is exactly what had been happening as I entered my thirties expecting (like many people) to finally meet my long held goals and aspirations to instead find them even more unreachable. Even if they aren't entirely familiar with astrology, many people have heard the term *"Saturn Return"* because it is such a universal and prominent transit, and refers to the time when Saturn returns to the place it was when we were born which, because one orbit by Saturn through the Zodiac takes about twenty-nine and one half years, occurs around the age of twenty-nine to thirty and is the first major life crisis that we all encounter. This does *not* necessarily occur in our first house of self, however—that happened to me because Saturn was also very near the Sun when I was born, so it just so happened that for me my Saturn return occurred *also* in my first zodiac house of self and conjunction shortly after with my natal Sun, to exaggerate even more than usual the disciplinary effects of Saturn on our life path, and for anyone else this may occur in any of the signs and zodiac houses which will likewise be concerned also with that area of the chart that it is also transiting (means, relationships, society, etc.)

Saturn is well known in astrology as the planet of structure and discipline, and traditionally rules the sign of Capricorn and the associated tenth house which governs profession, the public, government, rules, authority, and institution because these are all aspects of existence fundamentally based on structure as opposed to others based on such things as chaos, love, spontaneity, power, etc. Saturn is not the planet of fate, however much it is characterized as such, and while its imposition on our lives is great this does *not* occur not because of predestination or crisis but because Saturn is simply the planet of structure and discipline and demands that we demonstrate those qualities to even achieve stability let alone actual growth and progress, so when we lack those qualities in certain areas (God knows I did!) the lessons of Saturn are far more disruptive and thus appear more fateful than they really are, or have to be. The problem is that most people criticize those whom lack structure and discipline as if a person should just inherently know what that even means or even how to even do it, when most of us in truth are not often even aware of areas in our lives which require structure, discipline, and consistency, because each of us only has the life tools we were taught by our parents and general life experience, and we cannot learn those requisite skills from our caregivers if they also did not have them. Most of us also think we are disciplined when in fact we are simply controlling, obsessive, or willfully myopic, and controlling behaviors are absolutely *not* structure or discipline but the very *opposite* of discipline and nothing more than reactionary survival instincts. Although the age of thirty might also seem old to someone in their twenties or younger it is also in fact not quite old at all! And there is in fact very limited life experience up to that point which also inherently disadvantages us. As such, having seen some success in my life through work, finally secured a man to begin building a life, and adopted a regular schedule of diet and exercise I had entirely deluded myself into believing I had structure and discipline when in reality these were all simply reactionary coping and control mechanisms to handle insecurity, fear, stress, uncertainty, disappointment, unresolved trauma, and were decisions made in fear for my future and wellbeing and thus ironically became highly *ineffective,* as I settled for a disloyal and shallow partner (after the hot, sweet state trooper I was dating didn't work out because he was still trying to get over his own ex), abused and neglected my body, and deludedly believed financial success to be persistent.

Likewise, many people consider Saturn an authoritarian and in fact value its nature for the order, rules, and structure it demands which in turn helps us also delude ourselves that we have any control in our lives. But Saturn is also the planet of *honesty,* and the most perilous behavior during Saturn transits is the inability to be *self-honest.* Because of our past trauma and limited life skills most of us in fact employ *denial* as our primary coping mechanism, which is the exact opposite of self-honesty, pretending that we aren't killing ourselves through harmful dietary behaviors, that we aren't reliant on alcohol or drugs to get by, or avoid our bank account because we do not have psychological skills to handle the concept of money and finances. Many young people today especially believe that being medicated in treatment for their mental health disorders is a responsible behavior, but in fact the need to be medicated at all is a warning sign that we are not healthy, and reliance and dependence on pharmaceuticals is as much a dependence behavior as abusing cocaine or marijuana, and those who believe oppositely are merely adopting the propaganda of the pharmaceutical industry so they can make money off the chronic dependence on substances. But many people also refuse medications outright because of the same stupid delusions, believing that we can will ourselves to be well by simply believing it, and that too is nothing more than a coping mechanism designed to spare us the stress of reality which is that *we do not control biology,* and that instead our recovery is entirely dependent on learning about health and biology and adopting consistent, healthy behaviors which care for *all* our needs, most importantly the health of the body and the nutrients it requires as a mortal organism to be well since even the mind is a product of the body and biology. But many young people such myself at the end of my twenties obstinately refuse to concede that we cannot control our physical health, appearance, and mental health, desperately wanting to maintain sexual attraction because of our insecurity and desperation to be desired but feeling powerless to actually change things because we have never learned how.

But we are not only in denial or fail to be self-honest about problems of structure and discipline, but also in good and wonderful things such as that we are probably quite lovable, capable, or far more sexy than we give ourselves credit. All my life, for instance, I thought that being gay was something I did wrong because that was how my parents indoctrinated me to believe, but through my life falling apart found in fact that I had been an extremely (comparatively) behaved son and had never done *anything* amoral, and was a lot more intelligent, interesting, good, and capable than I had been self-honest about, since self-pity is also a control mechanism which seeks to spare us disappointment by anticipating any and all opportunities for failure (likewise mistaking failure as a bad thing). Contrary to its traditional assignation, Saturn in fact is not cruel or debilitating at all and instead is one of the most benefic planets in the entire Zodiac because the ultimate goal of Saturn is to *gift* us with *truth*, especially such truths that we are not our looks, bank account, or the estimation of other people but valuable simply for who we are as a person and the fact that we exist. Saturn transits for sure can be quite stressful if we are very invested in our control mechanisms, fear, and refusal to let go of trauma, but ultimately Saturn only wants us to literally have a better life, to actually succeed at the things we are trying to do, to find real stability, true love, good health, and genuine self worth, but which are only found through absolute personal honesty, structured behaviors, and healthy discipline, and its tradition-al association with authoritarianism, cruelty, and stress is itself borne from the dishonest coping mechanism of *defensiveness*, thinking that criticism is wrong or bad when in fact constructive feedback, especially when given by the Universe, is fully in our best interest.

Because of this, Saturn transits through the signs and Houses are always one of consequence which serve to emphasize parts of our lives which are not working well, lacking structure, or devoid of honesty, and can and does cause much frustration in direct proportion to our disparity of cognizance which may exist between our behaviors and that which is actually effective and useful. While they can be frustrating, Saturn cycles are some of the most important growth periods of our lives, and in every moment we walk this earth Saturn is forcing us to pay attention to things we would rather ignore, take for granted, neglect, or wish were easier to handle, because such opinions arise from nothing but insecurity, and we are capable of so much more than we understand and only need to learn the skills required to properly handle those areas of life, which Saturn actively teaches us through the law of causality. Even if we think we aren't capable of handling those things, such as was my experience when my life fell apart, Saturn does and believes in us, which is why it is putting those challenges in our path, and if Saturn believes we are capable of overcoming those challenges then just, maybe, we are?

To further underscore Saturn's truly benefic nature, it has also long been known by astrology that Saturn actually presents us with a literal gift at the end of each transit through a zodiac house, as if rewarding us for our perseverance in enduring its lessons. When Saturn departed my first zodiac house at the end of 2012, after having been there since mid 2010 I had also grown completely cynical toward my partner because of the way he treated me, where before I had been more permissive and patient, and felt listless and directionless living in a resort community far from my peers and work opportunities, and though at the time did not realize it found a strong sense of independence from him (even if it was in the form of resentment), and began to pursue interests as was possible in my limited situation such as working on the garden I had built or designing and programming my first video game. The following transit of Saturn through the zodiac second house (for Libras) was when our financial situation and stress peaked, losing everything I had (including the partner I had come to resent), and sleeping on friends' couches while walking or taking the bus or train to work (which was actually really fun and made going to work far more enjoyable), and at the end of that transit in August 2015 was unsolicitedly offered a full time job at the place I was freelancing for more money than I ever made in my entire life (though it was still about half of what my industry should have been paying us, accounting for inflation) which then allowed me to purchase a good, used car, secure my own apartment, and fully furnish it over just a few months time.

Saturn's next transit through the third zodiac house of knowledge and communication during which I even more ravenously consumed the work of Dr. Peat and many scientific studies in learning about biology and health, testing and experimenting on myself to see what might be useful for my recovery, and started my first website and blog to share that information with the increasing number of people asking for help and at the very end of that transit in late 2017 completed my first book, *Fuck Portion Control*, which was then supported by several hundred actual readers whom purchased copies. The fourth house concerns home and family and that transit was dominated by frustrating living situations in Los Angeles (exaggerated by the presence of Pluto here since 2008 as well) and missing out on my siblings and their growing children and at the end of that transit in early 2020 (Saturn retrograded in and out), moved to a very larger, cheaper, and peaceful apartment near my family (where I later wrote and finished *The Perfect Child*). Not only was the gift a new, larger, less expensive apartment nestled in the beautiful mountains but was also in literal proximity to my family as the fourth house governs. After its fifth house transit (creativity) I had produced an abundance of art for personal projects and learned how to be productive in terms of creativity like I had never been able to achieve before (I was also gifted a casual love affair but he was also a drug addict so I cut that off), and now as Saturn ends its transit of Libra's sixth house of work and health I have finally resolved many of the last health and research problems I was tackling while also finishing this book, my cook book, and have more clients than any other time coaching, which was all achieved through structured discipline which was easy this transit having long understood the role of Saturn and was already disciplined before it began (but Saturn helped me understand better where to apply and direct that discipline).

Considering all this, it is then very easy to see how in fact Saturn is primarily concerned with our wellbeing, *not* our undoing as is so often described in astrology, and it is only because of our attachments to control behaviors and toxic people and situations that we ever think otherwise. Of course, challenging aspects of Saturn transits such as opposition and square can be harder than trine and sextile, but that only means that the method by which we achieve the goals of Saturn in such transits are more requiring of structure, discipline, and personal honesty than other transits. Very often too life sends us on our correct path by literally destroying the one down which we were headed, and although I lost things and people that metaphorically tore out my heart, the life I have now is so much more beautiful, preferable, easier, peaceful, and fulfilling than the path I was mistakenly headed down in the first place. I could not see it at the time but Saturn fundamentally saved me from a life of mediocrity.

Because Saturn takes about thirty years to complete one orbit of the zodiac, it transits each individual sign for about three years—technically it's about two and half years total but there is usually a retrograde at the beginning or end, overlapping the neighboring sign, which then stretches the time to three years from first entering to final exit, so while its visit to certain areas may seem impossibly long in reality these spans are not nearly as exhausting as the other slow moving planets like Neptune and Pluto, and help us make more rapid progress along our life path that we sometimes realize, considering how different my life is and what I have achieved in such relatively short a time as ten years. But its transits are so consequential to our experience it can make us feel like we are treading water for a very long time, when in reality it is our refusal to listen and surrender to reality that makes it both longer and even stressful in the first place.

Saturn's full cycles with the natal planets and the Great Angles are long indeed, each taking about thirty years, representing specific parts of our journey such as with the Sun, which concerns our very identity and purpose, which comes back around for evaluation at ages thirty, sixty, and ninety (if we live that long). Interestingly, scientists recently reported finding two common periods of aging in most people which occurs at ages 45 and 60, the latter being the second Saturn return and the former exactly when Saturn is in opposition to natal Saturn in the second Saturn cycle. Because all aspects of the Zodiac are a cycle they continue throughout the entirety of our life and do not ever end but simply repeat over and over and over, and Saturn is thus the bookend of major life periods each of us will encounter during the course of our life. The most prominent aspects too being with the Great Angles, but the conjunction of Saturn with any other planet is equivalent also to the end and beginning of those cycles, which is why the Saturn return or its conjunct to the natal Sun can be so pivotal, because it is the ending of a new cycle and beginning of another, the outset of which is a reevaluation of our life goals and our effectiveness both in terms of our goals but also in the world around us. What went wrong? What went right? What are we good at? What do we want to be good at? What do we want from life? What even is the point of life? Choosing which questions to pursue is thus the entirety of the following cycle as we strive to meet our goals and how we handle the failures along the way. So where the Saturn return (conjunction to natal Saturn) is a revaluation of our structure, career, achievements, and discipline, the oppositions which occur around the ages of fifteen, forty-five, and seventy-five are a culmination in which our labors in turn bear fruit and show us what effect our choices and decisions made at the outset have had on our life path and whether they were effective or ineffective as reflected in what we experience as a consequence, with Saturn at all times in its transits demanding structure, consistency, discipline, and self-honesty. This transit of Saturn opposite Saturn has literally started for me at age forty-five mere weeks before publishing this book, so the upcoming year will for me be exactly this transit and its effects on our lives.

The most apparently auspicious transit of Saturn during this thirty-year cycle is to the Midheaven, which represents the outermost, material, public, and visible aspects of our life such as things we can see and quantify like occupation, finances, social status, possessions, wealth, etc., so when Saturn reaches 'the top' of our chart it often coincides with a material peak of our lives, when all our hard work and struggle to achieve, succeed, and find meaning and purpose not only pay off but is also accompanied by an increase in responsibility, obligation, and effectiveness in our lives which would also logically follow and facilitate that culmination. But because Saturn does challenge us to reveal weakness in our apparent discipline, structure, rules, and consistency (which are oftentimes simply delusional coping mechanisms and not in fact structure or discipline) some people can experience instead a time of crisis in which their profession and public life begins to fail instead of rewarding such as a recent failed candidate for President whom refused to address such major issues as genocide being committed by their own party in power. This occurs because no aspect of Saturn transits are inherently benefic or debilitating but instead simply insist on application of structure, realism, discipline, honesty, consistency, etc., to correct shortcomings and failures as they occur and learn from mistakes as a teaching guide, source of wisdom, and directions on where to instead orient our life path.

Because the Great Angles are determined by the time of birth and do not correlate with the Sun or Saturn itself, Saturn's transits to the angles are not consistent from person to person and the first time it ever crossed my Midheaven was at the age of twenty-four when I was not yet old enough nor had sufficient life experience as to achieve much in the way of obvious outward material success, but I had finally achieved some success in the early years of my design and animation career and was able to relocate to Los Angeles and find work in the industry there using that skillset. One of the reasons Saturn's transit over the Midheaven is such a point of emphasis for most people and astrologers is that most of us greatly value money, materialism, and outward success since these things bring more ease of life and reduce stress and fear, but in reality it is no more important than Saturn's transits over the other three points of the ascendency which are far more influential on our life and development, because transits to the Ascendant, Nadir, and Descendent decide the direction of our life while the Midheaven is more like a kind of reward (or lack of one) as a result of earlier transits. When Saturn leaves the Midheaven and passes next down to the Ascendent we begin Saturn's descent to the bottom of the chart, away from our outward life and toward the inner self at the Nadir, passing the Ascendent on the way. Some people report Saturn's transit across the Ascendent as difficult but remember that Saturn's influence is to engender structure, discipline, consistency, honesty, etc., so when those parts of our life are in fact working well the influence of Saturn can hardly be felt at all, and when they are not, as was my case, its influence will be quite profound.

My Ascendent is located in the very early degrees of Scorpio so Saturn passed my Ascendant *also* at the time that it was just finishing up its transit of my first zodiac house of self and during this time I was greatly struggling with how to achieve my professional and personal goals and the conflict of my current life with what I wanted for myself, but had several occasions through my partner to attend writing retreats and meet other writers and eventually learned from Saturn's path here how vile the publishing industry was and how little I truly wanted to do with it, especially since I was never going to be good enough or desperate enough to compete with other writers, and my disheartening experiences during that time were a lesson by Saturn about the lack of structure and discipline required to achieve that path, having never attended higher education. But I also learned how in fact it was not a necessary course to take, because my desire was not to be a writer, specifically, but to simply tell stories and create works of art, which I could also do through video games, animation, character creation, art, etc. But because of these experiences when later finding myself having to write about my life and health in my blog and books I instead put those immediately to the public instead of trying to go through the industry, so effective had been the instruction of Saturn to avoid that life path which, even if any publishers had been interested in such a work as *Fuck Portion Control,* would also not have facilitated the constant changes and updates to it as was required of my continual research that doing it on my own can and does achieve, and so the events which I had previously thought to be misfortune as what accompanied fateful Saturn transits were in fact preparing me for a different purpose, as it is with everyone.

It is important also to know that Saturn's cycle around the Great Angles is not like other transits and aspects, as while Saturn's crossing each point may bring specific events it does not lose its influence between them, and is instead one continuous influence at all times in our life in proportion to its proximity to each of the four great angles. So, as Saturn descends past the Ascendant to the bottom of the chart it is often a rough and bumpy ride, because Saturn is forcing the evaluation of our entire life path and not only one small portion as it does when transiting natal planets, to then find itself at the Nadir which entirely concerns our innermost self and things of a spiritual and homely nature, and because as humans we primarily value materialism and success and regard those without it as having failed, or at least not yet achieved success (and therefore imply failure), Saturn's transit through the bottom years of the birth chart can be the most distressing of all as outward progress and material success grinds to a crawl and we are deprived of things which otherwise gild a human life. Not to bring up my poor parents again but, since we often meet people with similarly aligned angles, because those are what determine the life path, our fortunes are often similarly aligned with a partner and my parents Saturn descent through the bottom of their charts were spent in voluntary exile in Hawaii when I was in my early twenties (and attempted suicide) in which they too struggled to put their own lives back together, not understanding that such periods in our lives is a purposeful lesson by Saturn meat to draw attention to our insufficient spirituality and inner selves, home, and family which in fact requires structure as any business, routine, or relationship that we may grow and nurture not only our bank account and professional reputation but our very soul and most intimate connections (and because institutions like religion not only deprive us of skills required to do so but also encourage delusion and dissociation we then entirely fail to grow and benefit from the experience).

Indeed Saturn's transit at the writing of these words is only just beginning to emerge from the ten or so years of this forced introspective period of my own life, during which I was not only brought low by cancer and ruin but learned new personal life skills such as how to resolve trauma through exposure to Alcoholics Anonymous (even if they don't recognize that's what their program does), to research, create knowledge, cure disease, and write my three primary books (including this one). Not everyone's experience will be so dramatic, as mine has many alignments that all occurred at once which are also a product of having Scorpio rising and a destiny to be transformed by difficult life experiences, yet even every single one of my family members shamed me personally during this time for not finding material success, so conditioned are they by materialistic religious conservatism as to entirely transgress their own religious creed in the condemning of people who do not worship money and instead serve others, considering my work and philosophy a folly for evidence of a yet meager (but wonderful and supportive) audience and small income even though I can pay all my bills as they also unknowingly host Saturn at the top of their own charts where they have sat upon superficial trappings of materialism, but now on his way down toward a portent future of the same kinds of personal trials, loss, and forced introspection as what brings all people through this period of life, to which they remain ignorant as the great irony of refusing even to hear me and decide after, not before, whether what I have to say is useful.

Before knowing about this specificity of the Saturn cycle I thought for sure that material success would come shortly after the publishing of my first book, having discovered the cure to diseases like alcoholism and depression surely should result in some notable success, right? I mean, researchers get picked up by the click-bait news industry if they so much as find a place to *start* a potential investigation. Then I would have the freedom to rebuild my home and personal life, start dating again and resume a normal existence. But Saturn passing the Nadir of my chart also coincided exactly with the 2020 coronavirus pandemic in which it was not even possible to pursue restoration of my social, personal, and material life, during which the truth of my dysfunctional relationship with my family also became intolerable and was, in truth, the Universe prodding me to finish my second book which before I had no idea how to construct. Because of the pandemic and my refusal also to be a fucking eugenicist like many pitiable colleagues which instead profited from exploitation and fearmongering, experienced yet further financial constraints that I could not even furnish the new apartment to which I had relocated. Since Saturn is the planet of discipline the combination of its influence at the bottom of our chart is one of obligatory rest and introspection as a consequence of our inexperience, mistakes, naivety, and selfishness of the previous years, to thus learn from those experiences and work on our inner lives in stark contrast to that which is outward and material (relative to our current life conditions), which will always include loss of things like material wealth, influence, power, prestige, regard, admiration, etc., the loss of which will only be painful if they are meaningful to us and rest, introspection, and concerns of a higher nature are not.

While some successes can and do occur during Saturn's transit through the Nadir this pattern is generally so reliable that anyone seen struggling in their effort to achieve life goals likely has Saturn transiting the bottom of their chart, while those seeing success from their efforts will have it transiting up through the Descendant and past the Midheaven. Even the two most recent Presidents of the United Sates *both* lost the office when Saturn was down near their birth-chart Nadir, confronted personally by failures of structure, discipline, and personal honesty of the past which then imposed upon them a long period of forced introspection. Of course, what we achieve during this time is also not guaranteed just as during other points in the cycle, and is entirely dependent on our ability to pay attention, be self-honest, and eschew material obsession in favor of that which is intangible. While my financial situation was more of a monastic choice since I had purposefully given up a career in design and animation that paid fairly well in order to take care of my own health do this work and research, at any time could have received karmic attention and reward for my achievements like any of the many other advocates for health and wellness, achieved even by those who are clearly duplicitous and opportunistic, but in spite of my great efforts to gain visibility and help others was denied at every single attempt, even people telling me to kill myself for trying to help them, until Saturn finally began to rise to my Descendant and will finally result in more public attention and engagement around 2028 and after when it reaches conjunct and heads to the top of my chart. Such is the Saturn cycle for all people.

I often repeat things ad nauseam but it is worth repeating plainly again that the bottom of the chart IS NOT a bad place for Saturn to be, and if this journey sounds stressful, especially because of my personal experience I have shared, take heart in knowing it can be one of the greatest periods of any person's life and is only stressful when we refuse to let go of material concerns and exist fully in the present and the more humble and restful years of life. In fact, now that this period of my life is ending I do not want it to at all, as the peace and relative quiet has been most wonderful, even if I mistook it for stagnation and frustration in early years, and I am already looking forward twenty-years in the future when I will be sixty-five and can once again have many more years of obscurity as Saturn again reaches back around to the bottom of my chart. The *only* reason we might loathe this transit of Saturn is because we in fact loathe ourselves, hate being alone, and constantly chase material solutions to our problems not of a material nature. If it helps, pretend you are a monk or some other benefic, monastic figure during this time, which will help frame the mind to understand what it is we should be doing now, to grow not in material wealth, but of things far more valuable, buoyed by plenty of self-care, writing practice, warm blankets, and cups of tea.

One reason for the significance of these transits is that whenever Saturn passes one of the angles (as does any planet) it is also always *opposite* the other of its pairing, the Midheaven and Nadir, or the Ascendent opposite the Descendant, and thus the reason for the more singular focus on that part of the chart, focusing on the area it is in conjunction while simultaneously opposing the area it is not. For instance the neglect of our inner, spiritual growth during transit of the Midheaven because life is good and we can more easily ignore our fears and spiritual deficit (to our later peril), and oppositely an inability to find material success when Saturn is at the Nadir and allows only the possibility of inner growth. During its Descendant conjunction we find it difficult to work on ourselves because of the time and energy demanded of others, and the opposite difficulty maintaining relationships as it crosses the Ascendant because of our deficient sense of self which in turn needs addressing, which only we can do for ourselves. In fact, many long-term relationships which break up very likely do so when Saturn is going through the bottom of the chart, as was my own personal experience or that of my parents when they divorced for three months, especially since our partners are also likely to have similar experiences and thus going through the same themes and challenges as ourselves, which in the case of the Ascendant is not necessarily compatible with partnership as Saturn opposes the Descendant. As was demonstrated by my own personal experience, the transit through the bottom of the chart still will contain much achievement and progress, just not of an expressly material nature, and we will definitely feel less visible, even when we try to be, which is again why it is so important during this time to instead practice self-care and stop resisting this obligatory course in existentialism.

When Saturn was exactly halfway to the Descendant in my birth chart (from the Nadir) I realized its transit across the bottom was not actually its first because I was forty-four and Saturn had arrived there first when I was ten years old in which, just as occurred in my early forties, I struggled to make friends and find any sense of outward success. It would not be until the age of eighteen when Saturn finally passed my Descendant that I was finally liberated from the confines of my lonely childhood and free to be the young, naive, inexperienced, autistic, gay boy I was, although at that time too Saturn demanded abandoning the few friends of my childhood which, at the time of coming out, proved them unreliable, shallow, and literally unavailable from moving away which instead required a search for others with whom I could associate, with varying degrees of success, including my very first boyfriend shortly after as what commonly occurs in Descendent transits. Being so young the transit across the Descendant was marked also by many mistakes and much naivety as Saturn challenged my absence of effective life skills, even being harmed by predators and opportunists, which are direct lessons of the disciplinary Saturn influence here that requires most of all personal honesty rather than simply hoping, wishing, and being naive, to invest time and effort in the building of relationships, which don't happen without investment, and into people and relationships which are clearly in our best interest and not simply those which are convenient.

This division of the chart in half, relevant to Saturn's slow march through the heavens and the "top" and "bottom" is the reason for those stereotypes of childhood as popular or unpopular kids—those growing up with Saturn in the top half of their chart are those who seem to peak in school, but then appear to stagnate in adulthood when Saturn is oppositely descending through the bottom, while those who have Saturn in the bottom of their chart while young are instead those whom are observant, shy, isolating, living mostly in their internal world, who then break out once they ascend into adulthood at the same time as Saturn transiting the top, all of us living the same cycles but at different times and seasons.

Because they are slow moving but also associated with harsh and unyielding influence, planets like Saturn, Pluto, and Neptune around the zodiac are also associated with those experiences of trauma during childhood in which we mistake the absence of close friends or an unhealthy home as products of our own inexperience, failure, or childhood vulnerability when in truth they are the same experiences that all persons eventually live at some point in their lives but which instead happen to us when young, and our inexperience deceives us into thinking we are alone in our pain and confusion and unique in our trauma but which in truth is also meant to be part of our life path on the long journey ahead. Indeed as a child when this transit of Saturn through the bottom of my chart prevented outward success and expression I became instead a talented artist with a vivid and active imagination which profited me greatly throughout the remainder of my life, not only professionally but also personally, being able to retreat into my imagination and create artistic works whenever I felt otherwise blocked or inhibited by the outside world, and would receive great comfort and inspiration from making characters, stories, and projects. Later life experience often enlightens us to the reality of these periods in life, if we are paying attention, to contextualize traumatic childhoods as not simply arbitrary suffering imposed by an indifferent Universe but instead one which is designed with our experience in mind and thus relevant to our entire life journey which sets the stage for all that comes after. Of course, knowing what decisions to make is learned by mistakes, experience, and wisdom, and anyone struggling with problems like making friends or finding stability in life must practice inventory therapy to in turn bring clarity to what we want and how to get it, as most of our frustrations in life come from the impediment of control and coping mechanisms upon not only our decision making but our very worldview and life perspective itself.

Next most important in the nature of Saturn are its transits through the individual houses themselves, both of the Zodiac and houses of the Ascendency. The first house is the house of self so Saturn transiting the first house brings challenges and crisis to our personal identity and purpose which, considering Saturn is the planet of structure and discipline, is not arbitrary or wanton but is a result of prior absence of structure which supports our personal identity and sense of purpose, meant to forge a newfound identity through consequences of past ineffectiveness. If for instance you think you can prepare for a first-house transit and avoid any difficult lessons that will probably one of the first which Saturn challenges, but often takes the form of wondering who we are or even what we want, being uncommitted or scattered in our thoughts, plans, or even entirely errant in our choices borne from our concept of ourselves. As this house is preceded by the twelfth, challenges to our spiritual nature and the ending of things we previously relied on set the stage for this reevaluation of our personal identity and purpose. Remember that Saturn always gives a gift when it departs a house, so at the end of this transit there should be a better sense of direction and self identity than before it began (which very literally shows up in the very last weeks of the transit).

Because the second house rules our means—finances, money, resources, and literally *how* we get them—this is often a time of financial problems arising from our ability or inability to earn money (means) from previous lack of discipline or personal honesty. This transit through the second zodiac house can be quite stressful if we are excessively reliant on money and material success, or believe that we are the only ones looking out for our best interests, or are especially miserly and self-centered, and this transit will illuminate how better we can not only earn money but how also to use it in practice. The gift Saturn brings here at the end is often quite auspicious, such as a new job or other opportunity to earn money, but will not be something like winning the lottery because this house is about *means*, not luck or fortune.

The third house is that of knowledge and encompasses both the gaining, application, and sharing of information, and as was my experience of Saturn's transit of this zodiac house from 2014 to 2017 first by ignorance of required knowledge, then its acquisition and application, and finally distribution of knowledge many people here will be challenged to learn, use, and then share important information. Resisting that lesson will instead cause consequences of ignorance which can, in some cases, even become so severe as to cause death if very ill, for instance, if we are especially arrogant and willful. Usually though this is not a stressful experience and simply involves reading, listening, exploring, researching, communicating, using that knowledge to test its validity, then sharing with others what we have learned. Not everyone at the end of this transit will publish a book, but some may, or have a social media account, or other means by which we may disseminate useful information to others. Many writers throughout history got established in this pattern as Saturn was transiting their third house—for instance, author Armistead Maupin published his first entry into the epic *Tales of the City* series literally at the end of Saturn's transit through his zodiac third house, May 24, 1976 and began conceiving of the series at the beginning of the transit in 1974, and likely worked on it and solicited advice from friends, family, and fellow writers and honing his craft during that transit, rewarded for his effort and discipline with publication at the end of it. Observing this transit in those I know personally, most recently witnessed those with the Sun sign of Capricorn whom either resisted or acquired such knowledge while similarly dealing with personal issues as I had from Saturn's transit of my first and second zodiac houses to varying degrees of success.

As I have mentioned before, astrology influences do not function in isolation from each other, and many transits stack onto others, alternately amplifying or muting the effects of other transits in a process which composes the composition of our lives, and since about the year 2008 Pluto has also transited my third ascendant house to cause similar themes as Saturn transiting there. Being the planet of death and rebirth, Pluto transformed my original desire to write fantasy novels, films, or video games by sidelining me with trauma, alcoholism, disease, and a failed relationship, to destroy my original motivation by destroying my confidence and ambitions to then reforge them in the fire of adversity into a greater purpose of ultimately writing to share my own personal story and useful information which could better materially benefit humanity. The last years of Pluto's transit through my third-ascendant house now in 2025-2026 as I finish this my third book, eleven years after first beginning the research which resulted in *Fuck Portion Control*, certainly conspired with Saturn's transit through Libra's third zodiac house in 2014-2017 to result in this life path. Saturn has a similar, disciplinarian quality as Pluto but is more instructional, and one of my young, Aquarius cousins who only ever spoke to me when they wanted something, like help getting published, will likely have a similar experience to mine in their upcoming Saturn transit of the same house but which will similarly be complicated by Neptune's position there for all Aquarians, (and so on for all signs and all transits which at any time may be modified or complicated by other planets).

Because the third house also rules siblings and spiritual siblings (people our age whom we grew up with), Saturn's transit through the third zodiac house also affects our relationship to people our age we knew when we were young, and can similarly expose our failures or inefficiencies in how we maintain those relationships, through whom we can also acquire or share information, and as Capricorns are the sign most recently with this transit have likewise also experienced increased tension, conflict, and communication of information with siblings and childhood friends. The scope of the third house can be as profound as publishing and research or as seemingly mundane as public relations and writing of emails, as all forms of speech, interaction, collaboration, communication, even that which is rote and boring, is still the conveyance of information ruled by this house, which even young children can and do experience in their own lives which help them cultivate their own ability to learn and communicate. During this transit for Geminis way back in 2006 the last of my siblings who is also Gemini was stuck at home alone with my parents as the others of us were all grown and off turning into adults. Lonely, she messaged me (her sibling, also ruled by the third house) an incomprehensible email which failed to use any punctuation or grammar which, because I had been raised by my father which taught me to be offended by poor grammar and expressions of intimacy, saw a mess of an email instead of a plea for help and embarrassingly and regrettably chided her for the email instead of being the kind of loving and helpful big brother I should have. Of course, she never emailed me again and I was so lost in my depression and alcoholism I couldn't even see how my own behavior isolated me from my family—but for her this confusing lesson of rote grammar and self-expression, as well as our relationships with our siblings— was part of her formative experience which, because she was only a child, reacted to with emotional isolation and self-protection from experiences of vulnerability which later impair our ability to connect to others, then later in life to learn self-compassion for such vulnerability, pain, and the disappointment they expose us to.

The nature of Saturn's transit through the zodiac fourth-house of home and family was previously related in my personal experience and how I was forced from apartment to apartment until finally finding a place in Utah near my own family which was larger and nicer than anything I could afford in Los Angeles at the time. Saturn's transit through the fourth house will always disturb issues with domicile and family, because the fourth house is naturally *squared* in aspect to the sign that contains the natal Sun, in which we are forced by life to associate with and deal with people whom are not necessarily naturally compatible with us by nature (because they have their own signs and planetary placements). Because Saturn is the planet of structure this can often look like simply having more structured behavior in how we relate to our family members, such as making an effort to call them and arrange get-togethers, or even to simply put in effort for the maintenance of those relationships. But for those whom are parents it can also be things like having to better coordinate the activities of children and our spouses, the cleaning and maintenance of the home, or even problems with the house itself. Two of my brothers in law are Scorpios which had this transit immediately after myself—one discovered the house they purchased years before had been built very poorly, with none of the electrical outlets or mounted lights having electrical boxes behind them, posing a significant fire hazard. It was also getting very small for their growing family and felt claustrophobic, especially after my erratic and unreliable parents moved in with them, while the other sister and husband moved to the East coast to start an entirely new life but rented out their old house and had similar issues with home maintenance and family stress. Knowing that is the point of the Saturn transit can then benefit from being especially responsible in our place of domicile, caring for its wellbeing and maintenance, not living outside our means, properly inspecting and vetting places of domicile, and being on time with related payments and obligations. The gift of Saturn's transit through the fourth house is often a literally larger and nicer place to live, which I found as well just down the street from my family. While fourth house transits are difficult they are also really quite transformative.

Saturn is not naturally associated with creativity, so its transit through the fifth house of creation is not, in my experience, especially remarkable. During my experience here I simply found it impossible to actualize any of the various creative projects I was engaged such as making video games or writing screenplays, but instead the actual practice of creativity in the initialization of art and projects was refined and perfected until I was regularly creating art simply in the capacity that I was able. For many people the fifth house encompasses the beginning of love affairs and the making of children, which are all acts of creation, and during this transit I met that drug addict and had a little bit of raucous sex, but a lesson I immediately recognized from this transit was how the absence of reliability, dependability, and stability in partners—even in casual sex and love affairs—can be potentially disastrous, and very quickly stopped it before there was any chance to become attached or infatuated. Since I was still doing my research I was not interested in beginning a relationship anyway, but anyone else with this transit may find similar themes requiring structure, responsibility, and self honesty concerning new loves, children, art, or anything we create including projects, new businesses, etc. Since I do not have my own children I do not have much insight to their relevance to this house, but don't be surprised if the care of children becomes more demanding during Saturn's transit through the fourth house (which my book, *The Perfect Child* is meant to address!).

Saturn's zodiac sixth house of labors transit is not necessarily inauspicious but often is because there is probably no aspect of life more requisite of structure and discipline than when it comes to our work and health in the daily labor required of daily life. All Libras had both Neptune and Saturn transit the sixth house at the time of writing this book and because astrological influences compound each other many Libras saw severe challenges to our health and wellness that also involved mysterious or unknown causes due to the presence of Neptune. Not all Saturn transits in the house of work and health are this perilous, but Saturn transits of the sixth house will *always* demand structure and discipline that otherwise result in consequential health problems and work problems. This can be especially stressful because work is also affected by this transit, with our ability to show up, be dependable, and have structure in the function of our employment being every bit as requisite also to our health, and at times people end up choosing one or the other, favoring work in neglect of diet and self-care, or favoring health in neglect of work, wherein the problem is that all our daily labors and responsibilities require attention and what must be achieved is balance and discipline which addresses all of those responsibilities and not only some. During this transit I even found it difficult to go to bed on time like a spoilt little child who loves sleep but hates going to bed. In the case that this transit is especially stressful, health should be prioritized over work because we cannot work if we are sick or dead, and paring down unnecessary or superficial responsibilities could be a useful strategy during this time, remembering that Saturn always leaves a gift at the end of its transits as a result of making an effort, such as the restoration of health or obtaining career advancements or stability. This house encompasses anything that requires daily attention, such as the care of pets, children, or even plants, going to bed on time, staying well fed, or oppositely the cessation of harmful behaviors or routines which are just as regular but destructive, and anything that demands our regular labor and attention will be affected by this transit. Before this most recent transit Virgos had Saturn visiting their sixth house but which was shortly after followed by Pluto's entry here, and those who handled Saturn's visit well likely have not immediately succumbed to Pluto's influence as rapidly as those who poorly handled Saturn's transit, Scorpios now just beginning this Saturn transit at time of publication.

Very often, Saturn's transit through the seventh zodiac house of relationships are stressful because it is naturally at opposition to our Sun sign, just as the Descendant is opposite the Ascendant. But all relationships require structure and maintenance anyway, being naturally in opposition in the zodiac to the self, so transits of Saturn through the seventh zodiac house are not necessarily stressful, although they can be if we are especially selfish or intransigent, because Saturn is the natural embodiment of structure and responsibility already required of relationships. As a Libra I am quite biased saying this because Libra is the very sign of relationships, but partnership is not only those which are erotic and also encompass business relationships and other close partnerships, and this transit may reveal areas in which we are neglecting or harming any relationship regardless of its nature, and instead required to show consistency, dependability, and structure. No one likes to feel neglected or marginalized, and having dependability in relationships means we in turn show others we value their presence in our lives, thus engendering love and appreciation which strengthen bonds, where neglect, abuse, and inconsistency instead communicates to others that we only prioritize ourself and therefore are a danger to their wellbeing rather than a bulwark. So simply being consistent and dependable during Saturn's seventh house transit, and having structure in how we contribute to and reinforce those relationships, can mean great success, and possibly the establishment of a solid relationship or the strengthening of ones already in existence when Saturn departs.

The ominous eighth house of power and control is a precarious position for Saturn because we do not in fact control anything but our own actions and behavior in life, and because of basic human survival instincts many of us desire control to feel a better sense of power and safety which can delude us into believing we do have control when in fact we do not. Being thus, Saturn's influence in the eighth house transit is to focus our discipline and responsibility on control of ourselves and our own behavior and not control of others or things outside the scope of our responsibility. This transit can help to elucidate those boundaries if it is not quite clear, and at the time of this writing Leos were the ones experiencing this transit, compounded also by Neptune, in which they experienced attempts by others to control them and their activities and to then use control behaviors in turn, or not (which is the proper way to handle this). A famous young content creator, Hasan Piker, who focused on politics is a Leo and was a major force in helping reform politics and promote socialist ideals across the United States in the younger generations, and during this Saturn transit was invited to attend the Democratic national convention but during which he was also summarily dismissed and asked to leave, while before, during, and after this transit was also commonly harangued by political actors who should have been aligned with him but were instead duplicitous and opportunistic. Though not privileged to know all the details of these events it appeared he behaved quite exemplary and simply did his best with what there was to work with, which is ultimately the best guide for this transit— to do your best and show up for opportunity rather than attempt to control others when they in turn do the same, least of all the nature of cause and consequence. It is seemingly ironic that when we stop trying to control we often end up having more control, but that is only because we end up controlling only the things we can, such as limited to our own behavior and choices, so our energy and efforts are directed only to where we can actually be effective, and Saturn's gift from this transit will be more power but only because we better recognize over what we truly have it and thus become more effective.

Although Saturn is naturally at home in the tenth house its ninth house transit is also quite amicable because the ninth house encompasses the mind and philosophy which directly very much benefit from any structure and discipline. This can, however, oppositely impair the effect of this transit if we are overly controlling in this transit because the natural ruler of the ninth house is Jupiter, the planet of luck, growth, and expansion, which does not necessarily like restriction or limitation, and fears of experiencing travel, other people or cultures, being xenophobic or racist, embracing ignorance, or other such moral and ideological stagnation during this transit can actually result in significant harm to ourselves and our interests through sheer ignorance, naivety, and willfulness. This recent transit of the ninth house (also compounded by the presence of Neptune) caused many Cancers to become insular, xenophobic, suspicious, and close-minded, which also robbed them of good will and even financial resources as a consequence of that behavior, especially after Pluto recently entered their eighth house of control. If these other planets were not affecting this Saturn transit for Cancers at this time it might have been much less perilous, but that does not mean that Saturn is permissive, because it never is, but is reasonable and does not ask more than what we are capable, and if were are humble, present, paying attention, and doing our best this transit can expose us to expansive conception of ourselves, others, and reality than what we knew before it, especially when Saturn departs and leaves its parting gift of greater enlightenment and perhaps even new philosophy or ideology which can help us achieve better success in the future.

In the tenth house Saturn is most at home, being the house of society and institution, but even here Saturn will still require its lessons be heard, and lacking structure, discipline, and self-honesty can and will be a detriment to our career, professional effectiveness, and even how the general public views us. Many Geminis had this recent transit (compounded by Neptune), which resulted even in the ruin of their public reputation if they were especially willful, obstinate, or amoral. But this does not occur as an evil influence of Saturn but instead as a direct consequence of our own behavior, where instead the responsible, structured, honest behavior required of Saturn transits can and does instead bestow honor, respect, and prestige, and during difficult Saturn transits we have only ourselves and our unresolved trauma to blame for its unpleasant side effects. It is never too late to change course, however, so if a Saturn transit such as this was met with a sorry fate, changing our behavior and resolving trauma and control mechanisms will only result in a better life moving forward, resolution of the past, and restoration of peace and prosperity, and when Saturn officially departs this house it can gift us with notable public recognition if we have handled the transit well and implemented Saturn's lessons.

Saturn's transit through the eleventh house of fortune concerns our wealth and friendships. Like all houses, the eleventh is the natural result of all the houses which came before it, but although it concerns wealth and fortune continues to require structure and responsibility as Saturn demands. If not, we risk losing not only wealth but also friends, as just like the house of partnerships our friendships also require maintenance and care. Many of us sometimes take our friendships for granted, and those friends will wander off and find new ones which are more reliable if we do not put in the effort required of them. I did not realize until the very last draft of this book that my experiences with Saturn transiting my eleventh House (where my natal Saturn is also located) was around my 29th to 30th year and Saturn return wherein I also tried cultivated new friendships and even threw myself a large, fun, 30th birthday party which was attended by many people I loved and admired in Los Angeles, but even then did not realize this was a karmic lesson by Saturn on how to build and maintain friendships, which is simply to put in effort, and in later years as life got harder I let those friendships go as insecurity, stress, and poor health destroyed my life. At the time I also had overcommitted on the expensive car lease and found myself unable to meet my obligations and had to surrender the vehicle, so while I had succeeded in the nature of this transit with friends had failed in terms of wealth. Similarly, loss of wealth through irresponsible, undisciplined, or reckless ventures or gambling will also especially result in loss during this time but structured, disciplined, and self-honest ventures will likely pay off well at the end of the transit.

As mentioned earlier, the twelfth house is the house of spirituality but also of endings, and these too still require structure and responsibility as spirituality does not simply come to us, but we must seek it out and engage its practice. During this transit preceding my last first-house transit was the time where I first learned yoga and meditation and sought out solutions to spirituality and peace, which worked quite well considering the degree of trauma I was dealing with. Many of us resist twelfth-house transits, however, as we are so focused on progress, growth, and success that we resent downtime, rest, restoration, and introspection because it is time we could otherwise spend in pursuit of stuff and shit. Primarily, however, this resentment is also rooted in the resentment of our mortal condition, that we do require rest and spirituality every bit as much as material resources, and acknowledging the need for time to reflect and rest means also acknowledging that we are not in control of our mortality. This stubborn resistance can and will then result in more anxiety, not less, as we run out of steam and become addled with mania and neurotic behaviors which will make the next cycle far more stressful as Saturn enters the first house again than if we had prepared for the next cycle appropriately with a period of rest and introspection. Indeed I drank quite excessively during this transit as another way to deal with such anxieties, not yet recognizing my problem of alcoholism, instead of taking time to rest and take care of my considerable need for resolution (which I had also not yet learned effective skills to do so), which then also made it much harder to be healthy in the next cycle.

Just as influential in our life path is Saturn's aspect also to natal planets, such as when it is aspected to natal Neptune, Jupiter, Uranus, Venus, Mars, Pluto, Chiron, the Moon, or the Nodes of the Moon. Remembering major aspects of conjunction, trine, sextile, square, and opposition, Saturn's influence to our natal planets occur regularly and concern our personal conception of discipline, structure, and responsibility relative to the themes of those planets just as it does the Sun and the angles. But these transits are likewise often mischaracterized by astrologers, and Saturn never causes harm for the sake of it but rather only as consequences of lacking discipline, responsibility, structure, realistic expectations, personal honesty, etc., and Saturn never intends us harm but instead to graduate to more effective life skills which will have auspicious effects on our life, even in the near future, which is why Saturn always bestows a gift at its exit if we have been paying attention and implement its lessons. One example of this disparity are harsh transits to Jupiter, such as Saturn opposite Jupiter that is often described as a period of struggle with insufficient resources—This transit for instance recently occurred for me and within days of beginning the sales of my book dropped to zero as economic anxiety and stress overtook the country. But all this transit means in truth is that growth (Jupiter) only occurs at this time through structure and responsibility, and even at the same time as my book sales (donations) dried up several new people signed up for personal coaching but which requires more time, effort, and structure in showing up and helping them. Such is Saturn's influence to all the natal placements, which can be interpreted through the structure and discipline nature of Saturn in the quality of its aspect (conjunct, semisextile, sextile, square, trine, inconjunct, opposition).

Because Saturn is a disciplinarian, even its supposedly auspicious transits can still be stressful depending on the degree of our delusion, ignorance, and lack of restraint, and dissatisfaction can and does occur even during conjunction, trine, and sextile influences. But Saturn is in fact a benefic planet whose sole purpose is our wellbeing and whom, like the truth-teller, we only hate if we in turn hate ourselves or the conditions of mortality. Otherwise Saturn is a lovely planet which in many cases cares for us more than our own parents and brings a great deal of incredible life experience and helps guide us toward the things we want most—prosperity, love, stability, peace, and contentment. Through Saturn we get those opportunities to achieve, and the guiding hand of a planet more wise and uncompromising than we can ever hope to be, so paying attention, doing our best, relying on responsibility, consistency, and self-honesty, and practicing inventory to let go of our control behaviors can turn Saturn transits into the most productive of our lives in which we resolve our problems and find progress and stability that we otherwise expect from sources that do not actually bring it. Saturn is a loving friend, be grateful for its guidance.

18. Auspicious Jupiter

Wonderful Jupiter is the planet of growth, expansion, and opportunity, but Jupiter is also greatly misunderstood and often considered a completely benefic planet when in fact its transits can still also be disappointing because expansion is not always inherently a good thing. Astrologers often lionize Jupiter with reckless abandon, but there is a reason for the saying *'be careful what you wish for,'* and nowhere is this more applicable than with Jupiter which very often can gift us with expansion and growth but which is not always magnanimous. Also, while we as humans easily recognize and resent unpleasant experiences we do not equally appreciate those which are auspicious unless they are truly exceptional, and many times the function of Jupiter helps us in ways which are extremely benefic but because it's not a giant dump of cash directly into our bank account without any effort on our part our human nature is to overlook its generous influence. My thought experiment which demonstrates our insecure, biased perception of life and the world around us, pretending we enter a room with ten people and one of them immediately calls us fat which, if it were to happen, most of us would think how cruel and shallow is the world and how unlucky we truly are even though nine of the people in the room did no such thing. Meaning that 90% of people are probably benevolent, if at least not ill-meaning, yet because of our evolutionary survival strategies which make us always vigilant for threats to our wellbeing we focus on negative influences and ignore the good, because good influences simply are not a threat to our survival, and this survival instinct mechanisms is even stronger if we are very traumatized and stressed so that we spend all our lives in hypervigilant expectation of harm and completely miss boons, joy, and growth even while spending the far majority of our time and days in prosperity and peace.

This is the same function we often hear concerning children—that many positive experiences are required to make up for the damage caused by even one that is negative, because this concerns our evolutionary survival biology and is not a product of conditioning or nurture (although abuse and trauma greatly enhance it). This is the reason for the mass discontent of the baby-boom generation, in spite of being the wealthiest to have existed in the entire history of mankind the collection of riches and fixation on wealth is a survival reaction to feeling in constant fear of danger. Even while enjoying some of the greatest stretches of peace, prosperity, and heretofore unseen life expectancy do not even see how lucky and blessed they actually are because their entire childhoods were spent in horrendous trauma, abuse, and neglect at the hands of their parents' and grandparents' warmongering generations which tormented them as children with promises of nuclear annihilation and false salvation in the adherence to authoritarian religious and secular ideology. Such are we also blind to the blessings of Jupiter (and other planets) when we are in constant vigilance for our survival even though life is by far a constantly wonderful experience merely peppered by occasional misfortune, which is why inventory therapy is so requisite to achieving wisdom and finding peace because even when peace holds us tight we do not even recognize it.

Beset by stress and deteriorating wellbeing many of us commit to making gratitude lists or other coping strategies like willing our attitude to change in attempts to be more positive or to reduce stress. Even therapists will often prescribe such behavior as a means to ostensibly improve depression, anger, and other emotional stresses, and in the recovery programs sponsors admonish their sponsees to do the same. The problem is that this function is not one of willpower but instinct and human evolutionary survival biology in which we are literally programmed by biology to become hypervigilant when beset by trauma as a means for helping us survive in a world in which (we believe) potential peril exists all around us. Making gratitude lists is thus a delusional behavior that serves to ignore the underlying problems which cause this conflict which are instead the unresolved trauma and destructive control and coping behaviors which absolutely must be addressed before we can make progress in life, because it is the absence of effective life skills which cause this conflict, not our mindset, requiring the learning of new life skills from which is then derived the confidence that eventually changes the mindset, feeling newly *empowered* to be more effective in our life, which cannot simply be willed into being and requires structured practice and dedication to resolve that unresolved trauma.

Most of us in fact lead entirely safe and prosperous lives, often in the very lap of luxury even when we think of ourselves as not-yet-rich, living as if we are constantly deprived of food, shelter, employment, hunted by wild animals alone in the wilderness even as we drive cars, use space-age smart phones, and order food pre-made for us through takeout windows by underpaid and exploited workers. Once when searching for a place to live I saw for sale many large, spacious apartments for sale in cities on the East coast in neighborhoods which were supposedly blighted, going for very low rates but still more than I could afford, supposedly because they were not in an ideal neighborhood but were full one and two bedroom apartments in small buildings with vintage moulding, wood floors, and a garden, and many people who similarly live in what are truly generous dwellings often believe they in fact live in squalor, and then treat the home as if it is in fact a place of squalor, so it does in fact become a place of squalor, but only because we made it that way through our own behavior and not because of anything real about it, which if I lived in such a place would feel like a member of the aristocracy.

This is why inventory therapy is required to benefit from the practice of astrology, because we will not even recognize when in fact life is taking care of us, even as it constantly delivers us from oppressive circumstances and the consequences of our own behavior, because our worldview and perception of ourselves is so traumatized from formative childhood experiences that we do not recognize good and constantly chase bad, and even mistake some bad things for good such as when we are duped into scams and schemes by maleficent predators. Even though it has been ten years since my life fell to pieces I am still overcome by feelings of gratitude whenever I slip into my too small king size bed at night (my feet hang off the edge), covered by two very well made, soft, and pretty blanket and comforter, grateful to have more in life than I need to be happy and very comfortable, believing all the while before even when I had more that God was not looking out for me nor interested in my wellbeing because of how I was raised to always expect misfortune and demand God deliver me from it. Now that I have many years of inventory practice which has liberated me from such trauma and control mechanisms nothing more than a cup of tea, a nice blanket, or a conversation with someone I admire is required to make me feel like the richest person in all of history. Progress, expansion, and luck are the domain of Jupiter, but we do not lose out on his influences or fail to capitalize on his transits because of ignorance of astrology but instead what truly is of value in life, which is achieved by resolving past experiences of trauma and the control and coping mechanisms which pervert our concept of prosperity, ourselves, and the nature of life which then mislead our sense of gratitude.

One primary peril which afflicts anyone with addiction and alcoholism for instance (even if you are a dry drunk!) is a fear that our lives and success are entirely up to us and that we alone are responsible for our very survival. Thus being insecure about our own abilities and effectiveness—having seen first hand how bad we are at life—live in constant fear of not having enough or failing portended misfortune because we know for a fact that we are not capable of securing our continued survival. In truth we are not, in fact, capable of doing this, and in turn this causes a great fear of loss and failure and we then chase after immense riches, fame, and power that we might have an inordinate quantity of control that through excess we might overcome our incompetence. But because money is not real and success is fleeting we also grow miserly and greedy, suspicious and defensive, and spend more time chasing and defending security than anything else in this short life, even when we already have it, and there is a great disparity between wealth and contentedness which belies the irrelevance one to the other.

Learning through the practice of inventory (as a consequence of actually losing most of what I had worked so long to build) realized I am not *at all* in fact in charge of whether my life becomes a success. I can try all I want—go to the best school, climb the highest corporate ladder, make the most connected and powerful friends—and still meet misfortune and ruin, and if that is still a common potential outcome even when possessed of such resources what is all that effort for anyway? The point of spending decades of our life growing wealth and return on investment only to die and the end and have it mean absolutely nothing?

This is true because in fact none of the reasons we do such things is part of this reality—we are not the ones which determine outcomes, and our only power in this life is to show up for opportunity and to do our best, and beyond that all outcomes are entirely up to fate and the will and workings of the Universe. While pessimists may say this limits our freedom of will this in fact *broadens* our freedom by removing from our burden the responsibility of things we cannot control anyway. Indeed men and women and nonbinary people spend decades of our lives dedicated to something as inane as investing in money *precisely* because it is something we do not control, knowing ourselves to be incompetent find comfort in avoiding things we do control, like our behavior, because addressing those would mean having to change, and we do not want change or believe we cannot, and so then waste our time in distractions, fearing all the while that no one is actually looking out for our best interests but ourselves and that we alone are responsible for the outcome of our lives.

If, however, we are not in charge of the outcome anyway then we can spend our time doing anything else but *wasting* it on that which we cannot control, and because we are no longer wasting our time and energy on things we cannot control anyway we then ironically become *more* effective because we spend more time and effort on what we actually *can* control, such as our behavior, or the resolution of our past trauma and control and coping mechanisms, or in repairing past mistakes instead of forcing people to pretend we didn't make them, or in *eating* healthy instead of trying to stay young and healthy, or in writing a book instead of trying to be a famous writer, or parenting instead of trying to be a perfect parent, or in being a good friend instead of amassing friends. Thus the auspicious influence of Jupiter is not dependent on our understanding of astrology or the transits of Jupiter but our ability to acknowledge reality, live in the present, and recognize when we are actually blessed as what is found in acts of self-care and introspective therapy as inventory practice.

I too thus never noticed any benefit of Jupiter transits even while looking for them after I had known quite a lot about astrology. But once I had been practicing inventory for several years began to see exactly how Jupiter's influence had been there all along, but I had mistaken its effects as everyday entitlements I expected, or was even disappointed because they were not sufficiently grand and life-changing, mistakenly thinking that auspicious life events should liberate me from my doom and stress but in fact only lacked effective tools to do so which I was later fortunate enough to learn. It was at the age of forty during one auspicious Jupiter transit *trine* my Ascendant that for the first time anyone with a large social media following (my now friend) publicly praised my book which was also shared by another person with a large following at a time when sales were only a handful each week and sold more than a hundred copies over two days which then funded a badly needed move back to the city. But even as this event unfolded I recognized a very human, reflexive disappointment that it was not greater than it was, failing to liberate me entirely from the shackles of poverty, and how much easier my life would be if the sales could just keep coming and coming, and unfortunately went back to normal after about a week when such things are buried in the algorithms of social media never again to see the light of day.

After even more inventory practice I was able to look back on that auspicious moment as a truly wonderful manifestation of the correlation of Jupiter with growth and progress, as a result of past choices and decisions, and analyzing other earlier Jupiter transits such as when it was last *conjunct* my Ascendant was also aligned with the beginning my first book and recovering from my breakup, which were Jupiter gifts I did not at the time recognize due to the immense stress and isolation which had resulted from that experience.

Entertainment astrology and predatory astrology are similarly rife with promises of windfalls, anticipated money, or unearned luck because we as humans so greatly desire the false security which money and resources can bring, which make them easy points of exploitation by unscrupulous and duplicitous astrologers in turn for *their* desire to also find security and stability through patronage. But that just isn't how life works and while astrology can be used to understand cycles and life experiences it cannot be used to obtain wealth and prosperity or any other desired control of reality, but instead to understand our life path and how we can be more *effective*.

Typically, however, Jupiter transits are in fact beneficial and auspicious, but in many ways which have nothing to do with money, the desire for which is merely the subversion of our fears and anxieties about life and our future which we instead must learn to accept though structured practice of inventory. Many of us regard our lives as hard and unpleasant or disappointing but in reality the far majority of our time on Earth is marked by nothing but pleasure, peace, and uneventful passing if measured for instance by the number of good days versus bad ones. I could not appreciate or experience any of them until I was able to heal that trauma, but that also did not even require the complete healing of physical disease, although my life was better once that occurred, and even while still dealing with physical limitations and illness was able to complete significant entries in inventory and heal my psyche and spirit, because the very act of practicing inventory is itself the practice of self-compassion and self-care. Most importantly, inventory therapy taught me how to live my life in the present rather than anticipating misfortune or dwelling on the past, and thus also began to see even the subtle influences of Jupiter, especially when they were not the fulfillment of my dreams and fantasies.

The expansionist nature of Jupiter is the greater of its qualities, and while most of its transits could be considered positive sometimes they can also be difficult since Jupiter merely serves to expand or grow anything it touches. For instance when transiting Jupiter is square a natal Moon it can cause us to be preoccupied with what we perceive as social injustices done toward those with whom we identify, expanding our conception of our inner selves to those in the world around us (since the Moon rules emotions), and become stressed or even fearful because of that transit. Sometimes even things we perceive as being desirable as wealth, power, and influence when touched by Jupiter can in fact bring peril due to entitlement, impunity, or ignoble behavior, and where we have mistaken Saturn as a harsh planet rather than one which is kind and loving, assisting us to get the things we want, we also mistake Jupiter as being entirely benefit when in fact Jupiter can be irresponsibly indulgent, enabling reckless behavior and even imperil our best interests. One of the most likely of this influence is when Jupiter transits our eleventh house of fortune and we find ourselves gifted with a great many friends but whom may not necessarily have our best interests in mind nor be good people, since Jupiter does not determine the quality of its gifts, only that they are given, and we later find ourselves misled, deceived, taken advantage of, and even abandoned by those very friendships later during harsher transits because they were founded in willful naivety, wishful thinking, and fulfillment of our ego.

Since Jupiter transits are slow but not as slow as other planets its transiting aspects to natal points can last anywhere from two weeks to half a year if they occur during its retrograde, while its transits through the signs and houses last about a full year or so, and its entire journey and cycle through the Zodiac is about twelve years. Like all the other planets Jupiter also interacts with its peers to influence (or represent) trends in life around us, and one particularly useful influence for understanding long term trends in prosperity and loss is the Jupiter-Uranus cycle. Many astrologers look to various planets to divine cycles of economic prosperity and bust and while the Jupiter-Uranus connection is usually afforded to themes of society, rebellions, etc., it is actually one of the primary combinations which reflect general material prosperity, and demonstrates a key problem in astrology which is that planetary relationships are not only important when aspects are exact but are in constant relationship. Just as the Sun has a constant, stabilizing effect on Earth as we orbit our star all objects in our system always have a connected, metaphysical relationship with each other, and exact aspects are only times when cycles have more apparent *changes* in influence with each other, but are at all times active and influential even when they are not in prominent aspect. Jupiter is the planet of growth and opportunity so it should naturally be connected to some economic trends, and Uranus is the planet of revolution, technology, and fortune (not always the benefic meaning of fortune, however, but merely concerning it), but this connection with Uranus has been overlooked because aspects of the economy are often grouped into one measure when in fact there are many different influences, each governed by different planetary bodies (this is discussed in greater detail in the upcoming chapter on world events and great cycles).

So much astrology practice does nothing but serve the human ego, where instead it is more useful for understanding the purpose of existence, which is not to accumulate riches or success but simply to have experiences, where material wealth is merely a control mechanism meant in practice to spare us uncomfortable life experiences but in fact more often causes them as we fight for wealth, influence, and status, even to the point of sacrificing at the altar of fear things we truly desire or which are actually more important such as our time, integrity, peace of mind, relationships, or our very humanity. For instance not one highly-wealthy person in the world can go wherever they want, and at all times must be guarded and protected from yet other money-hungry humans from harming them or taking what they have, where any regular person can go for a stroll, lay about in a park, or go to the grocery store as they please, and no matter how rich a person is they cannot ever buy that kind of freedom, and ironically spend their time wishing for such a life yet retain the wealth that prevents it in the first place due to fear and service of the ego, failing also to recognize the reality that if we lived in an egalitarian society in which the needs of all people were addressed there would be no one who feels compelled to take from others and the rich could by their own hand create the world in which they could have the freedom they desire but refuse to do that because it would then mean no longer having so much money.

Jupiter, however, has nothing to do directly with money and finances but instead simply grows and expands all that touches, so its supposed association with money and economies is of relevance only to those whom are interested in such things simply because we generally enjoy the expansion of money and abhor its contraction. But Jupiter can and also expand things such as relationships, sex, opportunity, wisdom, learning, authority, oppression, stability, and even the appetite (although many astrologers maintain outdated ideas of health and weight as a function of calorie intake, which it is not, and will nonetheless present misguided warnings about eating too much during certain Jupiter transits as if we have any control whatsoever over our own biology).

Jupiter also strolls through the Zodiac in hand with Saturn to participate in the structures of markets and economies due to their dependency on structure and institution, and during the early conjunction, semisextile, and sextile phases of this cycle very often cause unrestrained and unchecked growth which to those seeking riches is phenomenally intoxicating but which in reality merely sets the premise for the following crash as a result of undisciplined expansion which occurs at the squares such as is currently occurring right now in 2025 (which started in 2024). Jupiter's role is thus not to bestow wealth and prosperity but to expand in opposition to Saturn's role in contraction and discipline, not only of opportunity but also loss, the mind, or even delusion, and even expansion of things such as money can actually in practice be undesirable even if we choose not to acknowledge that for the service it provides our fear and control behaviors (again, more on these great cycles are in the later chapter on them).

By the nature of our birth time and place we are unable to choose which transits will affect us, but what we can choose is whether or not to listen to the lessons the Universe is trying to teach us, that we can be just fine without prosperity, or that prosperity must be responsibly managed and not used as a substitute for self-worth, a salve for fear, nor a means to control life, and Jupiter's path through our birth chart is most useful as one which helps us achieve and progress, especially at a more rapid pace than the slower planets, but only in proportion to the degree of our self-honesty.

When Jupiter transits the Sun we experience growth in our personal identity and purpose but also often improvements in our health and wellbeing which, if aspected in square or opposition might require more effort or is achieved through mechanisms which are not entirely auspicious but still helpful, meaning such that growth can be achieved but only by effort or the correcting of mistakes, the learning of new skills, or improving inefficiencies. Transiting to natal Saturn is often described as challenging times in which Saturn stifles the impulses of Jupiter, but in fact Saturn merely wants to provide structure to Jupiter influences and is highly auspicious to make Jupiter influences a higher quality through structure and discipline. Transits to Neptune can expand our spirituality but they can and do also expand our delusion (another thing we do not benefit from its expansion), and so transits to natal Neptune by Jupiter should be handled with much introspection, inventory therapy, and evaluation of our behavior as described in the upcoming chapter on Neptune. In transiting to Venus, Jupiter can improve our relationships and bring an abundance of love or beauty (such as my recent trine transit in which I purchased lots of organizers and set clean my closets and cupboards). With Mars, Jupiter brings action and initiative except in the square and opposition which can sometimes bring conflict and animosity if our methods are not considerate of others or we are not entirely clear about what is required of us. When Jupiter also transits our natal Moon this can bring wonderful emotional experiences, though care should be taken during the square and opposition not to let those get out of control. Transits to Chiron also bring much healing, but may trigger past wounds in order to help them heal and for us to make progress in their resolution (which is especially more likely in the square and opposition, and Chiron is further discussed in the chapter on karma and healing).

When transiting the first house Jupiter expands and grows our sense of self and life purpose, similar to its transit of the Sun, most commonly through increased opportunities to be effective in our life or grow and expand our sense of purpose. While this can indeed be auspicious it also comes with the risk of delusional arrogance (that's a bit of an oxymoron because arrogance is inherently delusional) or an inflated sense of self. Most often though this is simply the opportunity to do more and feel personally more effective, and if burdened by unresolved trauma or control behaviors we will be entirely unconscious of its effects.

In the second house transit Jupiter expands our *opportunity* to earn money. This is often mistaken as simply an increase in money such as from a larger tax return than expected or other increase in wealth but the second house, again, is the house of *means,* not simply money or wealth, so normally this influence is experienced as increased opportunity from which we can earn money. For instance when Jupiter transited the second zodiac house for Libras in 2017-2018 was when I began offering coaching services for people who needed help with their health. Although I already had a job this in turn allowed me to quit that one and begin transitioning to writing, research, and coaching full time instead of wasting my time and effort on a predatory, terrible industry earning money for a man who did not care about my wellbeing at all and in which there was no future for me anyway. But that job had also allowed me to accumulate quite a bit of savings which peaked at that time which in turn allowed me to transition out of that career without any real financial stress and do something instead which was far more rewarding even as it paid much, much less. So this transit strongly affects opportunity for means, not simply banal earning power.

The third house being of knowledge and communication Jupiter's transit here is a surge in exchange of information, learning, and opportunities for communication. For instance, being that my primary medium was then writing this was also when my first book was first published, in 2018, and the increase in chatter and traffic to my website and interest in my work that set the foundation for my work, research, and writing. While media such as writing and books is often an aspect of this transit it needn't always be the case, and can be any kind of exchange of information either from others to us or from us to others such as through social media, politics, journalism, teaching, school, trade schools, research, etc., and what is certain is that communication generally during this period will be greater than usual.

Jupiter transiting the fourth house is always auspicious, even though the forth house is at a natural square to self, because this concerns home and family which are always wonderful (okay, well, nearly always) in abundance, and sometimes literally correlate with moving to a bigger apartment or home but very often takes the form of simply having more time and interaction with family members. For Many Libras in 2019-2020 both Jupiter and Saturn transited this house together, which was when I personally moved to a bigger, nicer, less-expensive apartment nearer my family (but which I also then couldn't furnish due to the pandemic), which is exactly the type of influence that Jupiter has in this area even when Saturn is not also there.

The fifth house of creation in Jupiter transits often results in an explosion of creative inspiration, and sometimes even a new or casual love affair, and during this transit also in 2019-2020 many Virgos found themselves in a new romance or love affair, or creating works of art or with new children, as did Libras the year after (the drug addict I met), and then Scorpios in 2022, and so on, with Aquarius currently hosting Jupiter here and Pisces up next.

Sixth house Jupiter transits are likewise characteristic of improvements in health and work, but more in the context of daily labors and *opportunity* to make those improvements, since the sixth house is of our labors and not magically borne from the aether and requires action and effort, but will pay off well if this occurs (and if there are not more powerful opposition transits such as what Libras experienced with Neptune also here during this most recent transit).

When Jupiter transits the seventh house of partnership it is often thought as being a romantic influence, but partnership can come in *many* forms and rarely for me has this transit included romance, but one year while writing this book and having a near dearth of relationships an unexpected friendship developed as a consequence of my work and was with someone surprising and deeply refreshing after many years of failed or shallow relationships and involuntary isolation. Relationships always require that both parties be interested and active in maintaining them, however, which means that we are powerless on our own to get those things brought by any relationship whether it is romantic, plutonic, business, etc. We often dream of romantic love since it is so powerful and alone can often light the dark places of our lives, and indeed when this transit began I secretly hoped after eight years of being single somehow I might finally meet someone. But I was still working on *Fuck Portion Control* and trying to recover from cystic fibrosis and had put no real effort at all into anything remotely resembling romance, and realized toward the end of this transit that I had, in spite of my inattention, fallen into this friendship through no extra effort on my part, and it was neither showy nor the kind of friendship I would have anticipated and for that reason was profoundly appreciated. Transits through the seventh house typically do *not* initialize romantic relationships, that is more the domain of the fifth house of creation, and instead serves to more formalize the romance relationship structure.

Although the eighth house concerns power and control, Jupiter's transit here does in fact help us to find more control in our lives, especially in becoming more personally effective by controlling the things that we actually can control such as by increasing our understanding of something, making better decisions over shared resources (such as being able to come to agreeable terms on debts and the ability to pay them down), or to experience new sexual encounters or dynamics since sex is a primary domain of control and surrender. Sometimes Jupiter here can actually bring us opportunity to wield power over others, but that is a devil's bargain which is simply increasing our *desire* for power and often results in our downfall through later transits of other planets since we cannot actually control reality, whatever our ego may try to convince us otherwise.

In the ninth house of the mind is itself ruled by Jupiter, and its transit here naturally and greatly expands our philosophical understanding of our personal life and the workings of the Universe, often through exploration of philosophical subjects (even if we do not recognize it as such). For instance when I helped one of my siblings learn inventory and to resolve much of their stress at being a parent (which also informed the content of my book on psychology and parenting) Jupiter transited their zodiac ninth house during which they realized how their previous worldview and ideology had been the root of much of their life's problems and through my work learned a new, more effective worldview and life skills. While I began this very book on spirituality years before my own Jupiter transit to this house I was not sure exactly how to write it, but the bulk of it then came together during this ninth house Jupiter transit for Libras which then helped me greatly understand how to present the information and to better understand the concepts I was trying to elucidate such as the nature of time and consciousness which I had not before solved which then also helped me complete this book in the very last weeks of this transit. Twelve years ago when Jupiter transited this house in my early thirties was similarly a time when I was exploring how to actually be a writer (many of which generously tried to help me in turn!) and met many different writers which expanded my conception of the world through real experiences, as Jupiter also facilitates adventure and mind expansion through travel and real life experiences and not only study and intellectual investigation.

Jupiter in the tenth house is equally auspicious since this sector concerns our public image and professional success, so Jupiter can and does indeed expand our visibility when it transits here, normally for the best, as likely a consequence of our previous choices and life path which pay off in recognition or greater opportunity to be effective. New career options might open up, or a position of authority, or simply greater interfacing with society which can help us be more integrated into our communities and the structures which create stability, institution, etc. Since this transit for Libras is imminent at the time of this writing I expect an increase in public recognition, for good or ill, through to next year as a consequence of writing this and my other books. But this influence can be for any reason, for instance a Virgo I personally know had their public profile raised because of their cancer diagnosis and subsequent fundraising for treatment, which I suppose is better than not receiving help in the first place, but demonstrates how the nature of Jupiter is not always concerned with the context or quality of its transits, only that it does simply expand the areas it touches.

Jupiter is not the ruler of the eleventh house but the themes are quite similar as the eleventh house rules fortune and friendships, so Jupiter here can literally become a windfall of good things as we find our social circle expanding and perhaps some expansion of our wealth, if even slightly. When I had this transit last in 2015, however, it also occurred when my life was falling apart so I did not recognize the introduction of many new friends and acquaintances into my life from new places of employment or participation in recovery programs as the influence of Jupiter until many years later when looking back in analysis. Indeed I met many wonderful people who helped me get out of the rut into which my life had fallen, and was auspicious indeed. But, as previously mentioned, I also had this in my ascendent eleventh house in my late-twenties in which I cultivated a many friends and acquaintances but later found them absent when my life fell apart and neglected them, as Jupiter does not determine the quality of friendships (which is instead determined by our trauma and control behaviors), and its similar zodiac transit in my early thirties exposed me to many of the writers with whom I could have become friends, if I had not been so insecure. Sometimes this transit can result in growth of literal wealth, but that was not my experience so far, which is probably a consequence of having Neptune in my second natal House, and those without such modifying birth planets might also find this transit to be more financially oriented.

In the twelfth house Jupiter's transit can often be entirely mistaken or neglected because one of the primary concerns of the twelfth house is the conclusion and endings of past experiences. Many of us do not want endings to occur, as was my case in 2016 when I still did not want to move on from my old life. But Jupiter gave me experiences to help me step back from it, to rest and heal from those experiences, so that the new cycle moving forward could be something new. And again earlier in 2004 before my move to Los Angeles in which I recovered from the previous cycle and my suicide attempt in a period of inaction and (relative) stability that helped me put myself back together in preparation for new adventures and new experiences. Likewise this influence is not what we expect and do not always look forward to it, although we very much should as it is one of the best roles Jupiter can play in our life journey since it helps us to heal in preparation of beginning anew in the next first house transit but which for which we first must prepare through a period of reflection and rest and healing of past pain and trauma that will otherwise hold us back.

Jupiter is one of the most benefic planets of the zodiac, and its transits are always welcome opportunities for growth and progress, just don't cultivate unrealistic expectations or be deluded to its true, impartial nature, as doing so may be disappointed and in turn we miss out on realizing the true benefit of its transits. This is best achieved by understanding that Jupiter's function is not to shower us with gifts, but simply to expand whatever it touches, which can and does also include things that are not always ideal for such influence. Jupiter transits are also not always permanent, probably in fact because Jupiter moves too quickly around the zodiac for its influence to have inherent permanence, so we are obliged instead to keep its lessons and gifts, to make the most of what it has kindly bestowed.

19. The Love of Venus

Venus is the planet of love, beauty, balance, and partnership, which indeed are all concepts of balance, especially beauty because extremes which depart from balance end up being the opposite of beauty. Its domain of relationships are also not limited to romantic partnerships, and does also govern any relationship which requires balance such as those in business, because even those have close, emotional aspects to them which require equal give and take which makes relationships function which, when balance is voided then also void those relationships, and for this reason business and thus also sometimes money are often reflected in the state of Venus.

But Venus also has some great and mysterious influence over the course of human history, and people no longer know about Venus being the *Evening Star* or *Morning Star* which, before the advent of electricity (and knowledge that Venus was not a star), was a constant, magical sight for humans looking into the sky during the hours of dawn or dusk as one of the brightest celestial bodies, which also transited the sky on an almost nightly basis except for the short period where (in the Northern Hemisphere) it disappears below the horizon, and because of the motion of Venus relative to our Earth, for 263 days Venus appears brightly in the sky as the Morning Star, after which it will disappear below the horizon for 50 days, then reappear as the Evening Star, after which it will disappear again but only for 8 days, then reappear once more as the Morning Star to continue the cycle.

This cycle of Venus actually takes eight years to complete, for Venus to return exactly to the same place in the night sky, and when plotted on a geocentric graph this cycle forms a five-sided flower or pentagon called *the Petals of Venus* (as illustrated in the center of the back cover to this book), and for long periods of human history it was known that rival challengers to power would usually succeed in supplanting sitting rulers if challenging them during Venus as the Evening Star. It just so happens the eight year cycle of Venus currently correlates exactly with the United States Presidential election cycle, and Venus is always either at its peak or nadir when U.S. elections occur, and with few exceptions through-out the entirety of U.S. Presidential elections power balance has almost always switched every evening star election, which includes also the Revolutionary War against Britain and ascent to power of the first President, George Washington, in 1789 while instead power has mostly been retained by the party in control during Morning Star elections. This does not mean that those whom hold power are the embodiment of those themes nor legitimate in their control but rather the function of reality and such cycles which in the end is to achieve them regardless of anyone's personal goals or ambitions. This transit also concerns the position of the "king," or otherwise top leader, and does not necessarily concern lower posts such as congresspeople or senators (but maybe governors?), which sounds like a trite convention but laws of astrology are often quite specific and not generalized, a mistake that researchers or skeptics make in their amateur investigations, not fully understanding what they are investigating.

While politics and business may seem antithetical to the concepts of love and partnership which are represented by Venus, politics and business are nothing more than partnership dynamics in which we are obligated to partici-pate to achieve the things we need or desire, as Venus rules balance which is the fundamental law of all partnerships which, when they become imbalanced and participants take too much or give too much the partnerships dissolve and fall apart.

For this reason Venus does not only rule love but also its *absence*, and whenev-er Venus goes retrograde there is a marked increase in pure carnal desire for sex, which is not wrong in the least, but demonstrates how Venus influences love in all its forms, including when it is wanting, as its influence temporarily wanes and thus asserts less forces of balance over our existence which then has the effect of disrupt-ing plans, business, and romance, and the nature of Venus cycles influence all dynamics of partnership in our individual lives as well as world events, even when love and partnership are absent, as all cycles travel through periods of waxing and waning, culmination and contraposition, and in no other aspect of life is partner-ship or its absence more dichotomic than politics, so it makes sense that political fluctuations and dynamics would correlate more directly with the Venus cycle than most of the other planets.

For most of us Venus is a personal influence which directly influences our love, relationships, and business separate from larger, existential influences such as economies and political movements. Yet while all of us deserve love, sex and companionship these are some of the few things in life that we can *only* ever get from *others* and is not something that we can truly give ourselves, meaning while it is possible to love ourselves and even have sex with ourselves we literally cannot have sex and love from others *without* those others. Of course sex and love can be taken, and many people commit rape and other sexual assault or use threats and coercion in relationships in attempts to control others into doing what we want, but because there are in fact laws of causality there is also very serious consequences for taking love and sex from others without their consent, and in this context those are ruled instead by Pluto, the planet of control and power, not Venus, which instead concerned with the nature of balance which such heinous behaviors are the polar opposite. Many immature young men, especially those whom have yet to experience sex, have fantasies of sex with people without the requisite step of gaining consent because we don't know how to win such opportunities for romance and errantly believe it then outside of our reach. I know this personally because when I was a teenager I often dreamed of getting to touch a crush or object of desire but at the time had no idea this was a possibility for myself due to the hatred and oppression of the family, community, and religion in which I was raised, and that there were in fact other gay men who equally wanted to love to touch me back whom also led totally normal lives and had love affairs and lots and lots of sex, and in my conservative, hateful family and community the only future which was very presented to me was one in which I would never feel the warm embrace of another person who loved me (or just wanted to jump my bones).

Indeed many heterosexual young men believe wrongly that sex is something difficult to find consent and a willing partner, having been mislead into believing women both an enigma and not as interested in sex as men, which isn't true in the least, and thus will never get the chance to bed a woman, all rooted in the primary sense of insecurity and self-doubt, and this fear then leads them to lash out and treat women heinously which then obviously makes women dislike and distrust them and so greatly, greatly reduces their chance of getting to experience the touch of someone who wants them not because it is difficult, because it is not, but because our own destructive behavior prevents such opportunities.

The truth about sex and love is that we are all naturally motivated and inclined to want, desire, and seek companionship, sex, and intimacy, and all anyone ever has to do is be themselves, show up for opportunity, and someone designed by nature and biology will automatically want to take your cock or pound your pussy without you having do to anything more than be kind and allow another person to get to know you. Seriously, many men hide in their homes, terrified of romance and then blame others for their fear when getting laid is as simple as participating in life in complete opposition to what those with these conflicts are currently engaged. In truth rejection can be painful, however, and the primary motivation for isolating and avoiding opportunities for romance and sex is instead a juvenile fear of not being accepted by every single person on the planet, which is not only a delusional and ridiculous worldview but also a hypocritical one as we do not extend the same consideration to others that we reserve for ourselves, that we should be allowed to reject those as we wish but they in turn not have the same freedom?

This dual nature of love and sex, one which is incredibly violent and shallow and another which can literally plumb the depths of metaphysical existence with breathtaking experiences is not an accident of life but a purposeful dichotomy in which we are *required* to surrender the ego as the price for a voyage on the fleshy ecstasy of another person's body and soul. The irony unknown to those who take sex is they will never experience what it is to have *real* sex, because the price of entry is the ego, in order to expose our most vulnerable selves, and the act of taking is nothing but a pure expression of ego to which the doors of existence remain shut and locked, sometimes forever (at least until the person makes right their horrible wrongs).

Though many of us do not commit rape we still wield control and power in affairs of love and sex because it is one of the most vulnerable behaviors we do, and the fear of giving such control over to someone who can very well exploit our vulnerability and use it to hurt us fills us with a resentful sense of spite even in advance of experiencing sex, and become controlling preemptively, before our partner has even done anything to warrant it, ruining our own chance at fulfillment because of fear and desires for control at the root of all our behavior.

Even if our partners do act in ways which we feel are deserving of punishment, is that the nature of love? Are romantic films filled with plot lines where a woman didn't lose enough weight for her boyfriend so he continually brought it up and she unreservedly threw herself at him for being a controlling loser and the audience swooned at the romance of it all? Is there a film where the woman feels her fiancé doesn't make enough money and constantly shames him for his lack of ambition and he falls madly in love with her her ball-busting cruelty and all the women in the audience agree it's the most romantic thing they've ever seen? Of course not—There is a reason most romantic stories and art hardly reflect our actual experiences in romance and sex, because most of us do in fact use the vulernability of our partners as a way to exercise control over life which then comes at their expense and sabotage our interests. Even if our partner does not live up to our expectations no person wants to hear that the love of their life thinks they are a loser, and yet that is what most of us do all the time to the people we *supposedly* love, which in some sense we don't, truly, and they are instead just there to keep us company and fill space that would otherwise be empty and lonely.

Such delusions about romance are often caused by transits of Neptune or Pluto to our natal Venus or through the seventh house of partnership which Venus rules, since Neptune rules spirituality but also *delusion*, for instance believing that we can do things like sexually assault other people or hatefully control them without consequences, or Pluto which rules power and control in which we attempt to assert ourselves over relationships and control the lives of those on whom we depend for love or partnership, but since we are not in fact capable of control end up losing much through our behavior. When I once met a significant ex-boyfriend for the first time, transiting Mars and Venus were conjunct my natal Pluto, which then shortly progressed to conjunct my Ascendant when we began officially dating, but this also seemed to portend the controlling nature of the relationship (either me or him or both?). The Sun was also conjunct my natal Venus, and in fact I met two of my three lovers during Sun conjunct natal Venus which, since it occurs in the sign of Leo, then spent the Autumn and Halloween seasons falling in love with them as we made costumes and plans for the holiday.

Because one relationship occurred during my Saturn return as discussed in the chapter on Saturn, when nearing the age of thirty in which we all evaluate our life progress and make impulsive changes for good or ill according to the results of that assessment such as choosing to settle for someone beneath my standards since I had grown frustrated at a series of failed romances and yet to create my own family and even after suspecting from the start that he might not be a good person which was reflected in fact by transiting Neptune *exactly* opposite my natal Venus, directly pointing out my wishful ignorance and naivety and portending (along with the other influences) the tumultuous and ultimately doomed nature of that relationship. While there was certainly a choice to not be in that relationship, I still did. Even had I been armed with the useful knowledge of astrology probably would not have avoided it anyway since my choice was informed by a collected history of unresolved trauma and resulting insecurity and desire for control which makes us vulnerable to such poor decision making. As discussed in my book on psychology, our choices of romantic partners are entirely limited to others with matching control and coping mechanisms, which are often mistaken for excitement when we meet a new partner, and this occurs because our trauma and control and coping mechanisms inform every decision we make, including where we go, how we spend our time, and who and what are important to us which then serves to put us in the very situations and conditions in which we find ourselves wherein also there are those others whose control and coping mechanisms have also guided them there and then, and when we meet those others we recognize in them the same or similar worldviews and perspectives which inform those mechanisms, to feel kinship and relatability. A man who was traumatized about his intelligence and self worth as a child, for instance, will not meet the devastating, sweet math teacher at the house of friends he met at the local farmer's market he visited every weekend because he thinks cooking is too difficult and instead gets drunk every weekend at the local bar to instead meet women whom do the same and also hate cooking because they are also burdened with similar trauma and insecurity about their own self worth. So because our control and coping mechanisms limit what we do, where we go, and with whom we associate (or not) it becomes literally *impossible* to even meet the kinds of partners we might desire and fantasize because without resolving those trauma and coping mechanisms we cannot even begin to figure out how to meet them, let alone actually relate with and impress them when we do. A person that would have treated me with love and respect and been present in a relationship would oppositely not have even been in the places I was, with the people I related, so I never met any, and instead was attracted to the showy and demonstrative extroverts who in reality were simply unstable and narcissistic, searching for nothing but an empty vessel into which to dump all their expectations and fantasies.

This also brings up the point that many of us in fact mistake rejection as being all about us, because the experience triggers unresolved coping mechanisms such as insecurity or being controlling when in fact rejection from others is always about *them*, and the fact that we can be rejected even for reasons that we might appreciate or support, such as perhaps being too nice for them, or not shallow enough for their personal control mechanisms. Even when we are rejected for not being attractive enough it might hurt our feelings but we are also being spared from a relationship with a shallow asshole who would constantly remark on our appearance, compare us to other people, and likely end up cheating anyway. But such karmic arcs are quite difficult to consider when a devastating person tells us we're ugly if we do not have sufficient life skills to understand our higher purpose, where in fact all those people are doing is showing the world what awful human being they actually are, and doing us all a favor by providing information we can use to steer clear of them entirely. The first time I broke up with my ex was before we were very serious, and I specifically said we did not seem to be right for each other, and he was not as tall as I probably needed in a partner (since I'm uncommonly enormous), and I swear he took that as a challenge because our time together would prove his true disinterest in me and wasted both our lives for years for nothing more than the fear of rejection.

There were even times I actually rejected men that were *too* hot, especially one young man whom came home with me one night from *Akbar* in Los Angeles, because his beauty made me so insecure I would literally shake with anxiety, both feeling like I was not hot enough myself to be worthy of such a man and that I could not prevent someone hotter than me from leaving, because I did not yet understand that real love is stronger than transient concepts as physical appearance—I didn't outright reject him, but accidentally lost his number and didn't try to find him again, since I believed I was not worthy of love, since he could have any choice of lover (but apparently not since I was willing to refuse him anyway), and the concept that someone rejects us because they think we are too sexy or too good for them is not something we even consider when experiencing rejection, our insecurity making us think it is about us when in fact it is their insecurity which motivates their decision.

Because our choices of romantic partners are strictly limited to those with the same level of unresolved trauma and control mechanisms the most important behavior required to find healthy, stable, and fulfilling partnerships (remember, not only love but also plutonic such as in business) is to resolve those trauma and control mechanisms. But unfortunately these also influence our psyche and unconscious in ways which absolutely *cannot* be perceived by the conscious mind, and present in how we fundamentally view ourselves, the world around us, and how we navigate that world, and the belief that we can avoid making mistakes or unfortunate circumstances, events, and people by simply willing it or knowing about astrology is pure, irrational delusion and desire for control over life, which we absolutely do not have. Several times I have helped young, straight men get healthy and navigate difficult relationship problems at the same time, and those who actually did inventory therapy and faced their insecurities and problems leveled up in the quality of their partners, sometimes several tiers in the quality of woman they were able to secure, even to the point that they became insecure due to the quality of person the woman was and had to do yet more inventory in order to feel worthy and relaxed within the relationship, which of course they were they just didn't appreciate themselves as such due to childhood conditioning, fear, and insecurity by parents who have similarly woeful conceptions of their own self-worth (since our parents cannot teach us skills they do not have). Those who refused to do inventory instead simply jumped from the same kind of dysfunctional and manipulative woman to another, continuing in the same exact kinds of interpersonal drama and cycles of recrimination they themselves partook without any sense of self-awareness. Nearly all of the women I coached refused to do inventory therapy, probably because I was a man telling them what to do which triggered their deep trauma of being offended by men but, regardless, then similarly remain in cycles of self-destructive behavior and limited only to pitiful, desperate, child-men equally possessed of poor life skills in antithesis of their deepest dreams and desires, living in a prison where the door is in fact wide open but refusing to leave simply because they feel safer in the only world they have ever known.

Similarly, selfishness which often disturbs relationships is a human survival mechanism, but it is meant for organisms which have very little and are at great risk of failing should they not assert themselves by taking what they need from their environment, which can and does often include others of our species, and when we are in such extreme states of trauma and stress to trigger selfishness instead of selflessness the body and mind are not concerned with higher states of existence as found in fulfillment, satisfaction, and joy, never mind reaching the heights of human intimacy because the mind is simply concerned with not dying. But if we cannot see that such behavior is in fact nothing more than our shared human survival instincts which sacrifice our very wellbeing and fulfillment for the sake of the species in response to fear we willingly forfeit the opportunity to experience those things we most desire as ravenous lovemaking, adoring offspring, and respect and admiration from our community and relationships, and the difference between that life and one which is debased by our own choices and mistakes is nothing more than our worldview and perception of our own self-worth as a result of our childhood trauma, for instance the fact that I was in fact very attractive, talented, funny, and engaging when I was younger and could, like the hot men who devastated me, had my pick of many excellent options, but time and again self-sabotaged my own efforts without even knowing that was going on because our conscious mind *literally cannot be conscious of the unconscious mind*, which instead requires purposeful, self-care practices such as inventory to communicate with the unconscious to free it from fears, traumas, and insecurity.

While astrology is most often presented as a way to control our lives my own experience instead took me through the steps that were required to learn the truth about not only relationships and love but myself and many other facets of reality, and the position of Neptune opposite my natal Venus at that fateful juncture was not a mistake but simply one street on the long road of my life journey, and astrology is nothing if not a record of our lives, both past, present, and future. One practical manifestation of such delusion which can help us better identify it is whether we like our partner's friends or not. Whom we chose to have as friends is very similarly related to our trauma and control mechanisms, and someone who surrounds themselves with shallow, insincere, insecure, selfish people will themselves be that way as well, and not liking the friends of a potential romantic partner is a big red flag also about the quality of the person you are interested in and our need to better inventory our own control and coping behaviors in order to level up. While it is true that we cannot control reality we can make better informed decisions, learn from our mistakes, and apply knowledge to our actions so that we have a greater likelihood of effectiveness. Knowledge of astrology itself is also reflected in the influences we live through, so learning about astrology, especially in terms of romance, is not a hack or a cheat to trick life into giving us what we want but instead also part of the very intended life path on which we tread which was always going to happen in your, and the mistake is to believe that we have somehow pulled a fast one on life when in fact the trick is on us, and that knowing these things is all part of the workings of the Universe and our intended destiny.

When Venus transits the zodiac it indeed moves quite quickly, only spending about four to five weeks in every sign, and being a planet that is also mild in nature its daily effects are not usually so obviously noticeable, and specific events such as the meeting of a new partner or developments in our current relationships (such as their loss, or marriage, etc.) more often concern the transit of other influences to our natal Venus rather than transiting Venus. The one exception to this pattern is the occasional retrograde of Venus which briefly interrupts relationships, love, and business to cause much frustration and consternation. This is not to say those influences are not there, only that they are usually so mild as to be unnoticed, for instance transiting Venus was recently trine to my natal Moon during its recent retrograde, and since the Moon rules the home and the inner life I think it presented as simply enjoying my home and resupplying cleaning products to then Spring-clean my entire apartment (since Venus rules beauty, which also includes cleanliness). Venus is often attributed to creativity, but creativity is a function of creation, and Venus specifically governs balance and relationships but does not have dominion over creation. That instead is a function of the Sun and the fifth house, and transits to the Sun and the fifth house instead very often concern creativity and creative inspiration. But Venus does rule beauty, which can result from art, but not necessarily its creation.

Venus also has an intimate relationship with Mars, which is very interesting since we, on Earth, are often caught between them just as a child is between their parents, when they are in agreement it is great but when in tension our lives can be stressful, and when they meet in the sky many new love affairs begin because where Venus rules love Mars is also considered to sometimes be concerned with sex because Mars is the planet of action (pun very much intended) and initiative which are required to conquer someone in the bedroom, as opposed to Pluto which can and does also concern sex but fundamentally concerning its power dynamics of control and surrender, where Mars is instead the pure, noble form of conquering. Usually when Venus goes retrograde Mars is not, although the recent retrogrades of both did actually occur back-to-back, and the increase in sexual energy which usually occurs during Venus retrograde is probably because Mars is usually still powering ahead, leaving Venus behind for a short period of time and not taking her sensitivities into consideration. When Venus and Mars have challenging aspects such as squared or opposition there are also often eruptions in sex-associated conflicts and scandals such as what once recently occurred concerning nonconsensual, deep-fake pornography (and it is indeed highly immoral to create unauthorized, generated sexual content of people without their permission). This is not to say that Mars and Venus aspects are always necessarily public, although they can be, but instead are notable moments of disharmony in matters of love and lust, where at favorable aspects they are instead in harmony, and stressful aspects not. Whenever Mars and Venus challenge each other I personally feel a great deal of impatience and restlessness, specifically toward other people, and find my mind far more preoccupied with disappointments and frustrations during which I am compelled to practice more inventory than usual. It is possible this is more pronounced for myself since Venus rules the sign of Libra, and so would be more influenced by any aspects to transiting Venus, but since these are dynamics between Mars and Venus and not only natal aspects it is likely that everyone feels a little more restless and frustrated than normal.

Because Jupiter is the planet of growth it too creates a cycle with our natal Venus which bookends entire stories of love and romance in our lives. As discussed earlier, many people misunderstand the conjunction influence as one which generally grants boons to our life experience, and while it sometimes can it usually is specifically and more importantly the end of the last cycle and the beginning of a new one, and no less than seven months after the end of one serious relationship of four and a half years Jupiter came conjunct to my natal Venus. I remember at the time being confused because I had read that conjunct was an auspicious influence, but here the love of my life (clearly he wasn't) had abandoned me instead. But this actually fits in perfectly with the concept of conjunction being the end of a cycle and the beginning of a new one rather than specifically a moment for gain or blessings, ending perhaps the previous pattern in which I derived self-esteem from my partner rather than understanding my inherent self worth, or some other equally precious life lesson.

Do not misunderstand this as meaning relationships end at the end of Jupiter cycles to natal Venus—the point is that *cycles* end, not necessarily relationships, as obviously there are many relationships which last for many, many Jupiter cycles. But even those which do last for decades go through periods of evolution, which are likely encompassed by transits of the other planets to natal Venus. Indeed nearly the entire next Jupiter, Venus cycle I spent entirely alone, by choice, during my major research and work, never planning to be alone but finding that it was not possible to achieve some of the things I wanted to do while also paired in a relationship which in turn required me to be single for much longer than I anticipated (i.e. purposefully infecting myself with infectious oral pathogens to study), and as that research has now come to a close and as I finish this book have begun finally to put myself out there again. To return to the theme of rejection, it did not go well (lol), and my interest first caught by a man twelve years my junior (I'm 45, chill) on a dating app thought for sure I would be a type he found attractive and took my time gathering the courage just to say hi to him, taking pictures I thought were great and figuring out a nice thing to say, all the while dreaming how we would fall in love, how fun and charming he would be, the silly things we would do together, and we'd spend the rest of our lives building a home and family. Finally on a day of both courage and impulse months later I said hi, along with my photos and information. Completely and unceremoniously ignored. Not even a block. Just, nothing, as if I didn't even exist.

For about a day was extremely demoralizing, until I realized how silly I had been in building up the moment in my head after so many years out of practice. In truth I didn't really want to start by seriously dating again, but to just hook up and have some fun with lots of different guys until I happened to cross paths with one that might stand out from the rest, as had occurred so many spontaneous encounters in the past which I failed to recognize as potential love affairs, and where the last Jupiter cycle to my natal Venus ended in disaster perhaps this one will instead end in a climax. Or not. Maybe this one for me is all about being single, and maybe the one after is where I finally, finally meet someone. Or just remain single the rest of my life, having fun and freedom as I continue to create and build other things besides a relationship. Whatever the stars have in store for me, as is the case for everyone, is thusly reflected in the patterns we see in cycles such as by Jupiter transiting to Venus.

Compatibility in love astrology, especially for entertainment, usually fixates on Sun sign compatibility, and while it is true the Sun sign can be a factor in how we get along with others because our Sun nature is also our identity and purpose, Sun sign compatibility astrology and lay persons *incorrectly* assume that love and relationships depend on common interests, goals, and compatible personalities. If that were true we would in fact just be dating a clone of ourselves, which would be very boring indeed (not because we are boring, but because we are already familiar with ourselves) and no relationship is in fact ever like that because it is instead a person's *otherness* which excites and interests us, and the entire purpose and function of a relationship is to be with someone who has qualities, strengths, and weaknesses opposite to our own.

That is also why so many people are often with partners with whom they argue and argue and argue, because the conflict we find in their otherness is the very thing which also draws us to them, the dysfunctional excitement of control mechanisms triggered by someone who knows just how to push our buttons. Indeed if children understood intuitively that Mom and Dad arguing was the very point of why they are together (not that it is healthy, as it usually is not) none of us would be as traumatized as we are. Yet these relationship dynamics can also be very destructive because usually we date people because of their complimentary control and coping mechanisms, not because we are attracted to them (attraction can be a factor for initiation but not the reason we usually chose a partner). Instead, when we date or commit to someone whose strengths compliment our weaknesses rather than replicate them we in turn do the same for them and thereby through pairing become stronger and more effective together, and if we were instead involved with people who had the same goals, interests, and qualities as us ours our strengths would get stronger but our weaknesses *weaker*, so instead we mostly, instinctually choose someone very different than ourselves.

While Sun sign compatibility can thus describe how we relate to others it does *not* describe how we relate to a *relationship* and *relationships*. Even our Moon sign can be more influential than the Sun concerning our behavior in relationships and how we related to those experiences. Because my Moon is in Cancer, for instance, my emotional nature embodies the strengths and weaknesses of the Cancer sign, including being very sensitive but also loving to cook for and take care of a partner and family. So determining compatibility on only the Sun sign is a highly *ineffective* way to understand romantic compatibility. Primarily our love nature is instead, logically, the domain of the planet of romance, Venus and the placement of Venus in our birth chart. Like Sun signs this is very easy to determine too by the sign in which Venus is located and means our love nature takes on the quality of that sign. For instance my natal Venus is in the sign of Leo, which means my love nature takes on the qualities of the Leo Sun sign, specifically the desire to be the center of attention in a relationship—not in terms of being self-centered but in how the individuality of Leo can't help but shine brightly and attract attention like moths to a flame, as well other qualities of the sign as enthusiastic, optimistic, creative, and highly romantic in love (Leos *love* Love). In Aries our love nature is highly assured and independent (sometimes too independent), in Taurus means we greatly value relationships, in Gemini we greatly value communication in relationships, in Cancer is family and home, Virgo Venus values consistency, Libra Venus naturally values relationships, Scorpio is an interesting Venus placement which results in control behaviors that might even be inobvious such as avoiding relationships when we feel vulnerable to their control dynamics (which inventory can resolve), in Sagittarius Venus is adventurous, in Capricorn we tend to be married to rules and institutions like marriage, in Aquarius the love nature is unreliable or unpredictable, and in Pisces is deeply feeling and potentially either spiritual or delusional. As such, determining the Venus sign and understanding what it means we can better in turn understand our own love nature, including its strengths and weaknesses, and comparing ours to a partner's Venus can help us better understand relationship compatibility.w

There have in fact been many attempts to methodically categorize how people relate in relationships, and while there are some which are more effective than others (such as the so-called 'love languages') they are in fact simply accidental observations of Venus sign dynamics and astrology and the various divisions of signs into their elements and modalities. The good thing about Venus placement is that Venus spends nearly a month in every sign so unless we were born when Venus was on the extreme cusp of signs only the day of birth is really required to find this placement, not birth time. While compatibility can be as complex as we want to make it (mostly relationships just require empathy and as such do not need much analysis), in general it is determined by the element quality of each sign—*fire, earth, air,* and *water*. Those with Venus in a fire sign (Aries, Leo, Sagittarius) need passion and fun in their love affairs, and are not interested in anything but the most intimate of connections. When Venus is in an Air sign (Libra, Aquarius, Gemini) the priority in love is communication and dialogue. Earth sign Venus placement (Capricorn, Taurus, Virgo) prize stability and commitment. Venus in a Water sign (Cancer, Scorpio, and Pisces) means emotional experiences are most important. Looking at this, it is easy to then see how compatibility can get so chaotic so fast, as we each intuitively assume our partner wants the same thing we do since we relate to them and admire them but in fact we each have our own personal and unconscious expectation of what we should get from a relationship. Each element also has its weaknesses which become liabilities in relationships, for instance fire signs can be excitable and self-centered, air signs can be insubstantial and shallow, earth signs can be rigid and cold, and water signs can be histrionic and volatile.

Incompatibility is also only a problem if we are especially self-centered, however, and care more about what we get from a relationship than what we must give to it, which brings out the bad side of each element and sign, where we should instead be equally balanced in both give and take, and knowing ours and our partner's Venus placement and making an effort to understand our partner is a shortcut to understanding exactly what each of us need in a relationship and how to meet those needs. Whether or not what they need is something we can give them is then a function of whether the element of our Venus placement is compatible with theirs. Generally, having the same element as our partner is not always a good thing because too much of anything can sometimes be bad. For instance two fire sign placements can relate to each other's needs but fighting fire with fire just creates more fire and can get out of hand pretty quickly. Earth and Water signs go well together because stability and emotional experiences go together well, and Air and Fire signs go well together because passion and communication fuel each other. Fire and Water signs fundamentally do not understand each other as passion and emotionalism seem related but in fact are not, and Earth and Air are too disparately motivated to get along, one being grounded and the other a free spirit. Air and Water signs can be compatible but challenging, as is the same with Fire and Earth signs, and there are many other resources for specific opinions on compatibility that don't need to take up space in this book.

This of course does not mean that difficult placements preclude a relationship, and it is a mistake to use sign compatibility to determine the viability of any relationship—remember that the otherness of our partner is the very thing which interests us in the first place, and compatibility is not the same thing as fate, and most relationships simply require personal resolution of trauma and control and coping mechanisms (ours, not our partner's, although it's great if they can do that too), and doing that can also make ending of relationships more healthy since we then become empowered also by new life skills for endings and personal responsibility as well. Transits to our birth chart can tell if new relationship opportunities or potential relationship stress and breakup, but all compatibility does is describe the nature of the connection, and many connections can survive if we only learn to have compassion for ourselves and empathize with and understand our partner.

So Venus governs far more of our daily lives than only love and romantic relationships, but all relationships, including for those in power and the balance required of fate. Love astrology is not only fun and interesting but can be very useful if we properly understand how to use it, which mostly concerns the natal Venus placement, and other placements like the Sun and Moon describe instead our compatibility with all people separate from our love nature. As Venus too is a rapidly moving planet its influences can be exaggerated or muffled by the slower moving planets, and aspects to natal Venus are most associated with relationship events and developments which can be used to understand our love life and our romantic partners. But don't expect to be able to better control them with this information, which will only ever result in more frustration and ineffective choices, because we cannot control our partners any more than we can the chance of meeting them.

The Cowboy and the Rattlesnake

In the early days of the United States, before the West was invaded by settlers and the Native driven from their homeland, there was a lone Cowboy traveling on his horse through the deserts of New Mexico where the Sun was hot and the air drier than a rolling tumbleweed. One especially arid afternoon the Cowboy realized his water supply was running low, and his beloved horse was tiring, her gait slower than usual and her breathing heavy. So the Cowboy stopped his horse beneath the shadowed overhang of a great cliff, bordered at the base by a few stalwart cottonwood trees whose leaves spun gently in a lazy, roasting breeze and the Cowboy dropped his saddle, sat down against the cool cliff face, and quickly drifted off to sleep.

Hours later at the set of evening the Cowboy was waken suddenly to find a large diamondback rattlesnake coiled on his lap, staring at him angrily. The Cowboy, frozen with fear, wondered how such a large serpent could have climbed on him so stealthily.

"I am here to kill you," said the snake.

"Why did you not first warn me?" asked the Cowboy, "as is custom for your kind? Is your rattle not used so that others be made mercifully aware of your presence?"

The snake glared at him, "I was asleep down there in the shade of the cliff and a great hoof of your horse came down and crushed my rattle, nearly killing me in the process," and the serpent produced his tail which was freshly wounded where once a rattle was attached.

"I'm very sorry," said the Cowboy. "Surely there is something I can do in exchange for my life and repair the harm I have caused?"

The snake thought for a moment. "My rattle was part of me, and I can never have it back, and will be vulnerable to bears and wolves and badgers or other careless travelers paying no attention where they trod. It is only fair that your life also be so cut short."

The Cowboy cursed himself for having been so careless, and having stopped at this unfortunate place and encountered the snake at all. But quickly a plan formed in his mind.

"I am on my way to Albuquerque to join a cattle drive," he said. "Let me go and I will secure for you the most exquisite rattle you could ever desire, one gilded in silver and turquoise made by the Navajo women who craft the most beautiful of trinkets found in this great, endless West. I will bring it to you on my way back and fasten it to your tail and you will be the most charming diamondback which has ever lived."

The viper was intrigued by this proposal, and enchanted by a vision of himself so appointed. "Very well," said the snake. "I'll let you go on the condition you correct this wrong, as promised."

The Cowboy swore he would be true to his word, and as night fell mounted his horse and set off toward Albuquerque, and for several months he worked a cattle operation, roping and branding calves and herding the animals to lush mountain pastures. When his contract was up the Cowboy packed his things, collected his pay, and began his return journey home, all the while bothered by his yet unfulfilled obligation to the snake. "Surely the creature is dead by now," thought the Cowboy. "If not, I can simply take a different path and avoid him altogether, to be sure, and so have no need to secure such an expensive and useless gift."

The Cowboy traveled day and night as the weather permitted, more cautious to take with him sufficient water from stream to stream so as to never again encounter a situation as he did before. He planned a path far afield where he met the rattlesnake, and though he was sure of being a great distance from it slept only briefly and hurried quicker than he might otherwise.

One night long since passing through that part of New Mexico the Cowboy, sleeping beneath the stars, awoke suddenly to find sitting upon his chest the very snake to whom he was beholden. "Did you get me the gilded rattle you promised?" it asked expectantly.

"Where did you come from?" asked the Cowboy in surprise.

The snake smiled triumphantly. "A flash flood struck several days after we parted ways and carried me down a long riverbed until finally I was able to pull myself from the water in this place here where I have lived and hunted ever since." The Cowboy frowned and shook his head in shame. "I did not get your promised rattle," he said.

"Though I did not expect to see you again," said the snake "I was moved by your protestations of restitution. But seeing now how you have deceived me are no longer privileged to mercy."

Before the Cowboy could react, the serpent opened its broad mouth and bit the frightened Cowboy, then slithered off him and paused under the sage brush at the edge of the campsite.

"So I am to die out here, alone?" said the Cowboy mournfully. "What shall become of my horse?"

"No," said the snake. "I bit you but did not inject my venom. A warning instead I leave you, to never again tempt fate nor spurn those you have wronged, for the next time it may truly require more than what you are willing to part, and your adversary less merciful."

With that the snake disappeared into the bushes, his rattle-less tail stumpy and scarred, and the Cowboy felt shame for his deeds and deception, and having believed himself more cunning than fate.

20. Pluto, Power and Control

While control and power are mentioned a lot throughout this book I still believe many people have a difficult time understanding just what power and control means in our personal experience and behavior, because ultimately so many of the choices and decisions we make even on a moment to moment basis are rooted in desire for control, but because we might not think of ourselves as power-hungry or control seeking we then do not recognize that we are indeed so, and even our everyday behaviors are wholly founded in refusal to let go. Indeed one of the primary goals of my book on psychology is to help people recognize how our experience of being parented as a child was most often nothing more than an exercise in instinctual human power and control dynamics, with our parents trying to wrest literal free-will and choice from our young, inexperienced minds, to literally force us to have the feelings they wanted and to make the choices that accommodated their needs and not ours. Because even little children are their own persons with their own dreams, goals, wants, and needs we then resisted such unreasonable and unkind treatment only to find that many of our own parents were themselves a danger to our wellbeing and survival, purposefully or not, and thereby adapted as best we could in order to survive them.

Yet we are no longer children, but grow up not recognizing that the behaviors we adopted to survive childhood are the very control mechanisms which interfere with our adulthood, often learned from our parents since those are the adult animals our young, unconscious minds looked to for lessons on what life is and what is required to survive it, and then every word that escapes our lips, every slouch of the shoulders, every hand gesture, every averted eye contact is a control behavior that is, consciously or unconsciously (mostly the latter) intended to choreograph our moment to moment existence every day of every week of every year for the entirety of our waking lives. But then even our efforts to not be controlling ends up also being controlling! Because the entire problem is we believe our behavior a product of our conscious mind and will, or lack of it, when in fact the problem lies in the unconscious human instinct and biology which we cannot consciously be aware.

It is no wonder then that we are so loathe to give up control and power, because so much of our life and being is built upon it, and as a truly helpless mortal animal our vulnerability and helplessness is a liability and our evolutionary biological survival psychology tricks us into believing we have far more control over life than is truly the case, especially because control behaviors can and do help us often to get the resources and behaviors we require to survive, but not realizing we can get even more when we are more *effective* and less controlling. Control also *deprives* us of the very things which make life fulfilling, often the very things we are afraid of losing, and while control can help us survive it can never help us *thrive*, and many of us might not be lucky enough to be compelled to learn new life skills required to transcend baser human instincts meant only so we don't sit in place and die from exposure or starvation.

Because control is such a basic human instinct we then also do not realize when control behaviors of others people directed towards us, even when they are so plain they could be labeled with some tape and a marker. For instance every time a person pouts in a relationship expecting our partner to take notice of our mood, to force our partner to prove they are in fact interested in us even though we already know they are because why else would they be there? Which is a primary and common manipulation and control behavior. Or when someone inconveniences us and we then presume to lecture them about their behavior and why they should have been better, because in fact we are opportunistically exploiting their mistake and vulnerability to assert ourselves over them. Sometimes control is even tritely displayed in the very setup and organization of many institutions, for instance that in religious settings those in power are *literally* elevated above the congregation to impress upon them not spiritual enlightenment or useful counsel but *authority,* which is incidentally the reason teachers are usually *below* their students in higher learning settings because teachers are *not* authority figures but instead educators and instructors, but by raising themselves literally above their congregations religions even put control into the very performance of their claims to power, and followers then readily accept that control and authority as the price of being near what power they desire to be near but which is also entirely a facade, hollow and duplicitous.

Of course no person wields such power, which is instead the domain of God and the Universe, and of course people will claim to speak for God and others willingly believe it because we want so badly to be liberated from our fear and misery. Yet because man has no power no man can actually liberate us from fear, so those who do belong to religion continue fearing, continue submitting, continue giving away their money for nothing more than the *illusion* of fulfillment even until they have been swindled from it entirely. Instead, fear and the course of things of which we are afraid are ruled by the natural world and the Master of that world, which is the Universe and the laws of causality to which we are each beholden, and thus relief is found *only* through the Universe and what we experience to be God, to which each of us has total, unfettered access being that we are directly made from nature and possess a living spirit which is a direct conduit to the very energy from which all life springs, needing not third-party charlatans which claim to speak for God but instead nothing more than to sit in quiet reflection, practice self-care behaviors (like inventory therapy), and meditate on those experiences that we may listen to the song of the Universe.

Concepts as power and control being part of nature and the laws of causality are thusly also reflected in astrological observation, and are embodied primarily by the eighth house of power, of which Pluto is the natural ruler, and both Pluto and the eighth house are traditionally described as the house of death, sex, and power when in reality death and sex are simply associated with this house because they too are arenas for power and control or the relenting of it. I am even reminded of times in my life where brief romantic interest lived for less time than a mayfly due to extremely petty control behaviors like being admonished for texting a person during work hours (without even being asked not to), as if it was my fault they don't know how texting works and took no responsibility for their own time and priorities like, you know, turning off your phone if you can't handle that kind of thing, because my god if you are that controlling about *texting* just imagine how awful a relationship with you would be? That kind of behavior, for instance, probably resulted from Pluto transiting the fifth house of creation which governs causal sex and love affairs, since they are impetus of creation of love and relationships, and very often progresses to control even of what sexual positions and acts are performed during sex, which is just a control mechanism antithetical to and meant to prevent real sexual intimacy which instead requires letting go of that control.

Death is the ultimate surrender (as discussed in the upcoming chapter on death), and one reason we all die is that we all at some point must completely surrender to the laws of reality, an act we cannot do but through death, in order to finally experience the entirety of what it means to be alive and mortal. In the act of sex we similarly surrender ourselves to another or, like many people, find the vulnerability of surrender too frightening and instead use it as an opportunity to assert control over others. Thus, Pluto and the eighth house are not really about things like death or sex at all but about *consequences,* those parts in life where we experience issues surrounding power and powerlessness, often and usually against our will, or the immutable nature of causality which cannot be transgressed without the meting of consequences from the Universe, Karma, and fate.

Shared resources are also a focus of the eighth house because when we share resources with someone else we do not have total control over it, but instead must relent some control required of the act of sharing. While this can and does affect shared finances with romantic partners it also does with business partners as well, or also taxes because those are finances we share with society. When I was younger I knew a woman who spent a great deal of money on psychics and was likely experiencing the transit of Neptune through her 8th house of power and control which Pluto rules, delusionally believing she could anticipate misfortune by paying charlatans to tell her what she wanted to hear and thus maintain control over her life, which is not only impossible but completely fine that we don't have such responsibility. I mean, can you imagine if avoiding misfortune really *was* within our power? I certainly wouldn't want that—our entire life would be nothing but a terrible game of Frogger!

Pluto transits through the houses and signs are also extremely long, lasting anywhere from twelve to thirty-one years (even longer for our Houses if they are far stretched by extreme angles of the Midheaven to the Ascendant), its shortest being in Scorpio (the sign it also rules, appropriately) and its longest being in Taurus, which is opposite of Scorpio. Pluto transits are so long it has only completed about half of a cycle through the zodiac since it was even discovered nearly 100 years ago! The reason for such disparity of transit time through the signs is that Pluto's orbit is highly *eccentric*, which means that an orbit is not equidistant from the Sun along its entire path, with one side of its orbit traveling much farther away from the Sun and thus spending far more time in the signs which cover that stretch.

Pluto's transits thus encompass those themes which seem to be quite profound and dominating in each of our lives, but which unfold slowly over years and decades, and as no one person will experience even half of Pluto's influences in their lifetime also accounts for much of the uniqueness to each life path unshared by others. In fact, because Pluto is the slowest moving planet (apart from the likely Planet X which will have an orbital period of something like *ten thousand* years) it is in fact responsible for the mistaken, negative associations we may assign to specific signs. For instance in the sixteen years preceding its last transit Pluto was in Gemini's zodiac eighth house of power and control, and the Geminis in my life were highly controlling, intransigent, and manipulative, and for a time I thought it was due to the nature of Geminis when in fact it was simply the influence of Pluto and its stimulus for control and power dynamics.

This problem of the Pluto eighth house transit is also made all the worse since the house preceding power is the house of relationships, and before that is the house of labors, so the one-two-three punch of thirty-years or more transit through the most controlling houses can, in those who refuse to relent control the Universe, create the appearance of a highly anti-social, disagreeable, and destructive personality instead of the real reason which is they are severely challenged, temporarily (albeit a long temporary), by the Universe, and any person might do the same when having the same transits. This of course depends on the temperament of individuals, and I later made the acquaintance of other Gemini with whom I worked who strove very hard toward self-improvement and tried not to exert inappropriate or unkind control over others, demonstrating an abundance of magnanimity absent of the several Geminis I knew personally, and simply otherwise didn't want to do inventory, as the influence directs themes of power and control toward us and it is up to us how we choose to respond and handle them (informed in turn by our degree of unresolved trauma). Another Gemini I knew during this transit had contracted HIV many years previous and spent most of this transit longing to be relieved of the virus and spent many years searching for a cure, even traveling abroad to experiment with untested and experimental chemo-therapy, but ironically refused to read my work even though they were supposedly my friend, which discusses HIV and how to treat it (not cure, but treat, although it might help cure), because for many people accepting help is one of the most humiliating and submissive acts of relenting power, which those during a Pluto transit often find themselves, sadly, refusing to do.

Although we all bear responsibility for our actions even during the most stressful of situations and transits not one sign in the Zodiac is good or bad and instead the negative characteristics we often see in certain signs or that which are described by biased astrologers and consumers of astrology are instead triggered by the long-term transits of the slower planets simply because of the sheer length of time each spends in a sign, and we risk causing harm to others and being unhelpful if we let personal experiences and bias influence our characterization of signs and personalities, which is itself also ironically a control mechanism (even I will have unconscious bias in this book). One astrologer I really like, for instance, clearly had a negative experience with a Venus fire sign as their description of this placement fully criticizes those with it when in fact having a fire sign placement simply means a desire for passion and individuality within relationships, something some signs find appealing while others do not.

As with all transits, Pluto's journey through the natal zodiac houses concerns life events that happen to us, where its transit through the ascendant natal houses affects life goals and actions we choose to do or pursue. As mentioned earlier, at the same time Pluto was transiting Capricorn, which is Libra's zodiac fourth house of home and family, it was also transiting my natal third ascendant house which is the house of knowledge and communication such as acquiring information, using it, and sharing it as I did in my writing and publication of *Fuck Portion Control, The Perfect Child,* and now this book, *Under a Libra God,* which were all written during this transit which for me lasted from 2009 to 2026. Because of the nature of initiative such as what the ascendant Houses are concerned Pluto does not have an especially controlling effect there since initiative such as is required in these houses is already a form of power in which we are in fact required to take action, and Pluto here in turn helps us apply our power to those things which which actually can control, such as our choices and decisions and what we choose to do with our time and talents, for instance we can write a book but whether they are successful or find an audience is not up to us in the slightest. If our goal for success is defined by having readers and making money we are very likely doomed to failure since we cannot actually control such things, but if our terms are simply to write a book then we are the only factor which decides success or failure and in that case can easily find massive success every time we are productive. This dichotomy and irony surrounding success is also a major reason why so many creative industries like film, video games, restaurants, etc., often find failure, because their goal is most often to make money and win patronage which is not something anyone can actually will into being but we can, however, make our best film, our best video game, our best book, or our best food, which then has a greater chance of success because that is where all our energy, focus, and effort went instead of into things we can't actually control (which by the way is not at all served by delusion, and doing inventory to relieve ourselves of the liability of delusion which actually interferes with quality and creativity also then results in greater output). This control behavior can even be seen in popular media such as in Gordon Ramsey's restaurant shows in which people who are fixated on things over which they are powerless like attendance and patronage entirely freeze in their trauma because they know, instinctually, they can't control those things and thus feel rightly helpless, but all the while ignoring the things they actually *can* control like the quality of their food or the cleanliness of their establishment. Such are we also debilitated when attempting to wield the power of reality, which is not even our responsibility anyway, made worse by ignoring the things we actually do control, which the transits of Pluto help to elucidate.

Yet also it is usually lost on most people the solution to correcting our behavior and becoming more effective and letting go of control (which Gordon Ramsey accomplishes by yelling at them), because we still attempt to control our feelings and behavior and will ourselves to have a different attitude or make ourselves sit and be productive, but we cannot control even our emotions, which are hormones, nor our biology which is entirely beholden to nutrition, the environment, and susceptible to pathogens, toxicity, contamination, etc. The specific solution to these conflicts is in acts of compassion for ourselves like sitting down to do inventory, taking on less responsibility to make room for self-care behaviors like getting enough sleep, eating as healthy a diet as we can, and accommodating needs such as daily, generous Sun exposure. Even growing our own food to provide an increase in nutrition might be required, which even still is beholden to laws of nature such as from pests, drought, and plant diseases which likewise requires submission to those realities of nature such as through acquiring knowledge of farming and practice of regenerative agriculture in order to balance natural systems and support the soil microbiome. Essentially, rather than forcing ourselves to behave a certain way the practice of inventory, to compassionately resolve our trauma, fears, and insecurity, accomplishes that change because the fundamental conflict is will versus compassion, and those with the greatest behavioral problems have the least self-compassion (which you will remember is action, not feelings).

Pluto transits also do not lose *any* of their potency at the beginning or end of their transits even though they last for *decades,* and the effects start or end within a week of the transit beginning or ending. Not two weeks before Pluto exited Capricorn the first time (it retrograded back briefly) the literal brick facade also fell off our building and smashed into the metal awning outside my window at 2 a.m., and was so loud I was sure there would be a body laying there when I opened the curtains. In the literal last week of Pluto's first exit from Libra's fourth zodiac house my maternal grandmother died and my mother used it as an opportunity to invite me to dinner and attend the funeral with them even though I had long ago requested they stop communicating with me. But that side of my family is even more homophobic than the paternal side, and my mother had never even attempted to repair the damage of her behavior on our relationship as is characteristic of this transit, and I still continued to assert control by refusing to be part of any of it. I often talk like I know everything, but I honestly could not see any better way to handle this than avoiding them, even though it further alienated me from my family, as if there was a better choice to make I was absolutely done with anything and anyone homophobic and bigoted regardless if they were family or not, because that behavior corrupts the relationship anyway as they treated me with contempt, derision, and neglect, and primarily maintained the relationship from their side not because they wanted a relationship with me but because they didn't want to be seen as unkind, which is one of the most unkind things a person can do, and the only difference between having a relationship with them or not was a matter of several phone calls a year, many unreturned calls and texts, indoctrination of my nieces and nephews to similarly resent me, and patronizing, shallow interactions at Christmas, as all relationships require the participation of both sides. So, I don't know entirely how to handle those most difficult transits except to just do your best and try not to manipulate anyone (but setting necessary healthy boundaries is always fine, just try not to set only boundaries or get in your own way).

Pluto exited Capricorn for most of 2024 and I was immediately relieved of the constant noise and disruption that usually occurred from neighbors given to domestic violence and unruly homeless passersby on the street outside. I also secured a new apartment at a better rate and outdoor access and didn't hear from any of my family members, but then it transited back for just two months and sure enough I received yet more inconsiderate and controlling communication from my family after the 2024 election against my express wishes, as they too had Pluto retrograde back to their respective houses of control, relationships, and communication and literally told them to go fuck themselves. The city even early one morning outside my window began cutting down all the trees out front of our building, and my neighbor with whom I shared a wall apparently got a new boyfriend with whom they got high every night staying up till three in the morning laughing and fucking as if all our apartments weren't only ten feet wide with marginal soundproofing. At the very last week of the transit they finally seemed to become sleep deprived and spent of their fresh excitement and started settling into a more normal routine and being quiet, and after Pluto left for good I was finally relieved of the regular disruptions both of my place of residence and family dynamics entirely, even while there is an active construction site on the other side of our building (which is no doubt disturbing some Scorpios now with this transit over there) and noisy neighbors at the other end of our hall but far enough that it never wakes me or bothers me.

The effect of Pluto transiting our natal charts is the underlying mechanism for the experiences we have in our daily lives and interactions with others both individually and with the larger society. For instance, when we have Pluto transiting our house of partnerships and thus experience control and power issues from our partnerships (including business) those persons will in turn have Pluto (or Neptune or other consequential planet) in complimentary houses like their fourth zodiac house of home and family which in turn give them Pluto impulses that come from us. When I had Pluto transiting my fourth house various family members in turn had Pluto or Neptune in their first, third, sixth, seventh, and eighth houses, controlling identity, communication and siblings, health, relationships, and power, respectively. This is one reason why understanding how astrology works can be so productive because it allows us to in turn better understand and empathize with why and how things happen and people behave the way they do which can in turn make us more effective in our own behaviors and choices concerning them. We can never resist transits or the laws and influence of the Universe, but what we can do is act more *effectively*, to live in harmony with those laws and the nature of life and reality and thus become more effective in our actions, rather than acting in opposition to those laws and conditions which then makes us *ineffective*.

Because Pluto transits are so monumental, the end of Pluto transits through zodiac houses always also coincide with a glut of major world events and turmoil since human society is run by individuals whom are themselves subject to the same laws of reality as everyone else. Pluto's transit through Capricorn was the United States's seventh house of *relationships,* reflected in my country's support of others committing genocide, completely abdicating moral leadership (not that we ever had any) in the service of political relationships, even sabotaging opportunity for negotiation in the Russia and Ukraine conflict by the then Democrat President of the United States from a desire to reinforce NATO and control its member States while enriching the military-industrial complex as a lazy, destructive shortcut to sustaining the American economy which resulted in the continued deaths of hundreds of thousands of Ukrainian and Russian soldiers, the latter conscripted largely from minority populations to also genocide them in the process.

Genocidal world leaders at this time were also Libras and thus also had Pluto transiting their zodiac fourth house of home and family, so because they are also their countries' leaders viewed their country and citizens as their literal home and family which, trying desperately to control, then committed heinous crimes of racist, nationalistic, and genocidal intent, to the peril not only of their victims but the wellbeing, integrity, and stability of their own citizens since such behavior naturally interrupts economies, trade, productivity, and causes the threat of violent retaliation and recrimination. Generally it is thus a consistent rule that leaders with nationalistic abuses have Pluto transiting their fourth house of home and family either when they start or continue doing violent nationalism and xenophobia, because they view the country and nation as home which they try to control by determining literally who is allowed and who is not, with minority groups usually the convenient vehicle for their ambitions. Since there is in fact no such thing as race this behavior is wholly arbitrary to the whims and neuroses of those leaders and why they always end up also targeting their own people in supposedly privileged groups, imperiling even those whom supported them because there is no difference at all between peoples and anyone can be a target for their psychotic, hateful worldview since the point is not in fact 'racial purity,' which is simply a facade of excuse used to enable their personal ambitions for control over their 'home.' This was even true of Nazi Germany, whose world-famous, despotic, genocidal leader caused the need for a new word do describe what he did, *genocide*, but whom was hardly the first to do it, whom during his control of Germany likewise had Pluto transiting his fourth zodiac house of home and family.

While it is not predestined that world leaders with such transits will be heinous, their control impulses nonetheless become focused by wherever Pluto is transiting their birth chart, and thus it can be understood what their real motivations and interests are, such as in having power for the sake of it, or power over public perception and institutions, or their communication, of economies and finance, of the homeland, etc. The astrology of countries themselves is not always certain, however, as their official formation dates may not be the actual astrological point of 'birth.' Rather than the astrology of countries, which is not always accurate and since we cannot really do much about them anyway (God and the Universe do that), the astrology of their leaders is more relevant to our comprehensive ability and focus, and at any one time a great portion of politicians will always be people with Pluto transiting their gauntlet of the sixth, seventh, and eighth houses which inspires asserting control in the first place such as through institutions like government by entering politics and is a reason why most politicians and leaders are so often disappointing failures, as their motivations are not the wellbeing of the country but fulfillment of their personal desires for control. Because Pluto's transition to Aquarius means the sign of Scorpio is the next to experience its transit through the zodiac fourth house, the next world-stage campaigns of ethnonationalism and genocide will generally come from leaders with their Sun in Scorpio—in fact the previous President of the US was a Scorpio and we glimpsed this heinous behavior in support of the Israeli genocide of Palestinians in the early transition of Pluto to Aquarius, and had his health and poor performance in the office not intervened in his ability to keep it would have accelerated his authoritarianism in a second term, and so on and so on with the transition period of Pluto from one sign to the next being the most volatile of years as the consequences of ours and others' actions during its transits come due.

None of this is to mean that we must like trying circumstances or endure pain and loss with a smile, because even trying to pretend we are okay when things are very much not okay is itself a form of control which resents human emotions and our vulnerability to laws of reality. Yet while Pluto transits are long and often involve very upsetting and life altering events, Pluto is really very reasonable and straightforward and does not play games or act in secrecy or vagary, and the parameters of any Pluto transit are very plain to see and understand if we only know to look for them, and the point of Pluto transits are not to make us suffer but simply to demonstrate that the Universe, not ourselves, is the power in charge and that there are immutable rules for existence. Indeed Pluto transits can be *exceptionally* easy if we simply refrain from trying to control and let the Universe instead do its job, and submitting to Pluto's expectations and living within the real bounds of our mortality concerned only with our actual responsibilities can be productive, fulfilled, happy, and see much progress in our life path, especially since the Universe will take care of everything we are not in charge of anyway.

This is not to say that we surrender and give up, because giving up is in fact a form of control, and because we as humans loathe more than anything to relent control it can be difficult to truly acknowledge those limitations of a Pluto transit for those of us who have not resolved our control mechanisms and childhood trauma. To be clear, setting healthy boundaries with others is neither a form of control nor giving up because the act of control is oppositely the attempt to manipulate or force others into doing what we want, where setting healthy boundaries or even ending harmful relationships is an act of self-care (and sometimes a necessary one). For instance if I had continued to retain a relationship with my family but characterized them as bad people for having their beliefs that would be an act of control and manipulation. Many of us do this very behavior to our partners and lovers, telling them how disappointing they are to us or how they could better live up to our expectations of them, when if they really are such a disappointment you should not be dating them in the first place.

Similarly, control when Pluto transits our house of labors and affects our works takes the form of doing chemotherapy and fasting and dieting while having cancer, trying to force the body to be well and separate ourselves from the laws of cause and consequence, which is not possible. Advising a childhood friend diagnosed with stage three breast cancer at the outset of Pluto's transit through their sixth house of labors I assuaged them of their fear and sense of helplessness by educating them to such matters of spirituality and astrology, to show that both getting cancer and recovering from it is up to the Universe, and our only responsibility is in daily care and nurture of the body—*not* delusionally assuming we are doing so, but informed by enlightened knowledge (education) and self-honesty. People do in fact die during harsh Pluto transits as one unfortunate woman with cancer recently, due to our refusal to care for ourselves and do what is necessary to be healthy, mistaking our responsibilities for such impossible things as preventing death, made worse by that indoctrination from religion which tells us such lies, as none of us can do such things, but which are instead far simpler acts we are actually capable like simply staying fed, getting daily sunshine, supporting the microbiome, eating wholesome food and learning about nutrition and its various complexities such as in reading *Fuck Portion Control*.

Since none of us is capable of subverting the laws of the Universe anyway there is no reason anyway to challenge Pluto transits, and much peaceable harmony and satisfying experiences can and do occur during Pluto transits if we but simply submit to life and stop trying to control everyone and everything in service of our fear and insecurity. But, even doing inventory therapy so we have more control over life is still a control mechanism! And during the previous Pluto transit I worked with an Aries for many years who found inventory frustrating because, it turned out, their motivation for doing it was this very desire for more control as a product also of how I constantly spoke of the ironic increase in control which comes from doing inventory, but which comes as a *side effect* in redirecting our control mechanisms toward only the things we can control, such as our actions and behavior, which is very different than controlling how people feel about us, our waist size, or the timespan of our mortality. Finally making this realization at the end of Pluto's transit through their zodiac tenth house relieved them of that stress and were thereafter able to practice inventory with much enthusiasm and progress, finally being relieved of the responsibility for things we simply do not.

Likewise, Pluto has also been approaching the Nadir of my own chart (as if Saturn there wasn't enough), which concerns the inner life, spirituality, and home and family in the function of the great angles. For more than a year I anticipated the approaching aspect with some sense of dread, having read anecdotal experiences online of awful things like losing a parent or being involved in a very public feud with another celebrity (that's a joke—almost nobody knows who I am at the time of this writing). As the alignment got closer and closer and nothing seemed to happen I began to relax and believe maybe nothing awful would happen and I could be spared the awful effect typical of this influence, until one day realizing the entire absence of any relationships in my life and production of my work such as this book being how I spend my nights and weekends instead of socializing and romancing *was* the very transit of Pluto across my Nadir already underway, Pluto being the planet of control means I do not have the power to set the course of my inner life as I might prefer and even when I tried recently dipping a toe back in the dating pool was instead roundly ignored, obligated instead through circumstance to do this work and accomplish my life's apparent purpose, where a life and romance might otherwise distract me from the rumination which supplies these pages with content because, believe you me, while I am grateful for the chance to do this work and to help others if there was a choice I would absolutely not be doing it.

But all this grandiose language can mislead you, the reader, into thinking that all Pluto transits are equally grand and important, when in fact it is simple, common behaviors like simply being *defensive* which primarily drives conflicts of control and power. I say this as someone formerly very defensive that defensive people are the most fucking annoying of any kind of person, like, just shut the fuck up for a moment, you're not important and if you don't like what someone is saying you can just ignore them or don't take their advice. Why do you need to dominate relationships and force people into believing you're smart, nice, competent, not fat, very good at something, or not an asshole? Because I promise you they think worse when you're done defending yourself, going over and over the details of any argument ad nauseam like a weird little psycho with the most fragile of all egos and repulsive in your seriousness. The truth is that defensiveness *is* a control behavior meant to assert ourselves over others and is the very reason others dislike us for being defensive, regardless if we are right or not, because we're fucking annoying and nobody wants to be controlled or to have their interactions or participation in a relationship exploited so that we can feel domination over our environment and the people in it. Of course, defensiveness is one of the primary human coping instincts in response to unresolved trauma, in which we feel our person or status in relationships or society is under threat. Once when in the recovery programs I sponsored a person who, every single fucking time I tried to help them achieve recovery, vomited back at me all the reasons and excuses and justifications they could possibly think of to the point that I had to eventually fire them (I was also moving, though) and during which I gave one last bit of advice to shut the fuck up when someone is trying to help you—you can decide later if you want to use their advice or not, but you don't get to dominate every interaction every time and expect to retain that relationship (especially when it was you who asked for help). As was the case in this relationship, as it is in all others, behaving defensively *undermines* relationships and threatens structures and resources which would otherwise benefit us. Usually when I coach people they are more than willing to hear me out and, like normal people, afterward pick and choose what advice they want to integrate into their lives, but occasionally some asshole will come along and treat everything I tell them as a personal attack on their very character instead of my own personal insight into life and biology, one time so unwilling to refrain from trying to control the relationship between us I also had to end coaching with them and they in turn lost access to my knowledge and experience. As I mention in *The Perfect Child*, defensiveness is a sign of requiring inventory practice, otherwise we risk severe self-sabotage behaviors which are at the root of all of our life's problems, as goes the saying from twelve-step programs, '*our problems were of our own making.*'

One of the greatest strategies for surviving Pluto transits, besides fully completing inventory practice, is to just stop taking everything so fucking serious-ly. I mean, goddamn chill the fuck out and relax. Life is not that bad! Control is borne of fear and control is channeled through seriousness which is nothing more than a control mechanism, where levity and humility are instead the antidote to fear and seriousness, and one of the greatest effects of Pluto transits is in fact to gain a sense of awe and reverence for reality and the forces over which we have no control, to experience the true depths of life and existence and how truly remark-able it is that we are here in spite of the potential chaos and destruction inherent of reality which order and control otherwise restrain and regulate, to realize that almost nothing in life is our responsibility (only to show up for opportunity and do our best), to then ironically take ourselves and our lives with far more joy and appreciation and let Pluto and the Universe be in charge of those things which we are not. Pluto transits are not anything to be fearful of, but instead fear unwilling-ness to be humble in the face of life and its many wonders.

Unlike Saturn, which leaves a 'gift' at the end of each transit, Pluto's gift is to take from us those burdens we otherwise misunderstand as being ours, a lesson in empowerment to then be more effective in our life if we listen and heed its teachings and give up control of things *we don't control in the first place*, for instance it is not my responsibility to make my family (or anyone) love me, and we can find rewards at any time during its transits when we do relent, which will then allow us to progress and learn valuable new skills like patience, empathy, perseverance, awe, kindness, reverence, levity, confidence, etc. When Pluto entered my fifth house of creation immediately after the fourth house of home and family I began working furiously on many of the various art and video games I had started previ-ously and began producing very fun and satisfying work that made me both proud of my talents and entertained by the results (maybe one day you will get to play them?). As I also write this book in the beginning of that transit which will last twenty years there can be events which sill are unpleasant, since the fifth house rules other acts of creation like the making of children or the beginnings of sexual relationships, so I have no choice but to meet lovers under this transit and will be required not to control them. What Pluto brings to our life is a better under-standing of what life really is, not what we delude ourselves into thinking it is, and with this brings real growth and experience which will make us wiser, humbler, and ironically more effective in the long term as we are empowered with facts and information and no longer compromised by delusion, fear, and insecurity.

Ultimately, Pluto transits are about power and how our desire for power causes us pain and loss, and Pluto is unforgiving in its lessons should we fail to listen—The more we resist its influence the more it will insist, like a Chinese finger-trap, and the consequences can indeed be severe and extreme, even leading to imprisonment, homelessness, the loss of relationships, money, or even life itself. While that sounds ominous it is not at all fated, and such consequences of this transit are in direct proportion to the degree of our obstinance. As such, when Pluto transits the first house of self our sense of self, identity, ego, and purpose is challenged by circumstances and obligations. Plans we have made for our lives, based on who we think we are, will be upended by outside factors and redirected to paths we never even considered or thought appropriate for who we are. Often-times this transit involves a sense of feeling failed by our past sense of self, those pillars of ideology, purpose, and identity on which the self is built, to which many try to control by increasing our dependence and insistence on that identity, when in reality our sense of self must come from within and not from external forces such as institutions, groups, ideology, money, beliefs, etc. Being deeply disappointed by an inability to follow through on our dreams, because we are blocked by external circumstances or obligations, is a common theme of this transit, but that only occurs because we think that who we are is dependent on what we do, and since we cannot do or have as we wish in certain parts of our lives are challenged to find a different foundation on which to rebuild that identity.

At the outset of writing this book Capricorns had Pluto in their first house of self since 2008 and thus all experienced a crisis of who they are and what life means to them. Some Capricorns I knew during this time notably withdrew from family, relationships, and ambitions, and took shelter in obligations and mediocrity and they seemed unable to grow as people and individuals and relied more heavily on existing structures, even when doing so proved perilous to their wellbeing, relationships, and fulfillment. Capricorn is a more reserved and sometimes fearing sign who often naturally doubt their own sense of self (due to inherent reliance on institutions), so the transit of Pluto through their first house of self must have been especially challenging to middle-aged Capricorns who assumed they already had an established identity. No matter our sign, however, Pluto's transit through the first house will always disturb our sense of self, and other Capricorns which persevered in spite of their fear eventually found a security of purpose which this transit can bring if we relent control, for instance two very charismatic, young actors, Timothee Chalamet and Florence Pugh are both Capricorns who saw great success in the film industry during this transit and discovered their sense of identity when they were not yet famous which, because they were yet young, likely felt more natural than those who were more aged, and became famous *because* of their newfound sense of self and acceptance and *not* the other way around as might be assumed, as they were both only thirteen-years-old when this transit began. While hopes and dreams are a fuel for life they are simply goals and aspirations and do not actually define who we are, however, and identity and purpose must come *in spite* of them. Accepting these lessons from Pluto during a first house transit will result in a more effective and confident personal identity that is not based on superficial or shakable precepts, while refusing to accept these lessons will result in a weaker and more spurious personality less able to endure life's challenges and thus less effective in our lives in general and the upcoming second house transit. Fear of our own incompetence during this time can lead directly to highly self-destructive behavior such as committing sexual harassment or assault, refuge in alcoholism and drug addiction, or delusions of grandeur which later lead to ruin and remorse, and it is important to remember that *it is just fine to be incompetent*! You do not need to be a success, make people love you, or get rich—like both Chalamet and Pugh all you have to do is show up for opportunity and do your best!

Pluto transiting through the second house of means denies fulfillment of control toward money and other resources, especially in terms of *how* we earn them. Most recently it was the sign of Sagittarius which had the transit of Pluto through the second zodiac house for the sixteen years of Pluto in Capricorn, while

Capricorns will be the next to then experience this transit (and Aquarius after, and so on). One Sagittarius I know personally was quite constrained in their employment opportunities while establishing themselves as a young husband and father and, while excellent in those capacities, was never able to find a job that could provide much more than food and shelter for his growing family. Many astrologers present a simplistic view of this house as concerning only money because they do not understand that it actually encompasses means and *how* we actually earn money and not only money itself. Famously wealthy Sagittarius, Taylor Swift, could hardly be described as financially stressed during this transit, but instead endured the takeover of publishing rights to her work and an ensuing battle for the rights to own her music, which she never won until literally months after Pluto finally left that house but during which she re-recorded those albums under a new label. Like Chalamet and Pugh, Taylor experienced the Pluto transit of her first house prior to this where she worked to establish herself as a music artist, subjected to the themes of Pluto transiting the first house of identity, which experiences then inform the second house of means that will always concern control over the resources we build, which are not limited only to finances. During this transit I also worked with a Sagittarius experiencing many health problems and since health is a source of energy and is inherently a resource this transit can even encompass energy as means and thus themes of control surrounding that which seems even immaterial. Like all Pluto transits the point is not to give up and let life steamroll over us, but in the methods by which we try to assert ourselves which, if they are controlling and manipulative will end in destructive consequences, but if they are honorable, reasonable, and demonstrate integrity such as exemplified by Swift they can be very productive.

Since the third house deals in knowledge—specifically the acquisition, application, and disbursement of knowledge in acts of communication, a Pluto transit through the third house often presents as conditions which require knowledge, such as failing health or studies in school, and this transit occurred for Libras as myself in its transit through Sagittarius from 1995-2008, when my father bought our family our first computer when I was 14 which would lead to my becoming an avid and talented graphic artist through self-teaching by reading instructional manuals and then employing that knowledge in practice until eventually becoming a highly paid and sought after industry professional by the end. My two attempts at a formal education and application to the University of Utah at the ages of 19 and 22 were accepted, but due to a serious panic disorder never attended, which was an example of my own control mechanisms during this transit trying to control the manner and way in which I experienced the attaining of knowledge and its practice, but would have found a much more productive and satisfying career if I had instead studied film and writing as I had planned and been empowered to resolve the intense fears, anxiety, depression, and trauma which motivated these control mechanisms. For those who are older during a third house transit this could be even more consequential such as an inability to progress in our career due to absence of skills and knowledge we must first acquire, and resisting that requirement due to fear and insecurity can limit our progress, or it could be simply that work and effort are required to obtain knowledge and produce something as a product of that knowledge such as I have been doing during Pluto's transit of my third ascendent house, where Pluto requires that we first acquire information before we are able to use it. Or as was the case recently for the sign of Scorpio might find that our efforts to communicate to people, the public, government, etc., are not effective since we cannot *control* aspects of communication as we might try to do. One Scorpio I knew during this time actually yelled at me when I tried to help them with their health problems after they had been trauma dumping on me for months, refusing even to listen to me but also demanding I serve as an outlet to relieve their stress, insecurity, and anxiety. Because the third house encompasses all forms of communication this transit could deal with any form of learning and knowledge such as languages, professional skills, publishing and media, dialogue with people we know, etc., the primary point of which is to demonstrate that learning is an absolute requirement for progress and we do not get to pick and choose what knowledge we require.

As already abundantly described, the fourth house transit of Pluto is especially upsetting and difficult because it deals with family and our roots, which also happens to be a source of identity and purpose since it is from there whence we have come, and of course most of us love our families and are heartbroken by conflict and loss of those who are closest to us. This loss is hardly ever actual deaths, although that can happen such as it did with one fellow Libra friend of mine when his family lost their mother to cancer, but is more often the consequences of family dynamics and shared trauma as well as the actual place of domicile such as a house or apartment. Unfortunately because all relationships require both sides to participate, this transit is especially consequential because we essentially cannot control those whom are required to have a family and are absolutely and entirely dependent on their behaviors and choices in terms of whether those relationships function or even exist. At the outset of this transit I was for the first time in my life making significant income and tried even to buy the love of my family by gifting thoughtful and expensive presents every holiday and birthdays, but even these were regarded with contempt by people raised to hate and revile me as a person by our parents and religious institution, since they did not value me as a person did not value what I had to give.

This transit's effect on houses, apartments, and any other literal place of residence can occur in all terms such as environment, function, neighbors, landlords, etc. Right at the outset of this transit toward the end of 2008 before I ever knew much at all about astrology I excitedly moved from my stable, nice apartment in which I had lived for nearly four years in a misguided attempt at finding a social life in the city of West Hollywood. Little more than year later I met a new boyfriend and moved again to live with him, my first live-in relationship. About a year later we also relocated to Palm Springs, and a year after that moved into a home we had been remodeling for income. Our life together ended two years after and I found myself homeless and couch-surfing until finally securing a sublet and then months after my own place once again. A year after that my landlord evilly harassed me so I would vacate the apartment so he could get a higher rate in the rising market which precipitated the coming housing crisis. In the next apartment the manager lived next door to my unit and would go in and out of his apartment at all hours of the night slamming the giant, solid, mid-century oak door each time which shook paper-thin units as if an earthquake had occurred. He refused to stop when asked (well, demanded really, as I was still in control mode) and started harassing me and one day literally belittled me for working from home as he strangely resented anyone being around in their apartments during the day and even refused tenant applications of people who did not have an on-site job (I quit mine during this time to work on my books). So I move again and the apartment after that there was a merciful reprieve for about four months of blissful peace but then the neighbor who finally moved in next door which had NO sound proofing between our units took phone calls from two to five a.m. on the other side of my one room (not one bedroom, one *room*) apartment and I know there was no soundproofing after once night shouting and banging at her to be quiet put my hand through the drywall. After that I went back to Utah in a misguided attempt to rehabilitate my relationship with my family but whom immediately used the opportunity to try and control me and exert their bigoted beliefs and fight my work and writing, using my own nieces and nephews as ammunition in their chosen war and my in-laws demanding my siblings end their relationship with me because of their bigoted religious beliefs, which was the most disturbing thing that had happened to me all this long transit (but which also made me understand I had to write my book on child abuse). After that I moved back to the city with no family relationships surviving found during even the very final retrograde of Pluto from the fifth house back into my forth house my neighbor having the most spirited domestic abuse, the landlord surprising new, monthly charges not agreed to in the lease, and even the literal facade of the building falling off onto the awning outside my window. Really, the amount of literal disruptions to my family relationships and literal places of domicile over sixteen years was *uncanny*, and was no small reason for the conviction to write this book on the science of spirituality.

During this transit while in the city looking for yet another apartment in order to save money I finally learned in the last few weeks of Pluto's transit through the fourth house how not to exercise control over my life and just sat down with the apartment manager and asked for her help to lower my rent, which she kindly did, instead of trying to shame, lecture, or guilt the landlord into doing what I wanted such as in the apartment where I put my hand through the wall when it was discovered the landlord had been billing me for electricity of other units and I asked how he could possibly be the head of a bank with such poor math skills (lol wups).

In the midst of my last apartment search I even found for rent the very apartment in which I had attempted suicide—It had been completely remodeled, with brand new cabinets, a kitchen island, and all new appliances. I remembered my desperation to be loved and belong during my tenure there, and my little dog Angus whom I could not effectively care for, and was filled both with a great sorrow and regret for my many mistakes and failings, but also a profound sense of reverence for my mortal condition and the opportunities God has afforded me in healing and progress.

Not all Libras moved so often during this transit, though, and I'm sure my experience is an outlier in its severity due to my other many fateful transits, as each life is a product of all the different astrological influences which combine to construct any one moment of every life. Another Libra I know only moved four times during this transit, but each time to an entirely different state, and they also held onto a home in Hawaii after moving away from there and that property became more of a boondoggle than it was worth (in my opinion) and, although not to the degree I did they also struggled a great deal with their relationship to their immediate family. At the outset of this transit my Libra grandmother, in her eighties, was also duplicitously conned into marriage by her brother in law (her sister was long departed) who siphoned her life savings and social security and later divorced her when his theft was discovered, but was ruthlessly defended by his own children, her nieces and nephews, while my Grandmother and her children tried to recover the stolen money which also forced my Grandmother to relocate residences several times even in her old age.

One of the major lessons of this transit especially is that life is *inherently* transient, and that while it is wonderful to have a home and family and place to live and identify nothing in life is actually permanent, and we must know who we are and learn how to live even when we have no foundation to call home or others to call loved. While many of my losses were painful and seemingly senseless every one was also with a person or persons who were in some manner abusive, disloyal, controlling, resentful, or vindictive, and I came to learn how to enjoy my own company and what makes me interesting and valuable as an individual, doing my own thing, taking care of myself, and doing for myself what I used to think only came others, especially from parents, and at one point in this transit realized that I could now, as an adult, *be my own parent* and do for myself the things otherwise expected of parents since mine simply did not possess the skills and ability to do it. As the fourth house is in a square aspect to the first house this reflects the general tension and difficulty of this transit, which can take many forms but always will relate to the home and family, both literal place of dwelling and the people to whom we are bonded, including spouses and partners.

This transit especially touches on childhood trauma for which the inventory therapy is required, as the trauma from transits of the fourth house and learning the difference of what we can control (ourselves and our choices) and what we cannot control (other people especially) is something that children are not normally equipped. When this transit occurs when we are children, such as occurred to both my Gemini mother from ages 2 to 18, and my Leo sister from birth to the age of 12, it produces *profound* trauma since we are so utterly dependent as children on our family, and knowing what my mother must have gone through as a child and seeing her childhood pictures fills me with such profound sense of sorrow for what children like her have hd to experience in life and what might otherwise have been if each generation would take care of their own trauma to stop cycles of trauma and abuse as discussed in my book on psychology. Several of my nieces and nephews are Libras and Scorpios and no doubt will now have similar traumas due to this transit, and my sense of powerlessness in being unable to help them (except for writing these books which one day they may find and use) underscores even more how much we are not in charge of reality and are instead beholden to the mysteries of existence. Yet while we are young often think such trauma is unique to us because of our limited experience, but many of us share these same transits and experiences, and already being possessed of a personal identity its trauma was for me less formative and more salutary (in the end anyway), and what transits traumatize us as children might instead liberate us also from that trauma as adults if we are paying attention, and inventory can help us resolve those experiences and come to terms with the reality of life, to replace fear, insecurity, and pain with confidence, joy, and peace.

Pluto's transit through the fifth house can be far more fun and satisfying than others, aspecting the first house at a sextile rather than a difficult square as the previous. The act of creation such as is represented by the fifth house is inherently one of letting go of some power anyway, loosing into the world art, writing, love, and even children, all things which by their nature must be let go so they may exist outside ourselves. But Pluto will still require lessons in powerlessness and so those who try to control by force their creative visions, love affairs, or the creation of children (or the children themselves) during this period will still be rudely confounded. Many people achieve great creative feats during this transit but more so when not trying to control outcomes, and as the transit progresses will accumulate both failures and successes but never want for creative opportunities so long as we allow processes of creation to happen as they will and not so much for how we feel they should go. During this transit for my mother she got to experience some acting as a young model in Los Angeles, even landing a spot as one of the original Barker's Beauties on The Price is Right. But she also experienced a miscarriage after getting married, due to endometriosis, but then had four children before the transit ended including a set of twins immediately after I was born and was absolutely swamped with children. My Cancer father whose transits always follow those for my mother because of sign succession then experienced this transit as a boon in creativity and creation for his architecture and building as well as starting his own restaurant, but also challenges trying to control his children until I was fifteen years of age. At the writing of this book Pluto was ending its transit of this sign for Virgos and one famous Virgo streamer and content creator, Jerma, was enjoying the culmination of this transit and his previous sixteen years rising to the top of the first generation of creators as a passionate, fun, hilarious, creative, and inspiring influence whose work entertained and comforted millions before, during, and following the coronavirus pandemic due to his whimsical creations and enthusiasm for life and creativity. Jerma was notoriously affable and many of his projects were done in close collaboration with other creators and even leveraged the assistance of production crews, and while the larger projects took a lot of planning and resources the nature of his ideas were not so rigid and scrutable as to impair the creative process but could easily absorb or even benefit from things going awry, where someone who otherwise wielded unyielding control would, by their own behavior, likely sabotage such opportunities. At the outset of this writing Pluto is beginning its transit of the fifth zodiac house for Libras, so I hope my own experience follows at least a similar journey in creativity and collaborations. I think the fifth house Pluto transit is made so auspicious in part because the growth which occurs in the fourth house humbles us and make us appreciative of what we do have going into the fifth, which is an explosion of creativity previously stifled during the stress of the forth house transit.

Being the house of labors, of daily responsibilities like work, health, and even taking care of pets, Pluto's transit through the sixth house affects aspects of life that require consistent and unremarkable attention and will especially highlight themes of powerlessness in areas of wellness and workaday. Leos recently ended this sixth zodiac house transit and all of them experienced varying degrees of health and professional problems during the last sixteen years. One Leo with whom I worked contracted lyme disease in the early years of this transit when he was a teenager, then later as a young man tried to resolve his health problems through dieting trends like 'carnivore' and 'paleo' eating which made those issues far worse because, like all dieting behavior, are simply control mechanisms that try to force the body to do what we want. We are just as powerless over biology and nature as we are relationships and fate, but some people are so frightened by the realization that our bodies are not really our own we will often undertake herculean tasks trying to impose our desires onto biology and end up having very destructive effects on our lives and health, which the Universe will be sure to make us aware during this transit. This young man was also confused and misled by an overbearing and demanding parents, and after working with me at the very end of this transit learned from my experience how to have compassion for the body and life experiences and submission to reality leads to greater harmony with life, and immediately experienced improvements in his health that had so long been illusive but then later through the rest of the transit reverted back to those control behaviors to learn even more hard lessons that reality does not do our bidding until finally relenting control after the transit ended.

Another famous streamer, Hasan Piker, also a Leo, established himself early in this transit as a political pundit but also met challenges finding a stable venue for this profession until he began streaming which, when he did, required daily, persistent commitment in order to grow a fan base and success. At the same time he publicly struggled with health issues atypical for the young like weight gain, diet, and hair loss caused by our misguided ideals of health and fitness often at odds with how the body and reality actually work. Toward the end of the transit he achieved some greater stability in his health through daily, consistent dietary and exercise behaviors but thinks his progress was achieved by working out and taking prescription medications rather than the fact he got lots of sunshine every single day since his workout area is his own backyard (living in Southern California), since the Sun has one of the most profound effects on our health and wellbeing. The point of Pluto in the zodiac sixth house is absolutely *not* to overcome the issues it covers, which is instead simply the temptation of human control, to conquer that which makes us uncomfortable or fearful. Instead, Pluto's transits are lessons in the limits of mortality, that no matter how much we desire it we cannot actually bend life to fit our expectations and desires and that we must rather work in harmony with life and submit to its power and not our own.

Steve Jobs, the famous cofounder of Apple computers, died from pancreatic cancer during his transit of Pluto in the sixth house because transits of the sixth house encompass *all* mundane responsibilities and not only work, but workaholics (especially those which are wealthy) often neglect their diet or adopt restrictive eating behaviors for the very reason they are a workaholic, which is trying to control mortality, which can never be done (he was a fruitarian which is a particularly dangerous dietary behavior low in calories, fat, protein, and other nutrients we require). The body also requires sunlight and rest so it is not just diet which can lead to health problems during these kinds of transits, and refusal to listen to the lessons life is trying to teach us can have significant consequences. But again, the transits of Pluto are not lessons in overcoming adversity, but our ego. Another Leo I know during this transit was in a horrible bike accident and was nearly paralyzed, and while they recovered well the point of the experience was that beating up our body through excessive and risky exercise like biking at high speeds along a busy highway is not going to prolong your life. Both my Leo grandparents struggled with health issues and died during this transit, though nearly into their nineties it was not tragic nor unexpected but was still similarly rooted in our obligation to meet all our daily obligations, not only those which are convenient. Indeed all Libras will similarly experience this sixth house transit in two decades after this current transit, and after already having hosted Neptune and Saturn here Libras should be well familiar with the requirements of this Pluto transit, and if not will certainly have problems (and likely still will, especially considering I will be in my mid-sixties by then will yet have more health challenges in spite of my vast knowledge because the point is that *we do not control biology*).

Pluto's transit through the seventh house can be especially upsetting because the seventh house of relationships is always opposite our Sun sign, and as such relationships are always with an "other" who is, by definition, their own person with unique traits, behaviors, dreams, and trauma which may or may not be harmonious to our own but still being someone with whom we must cooperate, collaborate, and accept, which can be very difficult indeed if we are unwilling to relent control, because there is no greater stimulant for stress hormones than that of other human beings.

The way I stated that, however, might have implied we have a choice, which we absolutely do not during Pluto transits. We may choose how we respond to events which occur during this time but never what events occur nor how they play out, and it may be that even when accepting our partners or *not* trying to control them we still lose them or our relationships or they are somehow changed in opposition to our will. In astrology, relationships do not only mean those which are romantic, however, so this transit can also affect more than love affairs, and will influence business partnerships, collaborations, etc. Remember, this house is fundamentally the house of *balance*, so it is not so much about partnerships but instead doing what is required of us in service of balance which thus manifests through things like relationships, both giving and taking in equal amounts, contributing to relationships but also allowing others to contribute and not forcing them to do so or manipulating their contribution, which must instead come from their own volition and effort. In this sense I do sometimes see this transit affect parent, child relationships, which can then look like parents not contributing sufficiently to the relationship—meaning we continue to act as an authority figure, which we were never that anyway (we are only a guide and guardian to children) and is not appropriate to do to adult children anyway, which functions as the act of taking in excess in service of our own ego. Or we can use children as a replacement for friends, or even a spouse, since children are convenient targets for manipulation of emotional obligation, and we should instead fulfill those needs by establishing appropriate relationships for them and not requiring our children to do it for us, which will only result in the deterioration of our relationship with them.

While coaching someone during Pluto's transit of their seventh house they fell immediately into love with a psychologically traumatized person who was extremely reactive to some of their honestly minor control behaviors, whom themself probably had Pluto going through their eighth house of power and control, and while the other person was far more reactive and controlling Pluto is the kind of planet that does not care about such sentimentality, and means to exert his power over us without regard or consideration for our own sensitivities, deservedness, needs and wants, and desires and, as discussed in the chapter on love and Venus, it is our trauma which instead decides the quality of person we are exposed to or fall in love with, not our goals or intentions, and it is *impossible* to have a smooth and productive Pluto transit of the seventh house without first resolving our own trauma and control and coping mechanisms that will otherwise unconsciously compel us into such complimentary partnerships to our own unresolved control and coping mechanisms.

Even if are already in a committed relationship, the solution to conflict that might occur at this time is *never* about the other person, and probably the worst thing we can do is attend couples counseling which, as discussed in my book on psychology, is most often a device which serves only to empower Karpman 'helper' types and is nothing but an exercise in each partner telling the other everything about them they find disappointing. Even if we think the other person is the problem, all relationship problems in fact begin with *us*, such as the choice to be romantically involved with someone that does not treat us the way we should be in the first place, which is our decision, not theirs, and if that seems daunting or impossible it is only because we so far lack the life skills to successfully resolve problems and handle life which are in fact learned as an effect of practicing much inventory. Ultimately all Pluto transits do shape us into stronger and more capable people if we do not insist on resisting its transits, so though they may be painful its transits are not always in vain, and even difficult challenges such as this transit through the seventh house can result in more satisfaction, strength, and fulfillment through relationships if we listen to what life is trying to tell us.

The eighth house of power is the natural home for Pluto, but this does not mean we can get away with any of our own control and power antics. Quite the contrary, Pluto is most at home here and his power in this house is unmatched, and transits of the house of power deals directly with issues of power and control. During the writing of this book the sign of Gemini was finishing their transit of Pluto through the eighth house and boy was it full of fireworks. Two Gemini members of my family were the most domineering and cruel toward me in consideration of their religious ideology and dogma, and caused me the most harm and subsequently me needing absolutely to end those relationships. Repeatedly throughout this transit they refused to show compassion when I and others needed it, and attempted to use vulnerabilities in others to wield an iron fist over all their relationships, including significant child abuse and other amoral behaviors. The problem is that relationships can *never* be controlled and *should never* be controlled, and require equal give and take from all whom participate, of their own volition and autonomy, so any control behaviors during transit of the eighth house will absolutely be met with destructive consequences.

Abuse survivors often misguidedly believe that our trauma in childhood came from an absence of power and control, then as adults we seek to assert power and control as an antidote to the fear and insecurity which arose from abuse, but in reality it is this very behavior (as discussed in my book on psychology) which perpetuates intergenerational trauma, and while we think we are doing what is good more often are in fact submitting children or other adults to the same destructive, awful experiences we were subjected to as children (or worse). In truth the trauma and abuse we suffered was not a result of powerlessness, because even as adults we are all still fundamentally powerless, but was instead caused by others whom chose to do hurtful or even heinous things, and transits like this one will highlight such experiences, trying to teach us the reality of life and our true state of powerlessness over it and how the inappropriate assertion of power more often brings horrible consequences rather than security and peace of mind.

During this last transit many Geminis even did things that even brought legal consequences, prosecution by the law, and even imprisonment or loss of much personal resources and money so potentially destructive is this transit if we try to wrest control of life from the powers which rule it. Cancers are right now beginning this transit at the time of this writing and many Cancer politicians have already behaved severely in their quest to attain power and control which will in the long term bring them significant consequences, as the laws of causality *cannot* be transgressed no matter our motivations. In personal relationships these transits will doom us to losing them if we are intransigent, and while this sounds severe it needn't be difficult in the least, as earlier described of my friend and coachee who is a Gemini and was signed up for coaching for several years of this transit, because he was fully uninterested in trying to control others and thus fared very well during his eighth house transit, ended up even *gaining* a relationship and a child to start his own family. Having a sense of humor about life and not taking things so seriously is the primary key for success in any Pluto transit but works especially well for the eighth house (and if you lack that do much inventory that you may find more joy and levity in life!).

The ninth house of the mind is one of the less stressful Pluto transits, although it does still cause life-altering shifts in our conception of existence and reality, and can in fact sometimes produce some truly gnarly people when obscenely deluded. The house of the mind encompasses philosophy and learning, both in a literal sense and that which is intangible and experiential such as from travel or exposure to other people, ideas, foreigners, and concepts as well as ideology, zealotry, and law. Across their recent ninth house Pluto transit, a Taurus I knew was confronted or disturbed by the disparities of their religious ideation of the world and the reality which actually exists around them. They were one of the few people to directly ask me to teach them inventory, and after I did their experience being a parent became immensely less stressful. But then later their spouse demanded they stop having anything to do with me because their reevaluation of their hateful religious ideology had begun to make them doubt their religion, and for my efforts I lost an important relationship, but which then also underscored the control mechanisms inherent to Pluto influences in which they made that decision based on their ideology to also lose a relationship with me as a consequence.

The influence of Pluto in the ninth house is just as forceful as the other influences, but we are less upset by this influence since relenting control in terms of our own psychological wellbeing is an attractive relief from the tiring trauma and ideology lived through the gauntlet of the sixth, seventh, and eighth houses which causes us so much suffering, and by this time are usually ready to give up those control behaviors (and if not, Pluto will continue to throw stress at us util we are). Unlike transits through the fourth house which can literally disrupt our place of domicile, disruption to ideology occurs in nonmaterial ways, and so does not inconvenience us the way that loss of money, health, or resources can, but which means it is probably easier to maintain delusions of control during this transit since we are not always in direct conflict with others. During this transit we can expect our ideals and beliefs to be *challenged* by life experiences, and listening to the lessons Pluto is trying to teach us can empower us with greater insight into how life and the Universe truly work, and learn empathy both for ourselves and for others which can open our lives to richer experiences that encompass more people and opportunities, or close ourselves off to those things and lose those opportunities and then experience worse consequences in Pluto's upcoming transit of the tenth house.

The tenth house encompasses institution and our participation in public and social structures outside of our personal, intimate lives. This includes our public image, reputation, and our career since that is a function of how associates and professionals view us, but also membership in any group or organization, including even our own families but is not our relationship directly with them but instead how they view us as a member of society. Because Pluto transits are so long we often don't have many in a lifetime, and so we may not have wised up to the purpose of life and experiences to learn past lessons properly, and thus in this transit may experience the consequences of holding onto control mechanisms in terms of reputation that are diametrically opposite of what we expected or wished from our previous life choices. In fact, the person that demanded their spouse end their relationship with me because of their realization their religion was hateful and destructive was an Aries with this Pluto tenth house transit, whom also as I know experienced much stress and instability establishing themselves professionally, since at this time bosses, authority figures, coworkers, patrons, and other elements of institution and society will be sources of disappointment and frustration, and they likely demanded the end to our relationship not because they also believed in their religion but because they were embarrassed at being perceived by others in their religion has having a partner no longer interested or involved in it.

Again we meet issues of control and power in the transit of the tenth house in spite of our best efforts and showing up for opportunity, which is not a moment to further assert ourselves or work even harder as we might expect, but to recognize the futility of control and how the delusion of power makes us vulnerable to self-defeating behavior like trying to assert control during this influence, which can be especially destructive since our material security and productivity can be very impaired when we transgress laws of causality during this transit. It's always important to do our best and show up for opportunity, but the character weakness of *unrealistic expectations* could be an ideal descriptor for the challenge of this transit, thinking we might somehow look good when being a controlling asshole or bigot, or that we would get respect from people who are actively exploiting us. The good news is that we do not have to control others nor make them respect us, and when these lessons pop up during Pluto tenth zodiac house transits we can simply not give a fuck what others think, continue doing what is right and good, and make real amends, let go of control, and simply let go of people and organizations who don't respect us. More than any transit the ego is the only thing which makes this unpleasant here, so the greater the ego the harsher this influence will be, and inventorying unpleasant encounters during this transit can oppositely help the ego let go of pride, arrogance, and control behaviors which otherwise sabotage our public standing.

The eleventh house of fortune is a strange transit for Pluto which does not seem to bring obvious misfortune, which could be in part be because of the eleventh house's association with Uranus, the planet of spontaneity, which deftly outmaneuvers or compliments the lessons Pluto attempts to teach, and this sign is at a natural sextile to our Sun sign which is not so perilous as those at square, opposition, and inconjunct. But the eleventh house especially rules friendships, and most people appreciate their friends and try to treat them with respect and gratitude anyway, yet Pluto's transit here may in fact cause us to experience control behaviors which originate from our friends, or people that we mistake as friends, and which can also mislead us into life decisions which serve their selfish purposes at our expense. Most commonly we may in fact fear losing friends or fear their behavior concerning us, and assert inappropriate control over them, using the occasions of conflict, disappointment, mistakes, or misunderstanding as opportunities to manipulate, shame, abuse, and control them, and like all Pluto transits this will in fact result in consequences that we would rather not experience since this kind of behavior is antithetical to the nature of balance that such relationships require to be healthy and productive. Pisces are about the only sign I don't really know any, so my personal experience with Pluto's recent transit through Pisces' eleventh house is not well enlightened, but one Pisces I did know did not have very many friends except what he had through his partner, as Pisces are often so meek and sweet that the fear of losing friends or making them might have prevented him from making any. The eleventh house also rules wealth, so those with any kind of wealth might during this time experience desires to assert more control over concerns of inheritances and investments but which in doing can undermine those assets anyway through mistakes caused by decisions made in fear and insecurity. Especially since wealth and friendships are often private matters I don't have a lot of data or examples to go on here, but it will certainly affect those areas and the ability to do inventory instead of assert control can help bring more wisdom and effectiveness. Choices made in fear tend to always have negative consequences, while those made from confidence and wisdom are more likely to be effective.

While the twelfth house transit of Pluto is at an agreeable semisextile transit to the first house, it is very important because the twelfth house is where the current cycle comes to an end, and all twelfth house transits concern endings, spirituality, rest, regeneration, dreams, intuition, and secrets and the unknown, and refusing Pluto's influence here sets the stage for a very perilous first house transit in the next cycle which will be very rough indeed if we enter it not having listened to Pluto's lessons in the twelfth. During a relationship with someone who had this very transit I remember how they started bringing old or casual friendships into our life at this time and how these people were often unfriendly to me, for instance one time a person they invited to our house that they had known in college but had not seen for years actually teased me while in my home that I was not comfortable with their presence, which was an exceptionally weird thing to say and do to someone while in their house and which I most definitely was not after. It turned out my partner had started doing drugs with them a year before the end of our relationship, and I realized after they were addicts and enablers and I had not instinctually liked them because they were in fact secret enemies acting in opposition to my wellbeing and how sorry I was my partner did not protect our relationship from their influence. The twelfth house can equally represent *delusion* as spirituality, and the Pluto transit here is likely to cause us peril if we insist on being delusional rather than spiritual (the upcoming chapter on spirituality and Neptune clarifies what this means), and refusing to let things go as past resentments or life goals which have never been realized are likely to result in consequences. Pluto's presence here in fact may help show us what we need to let go of, and how to surrender to life and thus experience higher planes of existence that do not require the use of exogenous drugs and chemicals.

My presence in his life clearly triggered much unresolved trauma in my ex, and as healing can and should occur in twelfth house transits, as a period of rest and reflection, our refusal to actually allow experiences which would bring us healing is a likely experience during this transit, because we are afraid to let go of behaviors or the trauma which give us power. At one point I even taught him meditation such as described in the chapter on meditation, but he absolutely refused to do anything that might actually bring him peace and resolution and instead relied even more in activities and relationships that enabled denial and ego indulgence (hopefully through the remainder of the transit he eventually found how to do that). Another person with this transit with whom I was acquainted insisted that committing suicide was a selfless behavior, the very kind of absurd and contemptibly self-centered delusion common to many alcoholics and addicts but which poses a very interesting phenomenon of this power transit in which a person no longer wants to live but does not want their departure to cause harm to others, which it absolutely will. Trauma is indeed difficult to handle when we do not inherently possess tools for its resolution, which must instead be learned incidentally through opportunity, if we are lucky to be exposed to those methods, like learning the practice of inventory, but also required actually doing it, which many refuse since we are more interested in addressing our fears by getting rich, or fit. But Pluto demands trauma be acknowledged whether we want to or not, and refusal to face our trauma especially during a twelfth house transit will result in *much* psychological suffering and consequences. At the time these consequences may be what we desire—to be relieved of the reminder of our trauma through drug abuse or volatile interpersonal behavior, but this also comes with it the loss of opportunities and people which are actually good for us, the loss of which we may not appreciate until much later such as when my ex told me long we parted he had not realized how much I took care of him.

Because psychology, spirituality, and the unconscious are all the same thing this transit also causes conflict with our ideas of meaning and need for spirituality as something entirely separate from the institution of religion (which in truth has nothing to do with spirituality). During this transit most Aquarians also took a step back from public life, more or less, and willingly or not, resigning from public life or self-sabotaging opportunities because of psychological trauma and unfulfilled spiritual needs, as might Pisces in the upcoming transit and Aries in the one after, and so on and so on. During this transit the shelter of obscurity is comfortable since our internal psychological struggles make us feel vulnerable, and the worse our behavior or more shallow the friendships cultivated in the eleventh house transit the more guilt, remorse, and spiritual isolation we carry in into the twelfth, increasing the severity of the twelfth house transit, although we will carry also much trauma from our childhood which is no fault of our own and are only required to do something about it. Considering that inventory is not at all difficult to do, only requiring time and a paper and pencil, we are far more capable of handling this transit that we might at first believe, and the key to handling the twelfth house Pluto transit is contrite surrender to lessons in the twelfth house, have absolute compassion for ourselves, our trauma, our mistakes, and then cultivating spirituality—not religion or mythologizing, but true spirituality in order to heal our soul and prepare for the exciting new cycle when Pluto moves into our first house of self and we begin a whole new chapter in life.

Thus is Pluto (and its associated eighth house) concerned with themes of control and power which afflict all humanity and indeed the entire planet, since humans have spread across the globe. It is especially perilous to uplift a person to high office which has Pluto transiting significant houses whom has not done inventory and resolved their own trauma, for even if they are well meaning their fear and insecurity will compel them to act in their own best interest rather than that of the people, and taking major Pluto transits into account during elections would be very useful in electing leaders, avoiding those with potentially egomaniacal influences (*not* their birth chart interpretation, but active transits through their zodiac houses). All structures of power are eventually targeted by those who desire power, however, because they are afraid and the desire for power is the desire to subvert fear, but because fear is born of ignorance all structures of power are doomed to pass away in the hands of the ignorant. Because Pluto affects those in power it as such indirectly but consequentially influences world affairs as discussed in the upcoming chapter on great cycles.

21. World Events and Great Cycles

It is undeniable that humanity has forever changed the face of the Earth, even in ways which can never be repaired such as the total eradication of incredible ancient megafauna that were a regular sight for our ancestors like mammoths, giant slots, wooly rhinos, or glyptodonts (the giant armadillos). Yet human mythology and religion often characterizes man as entitled to rape the Earth and take from her anything they desire, excusing waste and exploitation, not only of our natural resources but other people, inherent to our divine placement above all others even those of our own species, which then robs later generations of resources as those currently living fail to consider the lives of their progeny in their actions.

The primary irony of this theology is that we are not in fact excepted from the laws of nature and causality, and as the climate destabilizes and our air and water are polluted with chemicals, plastics, and industrial gasses our position on this Earth becomes more and more precarious since, like all other organisms here, we are *entirely* dependent on the composition the biosphere and cannot survive without it. This should be obvious since none of us can breathe in the vacuum of space or survive poisoning by radiation and toxic waste, yet we consume and exploit and produce as if there is also no consequence in doing it. It is true that the powers which control industry and finance are those who make the chemicals and extract the resources but every one of us use them and blindly and willing participate in the systems which are destroying the only planet we have, and the doom which consumes us will be both entirely preventable *and* deserved, because there is no way we will ever move to a new planet as a solution to such a problem. Even at this very moment I type on a plastic keyboard made from fossil fuel, and I bought plastic bottles full of cleaning product (although the product itself is natural and sustainable), and the energy which powers my home are lit by a gas burning power plant which the conservative government under which I live refuses to replace with more efficient, cheaper, cleaner sources for no better reason than to spite their children and those whom wish for a better future.

Yet the reason we have failed to prevent this calamity is *not* for insufficient efforts for conservation or to protect the environment, but because we *misunderstand* what we really are as humans, an animal species which simply lives upon the face of this planet as any other of the many which also live here, and this misunderstanding then informs our decision making, based on a worldview which, like anyone with unresolved trauma and control mechanisms, is *ineffective* because it is also *misinformed*. Correcting this misinformation then begins with trying to understand what it is we really are, and the absurdity of humans denying our position as an animal creature might be understandable if we were entirely unique in our construction from other creatures on this planet, yet there are gorillas, chimpanzees, orangutans, bonobos, and other immediate biological relatives of ours which not only have nearly identical physical characteristics but even their emotional expressions we intuitively recognize as anger, sadness, joy, interest, or excitement, or how they interact with each other in nearly identical ways as we do amongst ourselves, throwing tantrums, stealing things from other members of their group, or oppositely sharing with those they love and trust, complete even with hugs, kisses, and cuddles.

Even animals which are less related but still related to us like lemurs have the same hand structure as ourselves and live in similarly composed tribes of related and emotionally connected members of their species, caring for their young, breast feeding as we do our young, and feeling joy from children or oppositely despair when they are lost, because emotions are not unique to human beings but instead is the role of hormones and neurotransmitters possessed of all animals on this planet that serve not so we can feel special but which direct and influence behavior ins creatures which otherwise would have no idea they need to eat, love, explore, fight, heal, or flee as required for survival.

One of the most interesting human traits is an inherent instinct to alter and order our environment. I love gardens and planting and one of the reasons a human mind is excited at the sign of a newly mown grass border next to a plot of perfectly cultivated wildflowers or other plantings is because of our real and innate instinct to organize nature as part of our evolutionary survival instinct. Yet as we plant fields of corn, raise cattle, or sow vegetables in well-tended gardens we assume this behavior is simply us doing what we want as a person rather than the obvious and irrefutable evidence our species *evolved* such instincts to garden, farm, and plant food as a primary evolutionary survival strategy the very same way beavers evolved to build dams, bees to build hives, and squirrels and birds to build nests. Around the time when I was first getting sick in my early thirties was the first time in my adult life ever having a garden, and I was more surprised by how much satisfaction I felt from seeing my tomatoes, flowers, and bamboo spring up from the ground as if by magic, doing almost nothing but putting them in the ground and watering regularly as the Earth repaid me for such little effort with an abundance of joy, awe, and the best tomatoes I have ever eaten. This behavior touched me to the core because it was in fact fulfilling a very primary evolutionary instinct to organize my environment, as an animal which came from other animals and our shared collective evolutionary history upon the Earth.

Our propensity as humans to organize the environment, to farm and build is truly an instinct and trait of our evolutionary composition and not a product of having culture and advanced technology, because *everything* we are as a species is entirely a product of our DNA and evolutionary history as a resident organism of this Earth, not simply wished into existence by any divine, humanoid deity but as a product of the construct of reality which also resulted in all the other innumerable plants, animals, and other lifeforms as made by the laws of the Universe. We cannot, in fact, have any features or traits as an organism which are not a product of our biology, which is why so-called debates as nature versus nurture and banal references to "biological sex" are so stupid because *there is no part of us which is not biological*, as the mind is the product of the brain which is also entirely a function of biology, and the energy which flows through us which is spirit is the very *point* of biology, the power which animates matter to life is not separate from biology, but one and the same.

The silliest irony humans can perform is admiring beavers and their productive instincts for dam-building without any sense of self-awareness that we are the very same in our desire to see nature ordered and managed, failing to recognize the sense of fulfillment which comes from a well-built house or pavered street is *not* because we are productive or ambitious but because it fulfills basic, animalistic instincts desiring order. So one of the greatest reasons for the failure to prevent the calamity of environmental collapse which is occurring is the very admonition to conserve and limit our use of resources, because the human animal instinct to survive is no less than any other creature on this Earth and just like any other creature on this Earth we *require* resources, so admonishing other people to conserve resources and use *less* of things we need to survive has the unintended function of telling people to just *die*, and since we also fear death as an animalistic instinct to promote the successful propagation of a species most humans thence respond to this admonition with anger and spite, not rational thought. Spite is also a primate characteristic observed in many species of primate, so other humans even double-down on exploitation of the environment just to prove to those assholes we don't need to listen to them, and by misunderstanding human nature we all inadvertently hasten our own demise by triggering very real animalistic instincts we fail to understand.

Yet it is also true that resources are finite, even more finite than most originally assumed. For instance one of the reasons humans used to dump sewage and pollution into the ocean was that the ocean seemed so immense as to never possibly be affected by human behavior and garbage. But then the population of humans exploded from a few tens of millions to nearly eight-billion, which is truly an unfathomable concentration of one megafauna species—even the mighty American bison at their height were only about 30-60 million before being wantonly slaughtered by genocidal European colonists down to just 1,000 remaining animals literally animals for no reason than spite. And since denial is also an adaptive survival trait of human animals we barrel headlong into our own destruction since in fact we are entirely dependent on resources which, because of our nature which has also caused this problem, also prevents us from recognizing it.

Throughout human history there have also been many great civilizations which have risen and fallen over centuries or even millennia, and the stories not only of their greatness but also their demise are the frequent subject of both academia and entertainment alike. What most people are entirely unaware, however, is that the fall of most civilizations from Ancient Babylon, to Ancient Egypt, Ancient Rome, and even the Ancient Maya Civilization in the Americas were *not* products of sociopolitical conflict as is so often cited but of failure of the *agricultural soil* around those civilizations. Economic centers like great cities consistently rise throughout history due to productive and effective farming which produced plenty of food to not only support a growing population but also to enable division of labor into trade and services which are entirely impossible *without* an excess of food production, otherwise civilizations are limited to farming or hunting and gathering in order to sustain their members which then have no other time to invest in art, writing, pottery, massage therapy, priestcraft, etc. But an unknown danger grew beneath the literal ground around all great, ancient, populous civilizations which is that intensive human agriculture, *especially* the tilling of soil, ends up killing the soil microorganisms required by plants for their own survival, so crops fail and the people starve and food supply crises erupt into corruption and civil unrest and as the population suffers and power-hungry despots take power to crush dissent and the population finally rebels or flees to safer, more productive lands. I mean fuck's sake the area now called Iraq and Syria used to be called the *Fertile* fucking *Crescent* which it has not been for at least a thousand years due to the destructive effect of intensive agriculture which kills the soil microbiome on which all plants are dependent, with famers believing their soil is "depleted" when in fact it is only missing beneficial microorganisms which could be easily returned by transplant and regenerative agricultural practices, since plain dirt is nothing but pure nutrition for plants and far more dense in minerals than soil.

The solution for solving such a crisis as what we face as a species is thus not directly conservation but in *providing* for the needs of *all* humans, and to help all people succeed and survive in ways which *are also* in harmony with protecting nature. This is not only possible in a way that can save our planet but will also help restore many systems and natural resources in the process, for instance principles of regenerative agriculture which not only restore the quality of the soil produces far more food and far more nutritious food than what is achieved by industrial agriculture while also achieving things such as weed and pest suppression without expensive, toxic chemicals and amendments. Pests that destroy crops are in turn predated upon by other predatory insects like ladybugs, wasps, and lacewings which lay their eggs in plant litter like fallen leaves which normally carpet the ground and also suppress weed growth, so the act of literally helping to promote an organized but natural system not only produces more and better food for us but also does so through natural cycles that in turn sustain and promote the natural cycles of the Earth. As discussed in *Fuck Portion Control* the plant polyphe-nols called tannins also condition the soil to promote healthy microbes and kill those which are pathogenic, so growing high tannin plants in regenerative systems protects plants without tannin from harmful mold, fungi, and viruses.

And of course this would be the case, because we are not an exception the natural cycles of the planet, but a product of them, and so restoring, protecting, and maintaining those cycles *also directly benefits human animals* since we are an animal also directly dependent on those cycles. In order to convert conventional agriculture to systems which do not depend on constant synthetic and industrial amendments (which are also costly anyway) fields should be planted with dense and productive cover crop which can then be flattened or mowed into mulch over which should immediately be planted pre-sprouted crop seed such as by briefly soaking in water for about 8-10 hours which will shoot up before weeds can get through the mulch layer and benefit from the protection also of predatory insects living already in the mulch and soil. If mulch is composed of high-tannin sources it not only suppress microbial pathogens in the soil but also *encourages* commensal microbes which *help* plants grow, and safe tannin crops can also be grown to be fed to cattle and other livestock to naturally also inhibit gastroinestinal pathogens and suppress production of toxic hydrogen sulfide, methane, ammonia, etc., and give them drinking water in which sources of tannin has been allowed to steep (again, *do not* use toxic species, which in includes also any very high in nitrate), greatly reducing the need for antibiotic use, and these animals also *love* tannin sources anyway for this very reason and will also be fulfilled by its availability.

Weeds are plants which have developed to colonize microbe-poor soil, having the ability to exist in biomes and soils so bare of microbes and organic matter that other plants would not survive, so constantly tilling the soil or using herbicides and pesticides actually *increases* weed pressure (because these chemicals also kill microbes, not just weeds or insects), because it cultivates the *exact* conditions that weeds and pests are designed by nature to colonize, so the more intensive a farmer farms their land the more weeds appear and take root or plague insects appear to decimate crops or the more widespread microbial diseases which cause plant rot, driving farmers insane and thinking it is the fault of the weeds and nature and not their own naive abuse of the land which is causing it. Oppositely, when soil micro-bial life is rich (which is a function also entirely dependent on plants) the microbes produce so much acid to dissolve nutrients out of dirt and rock that weeds cannot easily survive and naturally die off. But the weed seed bank also often descends many feet into the soil and tilling also brings weed seeds to the surface where they rapidly germinate, which can be great if you are planning a wildflower meadow or restoring degraded lands by simply turning over the soil but completely antithesis to farming unless your purpose is to literally farm weeds.

People often discuss the nutrient of soil as if there is somehow a limited supply of nutrients. This is not only wrong, it's completely stupid because dirt and rock are nothing but pure nutrients like iron, calcium, magnesium, copper, silicon, boron, selenium, and other elements which extends for miles and miles beneath our feet, and it would be *impossible* for humanity to ever entirely deplete the nutri-ents required of plants and agriculture because that would require removing the entire layer of Earth's crust to outer space. The way plants extract nutrients from dirt and soil is by capturing the Sun's rays, transferring that energy into sugars synthesized in their leaves, then sending some of that sugar down to the roots to feed soil microbial life which in turn eat that sugar as energy and as a byproduct produce acids like acetic acid, butyric acid, and lactic acid which dissolve the elements like iron, copper, phosphate, molybdenum, and boron out of minerals to complex with these organic acids which the plant can then effectively absorb those elements into the roots to meet its nutritional needs. In this sense, roots are simply inverted intestines in which plants cultivate a microbiome in order to process nutrients in the same way our bodies do in our intestines, so practices which destroy the soil microbiome also destroy plants. Even the use of synthetic fertilizer is an overcorrection for this problem, and the reason fertilizer is even needed is to more rapidly regrow the lost microbiome which is only lost in the first place due to farming practices like tilling the soil or use of herbicides and pesticides.

But elements are also never destroyed—so they too cannot ever be depleted—such as iron and copper, even when taken up by plants or animals or humans they remain as iron and copper and are eventually excreted in poop after use by the body, return to the environment and continue participating in the cycle, so long as the cycle is active. When human excrement waste is used as fertilizer it returns elements like boron, iron, iodine, copper, etc., back to the fields from which they were taken. One of the problems of current industrial agriculture is that our wastewater also contains toxic chemicals in the form of medications, cleaning chemicals, oral care products, skin and haircare products and toxic chemicals in makeup, and if human waste was instead composted back to agricultural fertilizer without these contaminants there would never be any need for industrial fertilizer since the sheer amount of human animals constantly pooping on this planet provides more than enough fertilizer to in turn grow the food we need since the amount of poop produced is the same as the amount of food eaten.

While many people might be worried about the practice of human waste composting due to the spread of disease it's already a widespread practice, and the act of composting actively destroys pathogens, hormones, and sometimes even some toxic chemicals due to the action of heat, microbes, and decomposition which occurs in the composting process, and because of our advanced technology can also easily be tested and monitored. But it is also silly to be worried about those kinds of dangers because they already exist and occur in the regular contamination of the food supply because these pathogens are ubiquitous, natural opportunistic microbes that can come from animals just as they do from humans, and it is usually not lack of exposure in people which is the reason for the absence of transmission but healthy and working immune systems, quality food production, and protective commensal microbiomes which primarily protect us from infection. Any time you get food poisoning at a restaurant, or even at home, that is exposure to these pathogenic microbes. The only reason any person believes, wrongly, that we will run out of resources like food is because we already misunderstand that food scarcity is even now entirely a fake crisis, with food systems and governments purposefully destroying food excess and withholding from impoverished people in order to force them into menial labor, and if we instead maintain effective systems of food production and include those who are less fortunate there never again must be such things as food insecurity for any person on this entire planet even should the population explode to 100 billion because of the properties of elements and the cycles of nature the nutrients that both we and plants require for our survival exist effectively for eternity, and combination of regenerative agriculture principles with industrial food production in vertical farming would be even more resistant to environmental disruption and provide a stable, sustainable, abundant, and safe food supply.

Greater than the misunderstanding of ourselves as an animal species which impairs our effectiveness in solving our problems is that *all things* in existence operate on long-term cycles of karma and cause and consequence, which are not only expected but very often predictable, and not even because we understand astrology but simply because we are not fucking stupid and willfully naive to them. For some reason it does not matter how many times capitalist markets nosedive into recession most investors, government officials, and workers who like to pretend they are capitalists (you need CAPITAL to be a capitalist) pretend like those cycles are not *inevitable*, gleefully shredding their money into bubbles of speculation only to be "surprised" when it vanishes into the aether as if they weren't the ones actively and voluntarily burning their own money. Cycles exist in every part of our lives, from the years of childhood spent in school, each year beginning, progressing, and ending in nearly identical ways but with some progress achieved in graduating to the next year each time, or cycles of birth, marriage, reproduction, and death by which all human lives are bookended. But as demonstrated by cycles of things like the financial markets there too exists cycles which encompass the greater structures of our world such as countries, empires, governments, markets, movements, and human society, but not because any of inherent value but because they are also constructs that result from fundamental principles of existence as organization, power, and dependency. For instance, money is not real, but we *are* dependent on other humans for our survival, and lacking especially large claws, speed, or strength (hell we don't even have our own fur anymore) we have evolved our dependence on one another, and our systems of cooperation are more structured and organized systems such as reflected in the use of money as a contractual construct which implies and communicates the terms of dependency and cooperation but on its own is entirely meaningless and without value. It is true that many humans can be duplicitous and exploitative, taking more than they are willing to give (which is also a function of natural laws and thusly also correlates with astrological phenomena), so money helps us effectively tabulate and document such disparities, such as recognizing when self-centered capitalists accrue far more wealth than other people in obscene disproportion to their contribution to humanity.

One of the most common of the great cycles occurs between Saturn and Jupiter which, when they meet every twenty years, is called *The Great Conjunction*, and are apparently called that because they were, before the discovery of the further planets (Uranus, Neptune, and Pluto) the rarest conjunction between all the planets seen by the naked eye. So the term is a bit antiquated and not even close to the longest-term cycle in astrology but during the time of writing this book the Jupiter-Saturn cycle is right now approaching the first *square* in the current cycle, an alignment often associated with instability in markets and will likely result in a massive market correction that this time might be comparable to the Great Depression. The Jupiter-Saturn cycle influences things like the market and economies not because either directly governs those themes but because Saturn rules order and institution, on which economies depend, while Jupiter rules expansion and growth, which is extremely consequential to them. When these two planets meet in conjunction nearly every twenty years (19.85) such as recently occurred around 2020 there is thus a marriage of both expansion and institution which appears to directly benefit systems like economies but not necessarily workers or people since Saturn rules institution. But we also are guilty of filtering such concepts as growth through the human filter of insecurity and greed and while growth and expansion can often be good it can also often be *not good*, which occurs regularly during the Jupiter and Saturn conjunction because Jupiter expands all that it touches regardless of what it is, and while Saturn can stabilize that expansion it ends up often expanding institutions which in turn believe their own shit and see validation in their growth rather than the danger of getting what we asked for, which during the more challenging aspects between Jupiter and Saturn then come crashing down due to the very absence of discipline and structure.

figure 4.

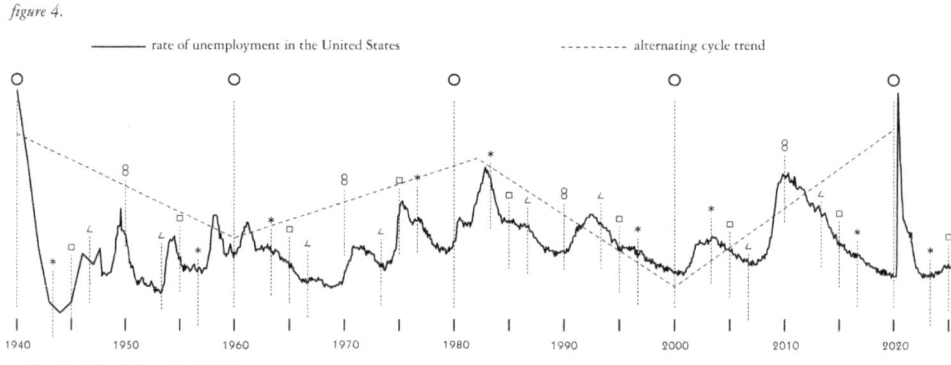

*employment data from Bureau of Labor Statistics
*employment graph between 1940 and 1947 was hand drawn
*trend lines are not exact data, meant to show patterns

   The most recent conjunction between Jupiter and Saturn in 2020 was thus the beginning of the current cycle and end of the previous, when financial inventions and the stock market exploded in growth and companies embarked on massive stock buybacks and obscene amounts of money was invested in new ventures and speculation. This was even more amazing because of the ongoing coronavirus pandemic which stalled productivity and interrupted trade and commerce which otherwise should have caused a far more long term financial stress as worldwide pandemics usually cause, and yet economies rode the pandemic relatively well, especially many institutions exploiting public bailout money to enrich themselves and engage in market manipulation like stock-buybacks (which, by the way, used to be illegal because of the artificial inflation of stock value it causes) with one of the most breathtaking rises in markets in modern history. Of course nothing lasts forever, which is the entire point of cycles, and as the current Jupiter, Saturn square (which began in 2024 and comes exact at the same time this book will release) finally reaches its fullness the economy, markets, and employment began once again to become unstable and eventually crashing back down as Saturn demands discipline and responsibility from growth.

One of the reasons economic cycles have not been expressly identified with the Jupiter-Saturn cycle is because no one has heretofore recognized it is an *alternating* cycle, where the same pattern does not repeat from beginning to end but is instead *mirrored* from beginning to end, so that at one conjunction unemployment is generally lower while at the next it is higher as shown in *figure 4*. As can be clearly observed there is thus a trend of high unemployment at *every other* conjunction of Saturn and Jupiter, and within every cycle is an absolutely undeniable trend in the opposite direction from where the cycle began, which then represents a 40-year cycle and one reason why this pattern was not so readily apparent, since we first needed this data and that every person may only see this cycle only twice in an entire lifetime. Anyone familiar with recent US history will see in this graph how the economic strain at the Jupiter and Saturn *oppositions* occurred with notable periods of economic turmoil such as after World War 2 when troops returned from war needing work (1950), the election of Nixon to office (1970) which undid low post-war unemployment, a similar problem with Bush Senior (1990), and the later catastrophe that his son triggered in the 2008 economic crisis that peaked at the opposition alignment in 2010.

The problem with this data, however (the unemployment data), is that it does not fully reflect the actual influence of the Jupiter, Saturn cycle. For instance since around 2015 economic stress on the working class began to increase immensely, with also inflationary housing, rent, and consumer goods putting economic pressure everywhere even if the elite or government officials were uninterested, which in fact was another primary reason the previous President lost reelection (well, his sloppy surrogate losing), so even though the *absolute* unemployment number was low before and after the 2020 unemployment crisis that data did not reflect the real-world conditions wherein workers were greatly squeezed by the economy even if they technically had jobs.

Of course there are other influences which also affect such cycles, which is why they are not even, featureless slopes and clean data but instead complex and dynamic, as most other planets will *also* be affecting any system or set of data such as economies and unemployment, since all reality is affected by the same forces and none exempt from any which account for the nature of variety and variation in any dataset, yet looking at this graph it is still wild to see the apparent correlative patterns such as the fact that there are clear changes in activity at the conjunctions and oppositions (halfway between conjunctions), without exception. Eventually when metaphysical sciences fully understand these laws it will be possible to reproduce this graph exactly but by using specific combinations of planetary or other astrological influences, as the reason for the fractal nature of such cycles like trends in economy, employment, and our insular, personal lives is that the underlying metaphysical laws of reality move throughout the Universe in *waves* and wave behavior just like all other forces or influences such as gravitational waves or the waves of an ocean or earthquake. Each individual influence being an influence wave also then builds upon and combines in constructive and destructive patterns with other waves of influence, with longer waves providing long term trends and shorter waves adding higher frequency, complexity, and detail, to thusly create the fractal nature of metaphysical laws like cause and effect seen in *figure 4* and not only in physical matter. This might sound like I'm making a metaphor but in fact the waves are *literal*, but exist in the dimensions of metaphysical reality that determine laws such as causality, and the mathematical modeling of such complex waves is itself a fractal model which is why the graph of things as economic data shown here for unemployment is also a mathematical fractal and not a homogenous, uninteresting line, while also seemingly random and chaotic actually demonstrates clear, repeating fractals and yet people will stare directly at this plain evidence of fractal patterns in metaphysical reality and claim there are no laws which underpin it when they are clearly visible in the data and even though fractal models are a fundamental organizing principle of all existence.

Because of this, if specific influences could be correlated and identified with specific trends such as the different patterns in unemployment or other economic data it would be possible to then strip each influence out of the overall data to isolate and understand each fundamental underlying influence wave and thus how they combine to create patterns, trends, and cycles in any system, to better understand how reality is constructed and correlates with other visible phenomena as the transits of the planets. This of course could be used to predict things like the cycles of the stock market but that is such a banal and obvious use of such knowledge it would also simultaneously destroy such shallow systems of greed due to the destabilizing effect such "cheating" would cause and could instead provide greater insight into the march of history, world events, and general future history much like what occurs in the science fiction fantasy, *Foundation* by Asimov. A fundamental law of astrology is that it can never be used to *alter* the course of the future, as reality is always fulfilled regardless of our expectations or insights such as when I avoided my grandfather's funeral in fear of my homophobic relatives but instead had an unpleasant encounter with my sponsor, and at best this information could likely only be used to *understand* reality, never to alter it.

It is not only the large planets which also influence long term, world events, and as is true of all astrology there are often greater patterns and cycles most people are not aware of which in fact have greater and more far-reaching influences over the course of history. While Venus is often considered marginally important in entertainment astrology Venus also has a 243-year perigean cycle (concerning its perigee, or the closest point) with Earth itself, which also likely has similar characteristics to other Venus cycles which have not yet been recognized by astrologers. For instance in 2016 it was exactly 243 years since the cataclysmic rise of the United States of America with the catalytic Boston Tea Party when the colonies first began to rebel against Great Britain and the events of the Revolutionary War, which at the same time heralded a new world order from that of monarchies to representative democracy. Exactly at the end of that cycle was the first term of the current President now dismantling the American experiment one 8-year Venus cycle later into the new 243-cycle after his shock second election to the office (shock only if you weren't paying attention). The British Empire also began rising immediately at the beginning of the previous 243 perigean cycle, which started in 1530 when King Henry VIII literally declared in 1533 in the *Statue In Restraint of Appeals* that that the realm of England was an Empire and began the Protestant Reformation.

Yet, because the 8-year cycle (8 years plus 2.4 days) is not evenly distributed into the 243 year cycle, the larger is thus halved not equally but instead the first half being 129.5 years in length while the other is 113.5 years. As such, the British Empire finally began to decline at the end of the 129.5-year cycle in the early days of the 20th century and the start of the World Wars. Then, in the exact middle (opposition) of the latter half in the 1960s was the counterculture, anti-war revolution which swept both the United States and Britain. Even the Ottoman Empire began at the beginning of a Venus perigean cycle three cycles ago just after 1287, while the Mongol Empire was at the very same time ending along with the yet preceding cycle, where the Mongol Empire in turn seems to have begun with the sociopolitical events which birthed the Khamag Mongol in the beginning of that cycle in 1044. It becomes more difficult to establish definite dates the more ancient the history (especially filtered through academic bias which includes other data that might shift the real dates of such periods), but even the Viking age is officially reported to have begun around the year 800 which was yet the yet previous Venus perigean cycle beginning 801.

But because the larger cycle is itself made up of shorter cycles there is not *only* one linear, end-to-end 243 cycle but instead patterns of smaller cycles throughout history fit into larger cycles which thus help elucidate the metaphysical bounds of things like the rise and fall of empires and other large-scale world events themselves divided up into smaller events and developments within their larger cycle. In fact, the *entire* length of the Spanish Empire from 1492 to 1976 is EXACTLY two Venus perigean cycles, the officially stated dates being just 2 days shy of 486 years (so it is in fact, exact). These kinds of correlations *cannot* be coincidence, it's simply too uncanny and wild, but occurs *because* the Universe operates on laws of mathematics and order, patterns and cycles, and it is therefore much more likely that metaphysical laws construct *all* of reality including such things as the rise and fall of Empires and the sociopolitical and geopolitical forces which shape history, which are not random and chaotic but in fact evidence of how reality is constructed.

More important to our experience of recognizing these correlations of history and events with astrology than attempting to predict the future is to recognize that life and the events of the Universe on both a cosmic and micro scale are always *cyclical*, that no one event is ever a beginning or an end, or the last or the first, or persistent and unchanging, but instead the continuation of unending cycles and intervals and that life and geopolitical dynamics do not move in a straight line but wrap around on themselves in repeating, fractal patterns with the same kinds of events occurring over and over and over again and again and again, and as what always occurs in fractal patterns there are smaller cycles which exist within larger ones which exist in yet larger ones such as the Venus cycles of 1.6 years, 8, 113.5, 129.5, and 243, which is why every generation of humans experiences the same types of life experiences and why larger human institutions like governments rise and fall with predictable repetition in quantized intervals which also correlate with the movements and aspects of celestial bodies, because everything in the Universe, not only physical bodies, is governed by the same, unifying laws.

It's possible that before even being familiar with astrology many people have heard through artistic media like film and music (particularly the Broadway rock musical, *Hair*) that we are in the *Age of Aquarius*. Ages are one of the largest divisions of astrology, marked by the *precession* of the *axial tilt* of the Earth through the zodiac belt. This occurs because the axis about which the Earth rotates is in fact tilted about 23.5 degrees from perpendicular to its orbital plane (the plane of Earth's orbit around the Sun), but this tilt itself rotates (precession) around the imagined perpendicular axis, thus also creating a cyclical path of precession through the zodiac which takes about 25,772 solar years to complete (solar years being different than the 25,900 sidereal years cited by astronomers), which also means that the axis stays in each sign for about 2,147 years, and thus also the length of Ages. Currently the Earth's North Pole points toward the star Polaris but, because of precession, by the year 14,000 it will instead point toward the star Vega (and back again as the cycle continues). One 25,772-year orbit through the entire zodiac by the axial precession is called *The Great Year*, and this shift in the Earth's axis and thus the apparent 'wobble' of all the stars in the sky was first observed by the Greek astrologer and mathematician, Hipparchus, whom also discovered trigonometry, whom is often entirely, biasedly, prejudicially, dishonestly mischaracterized as not being an astrologer by modern historians and citations even though he very much *was* an astrologer and that the entire purpose of ancient astronomy was in service of astrology, to understand metaphysical sciences and not only those which are physical. Axial precession also directly affects seasons, where the current state of axial precession makes winters in the Northern hemisphere shorter, in the Southern longer, and by the year 14,000 this will be the opposite and Northern hemisphere winters will instead be longer and but Southern hemisphere winters shorter. These changes to the Earth's climate due to dynamics in the movement of the Earth are described in *Milankovitch cycles* and have direct implications for human civilization, as by about year 10,000 it may become increasingly impossible for the many societies between the 40th and 50th Northern parallel to exist due to such long winters.

Like the nodes of the Moon the nodes of the Earth's axis tilt also move retrograde through the Zodiac. But exactly where the axial tilt is located in the Zodiac is apparently not a point of agreement among astrologers, with dates ranging quite far and wide. Most influences in astrology also have transition periods at the cusp, where both the previous and upcoming signs influence the quality of what they are concerned, and since Ages are so freaking long this transition period will also be at least several hundred years anyway, but it is very clear from the events of recent history that we are *well* into the Age of Aquarius, the sign of humanity associated with the eleventh house of fortune and friendship, and as reflected in its rulership by Uranus governs technology, revolution, independence, progress, unpredictability, sharing, and wealth. Considering the nature of this Age it likely began before the dawn of the Industrial Revolution and the foundation of Democracies as the United States and modern France, with the shadow period probably encompassing the signing of the Magna Carta, and the advent not only of technology and the information age but of representative Democracy and revolution by the masses, when we also see an explosion of the human population as a product of new and revolutionary technologies like plumbing, sanitation, mechanized farming, and novel medicine which also in turn leads to the greatest accumulation of wealth and friendships ever seen in the history of this Earth.

One reason the exact position of the axial tilt in the zodiac is not confirmed is that some people mistake this precession of the Earth's axial tilt as affecting the system of astrology itself, because of how our progression through these cycles also cause the geocentric location of the constellations to change location. But as mentioned in the chapter on how to use astrology, astrology is *not* based on the *sidereal Zodiac* (coordinates of the constellations), but instead on the the *tropical zodiac* (coordinates based on Earth's seasonal markers like equinoxes and solstices), and the naming conventions of the zodiac signs in astrology are simply derived from the constellations which appeared in them at the time of conception *but are not based upon the coordinates of the constellations*. So because the axial precession then occupies zodiac signs, the Ages thusly also coincide with various and distinct phases of history and Earth's periods defined by strong, consistent themes and changes in world order, history, and even the environment. Because the axial tilt retrogrades through the Zodiac the age which preceded Aquarius was not Capricorn, but Pisces, which is the sign of spirituality, illusion, and delusion, and indeed world history immediately preceding the modern era was dominated by religion and man's obsession with both discovering the divine and the delusion of trying to define and even control God and reality, even in the dismissal of astrology mischaracterized as something superfluous and unreal even though it very much contains math, science, and is based on the physical world around us while something entirely manmade and based in fantasy and superstition as religion was instead regarded as real. As Pisces is not a particularly intellectual sign, what progress in technology and science that was occurring through the previous axial transit through Aries was paused or abandoned in favor of religion, and in fact the Karmic arc of history which birthed the infamous Dark Ages was due to the Earth's transit through the Age of Pisces, not random accident or chaos, as even before this age the invention of the earliest computers and machines as demonstrated in the Antikythera Mechanism and the first automatons (primitive machines that performed tasks as pouring wine or opening doors) may have otherwise led to an earlier technological revolution two thousand years earlier.

Because Aries is the sign of action, aggression, initiative, war, etc., the Age of Aries preceding the Age of Pisces was in turn a period of ceaseless conquest and domination such as the rise of the Roman Republic, and its transition eventually to a Theocratic Empire and usurpation by the Christian religion marks the transition into the age of Pisces. This is not to say the Age of Pisces was free of conflict and domination—obviously some of the most heinous campaigns of conquest occurred during that time, but were motivated through and justified by religious preoccupation and delusions such that all people should hold the same religious beliefs instead of the generally accepted and recognized regional nature of belief systems before that time. Before the Age of Aries was the age of Taurus which represents materialism, resources, and the literal resources of the Earth like minerals and precious metals and that period was likewise also predominated by building with earth and stone and adornment with gold and other precious metals (Egypt, Babylon, Persia, Greece). Before that was the age of Gemini, the sign of knowledge and communication and instead the preoccupation with technological advancement, writing, and the development and implementation of agriculture and better production of tools and systems. Before that was still the age of Cancer which is the sign of home and family which includes not only the cooking of food for our families but also the literal place of domicile which likely coincided with the increased building of shelters (rather than dwelling in caves), and can in fact likely be seem in the archaeological record with the population explosion of humans that began around 8,000 BC as a result of increased family care and the literal building of homes. Leo before that is the sign of creation which rules art, casual sex, and children, and is a period from which a great deal of prehistoric art originates. Actually, prehistoric art appears to begin in the previous Leo cycle one full Great Year earlier, as also demonstrated in the archeological record, from about 30,000 B.C. Leo is also ruled by the Sun, and the end of the last ice age and retreating glaciers dates to the last Age of Leo at the end of that Great Year. Before the Leo Age was also the Age of Virgo and the Neolithic Revolution in which agriculture first became common because Virgo is the house of labors which concerns our daily obligations required of life such as not only the growing and producing of food but also eating it, and taking care of other responsibilities including the care of pets and livestock. While the evolution of dogs occurred *before* domestication, as wild canines first adapted to living off human populations and not the other way around, the first indisputable evidence of dog domestication is the Bonn-Oberkassel dog from Germany in 14,200 BC which was likely right smack dab in the middle of the Age of Libra, the sign of partnership. Because the evolution of dogs was unlikely to have occurred so recently that likely occurred many generations earlier and it just so happens the disputed origin of dogs from East Asian wolves in 39,000 BC is also squarely in the previous Age of Libra one Great Year earlier.

As is now underway in the current Age of Aquarius the near future promises continued focus on humanity rather than institutions such as religions and government, and continued progress in technology and science, but also the revolutionary spirit and defiance of norms and institutions, perhaps even some greater degree of chaos and disorder as what typifies the domain of Uranus. If the earliest estimates for the start of the Age of Aquarius are correct (about 1447 C.E.) it means we are about in the middle of this transit (which would make sense concerning the state of everything), and have about 1500 years remaining. Obviously, every last one of us will have long passed on by then, and these are lessons for the future, or in understanding our past, as the dynamism of Ages is beyond the experience of normal human lives, but following the Age of Aquarius will be the upcoming Age of Capricorn which *does* concern institution, order, and rules, and will likely be a response to the disorder of this Age, dominated by the return and rise of institutions, even possibly to severe authoritarianism and conformity, as the march of history and time move ever forward.

If there are any lessons which can be applied by humanity in the possession of this knowledge it is that all signs have equal potential for good or ill, to be productive or unproductive, helpful or unhelpful, and we risk the negative when acting on fear and desire for control wherein the humble recognition that the Universe is in charge of such things as fate, destiny, and the workings of the Universe we may simply show up for opportunity and do our best, recognizing that there is nothing inherently wrong with either order or chaos, since both are required together to construct reality, but in the attempt of human beings to impose our desires upon reality and other people, which we cannot do without consequences, and the successful existence in the age of Capricorn would be the rightful and healthy use of institution, to *help* humans achieve order, not to impose it upon them (but considering the nature of reality and Capricorn it is more likely to have much authoritarianism).

The transit of other planets to each other similarly mark other cycles, and every planet also has repeating cycles with every other planet, so there are in fact quite a few other combinations of binary planet cycles like the Jupiter, Saturn cycle which govern specific aspects of life and reality which concern their composition. This time in which we exist as I write this book is indeed more fateful than usual because there is also about to occur the fated Saturn conjunction with Neptune which happens only every 36-37 years. As I'm almost forty-five at the time of this writing the last conjunction occurred when I was almost nine-years old, which occurred with the fall of the Berlin Wall which had divided Germany since the end of World War II. Many astrologers incorrectly characterize the relationship between Saturn and Neptune because each is quite disparate in its domain from the other, the former being the planet of structure and discipline and the latter being spirituality, delusion, and ideology. I think this error comes in part by not understanding what spirituality truly is, which is not some ethereal, extradimensional abstraction but in fact a real part of our daily lives and the energy which makes us living which can and does also benefit from structure and is not inherently in conflict with Saturn influences. While it is true the domains of these planets are not obviously cohesive the conjunction influence is also a marriage of those energies, not a conflict, and at all occurrences of this influence it appears to be a resolution of past delusions and conflicts in which people come to starkly recognize the disparities between their ambitions and designs for the world and the awful results and consequences which have arisen instead. In addition to the fall of the Berlin Wall, apartheid South Africa also finally started to fall at the last conjunction with the release of Nelson Mandela from prison followed shortly after by the repeal of apartheid legislation. At the conjunction previous to 1989, in 1953, was the end of the Korean War in which 3 million people, mostly civilians, had been slaughtered by the United Sates, Russia, and China as they sought to impose upon the world through force their respective economic and political ideologies. What actually occurs in the Saturn cycle with Neptune is that Saturn demands structure of ideologies rather than abstractions, contrivances, or mere conceptions, and they must prove themselves to be practical, effective, and applicable. Part of the chaos which accompanies and follows the square or other stressful aspects is due to persons and movements believing their ideology is inherently structured for no more a reason than believing it, but all systems require effectiveness to persist, and when attempting to implement those ideologies their inherent weaknesses are laid bare through failures or consequences such as when attempting to *force* people to participate in socialism, which is antithetical to its ideology (achieving egalitarianism through murder? lol), or to oppositely prevent the spread of socialism because market capitalism exploits people and always ends either in fascism or socialism anyway and is nothing more than trying to temporarily plug a leaky, aging dam.

Each cycle thus bookends periods of ideological conflict in the course of human history, the conjunction ending the previous cycle while at the same time birthing another, each likely nestled in even larger cycles we have yet to identify. At a prior conjunction of Saturn and Neptune in 1846 Marx and Engles had begun their philosophical revolution and the Communist Manifesto was published shortly after. The German Empire also transitioned to the German Republic with the end of the first World War at the 1917 conjunction and the German Revolution which was itself famously prevented from becoming socialist by the establishment which then led to the rise of Nazi Germany and the Second World War in 1939 when Saturn and Neptune were inconjunct because, again, capitalism always ends either in socialism or fascism—because people generally do not like being exploited and if they are not given the opportunity to resolve their oppression through egalitarian means (socialism) they will reflexively resort to force (fascism). We likewise find ourselves at this familiar crossroads where the failed ideological structures of the past now give way and allow a new ideological cycle to take its place. The great astrologer, Andre Barbault who died in 2019 at the impressive age of 98, described this upcoming conjunction as a potentially "splendid relaunch of civilization," because not only is Saturn conjunct Neptune but this conjunction is in sextile with *both* Uranus and Pluto, which are *also* in trine to each other, indicating favorable forces of surprise, revolution, humanity, and power to accomplish it. Whether or not that finally produces something positive remains to be seen but what can be predicted with accuracy is that something *will occur,* and it is still likely that antagonistic influences will also affect this process, then the consequences of those delusional and duplicitous actors will spawn other conflict later at the stressful aspects of Saturn and Neptune as the cycle continues, as all cycles do, for instance Andre also pointing out the following conjunction in 2061 will instead be squared Pluto, promising open conflict and going so far as to identify North America as the primary staging ground.

Remembering the point of De La Fontaine's most famous fable, knowing such information and attempting to prevent it can be the very thing which brings it into existence, and Neptune cycles thusly highlight a particularly unproductive human coping behavior which is to believe and hope that after we achieve our goals and dreams we no longer afterward struggle or meet misfortune, a worldview which is entirely antithesis to the concept of cycles. Indeed dreams of utopia, heaven, peace, and other such absence of conflict are inherently delusional coping behaviors with no basis in reality because there is no future where we achieve total social cohesion and afterward are forever free of conflict, pain, tragedy, loss, and misfortune because all life is a series of interwoven cycles and will always return to times which are full of conflict and tension just as it does then to periods of peace and prosperity. Part of the delusion of disarmament is a desire to avoid violence and conflict, then the unarmed and unprepared are conquered by the ruthless lovers of violence, though more often we are actually actively harming others, not having true peace, and denying it which is what sparks such conflict, such as in Americans acting surprised by the events of 9-11 even though they were a direct consequence of our forceful occupation of oil-rich countries and the moral and societal crimes committed by our government in service of industry and greed. This particular delusion prevents us from being effective because when we do experience periods of prosperity we behave and make choices as if they will always last, which they will not, spending money and resources as if they are permanent rather than fleeting, or oppositely refusing to spend resources to build quality and stable systems in fear of later declines and loss even though they have not yet occurred.

Equally delusional, however, is when things are going badly we likewise think they will never get better, and succumb to nihilism and fail to enjoy what we do have or take effective steps which can help us achieve progress or prepare for when progress does occur, much like what occurs during the Nadir transit of Saturn through our birth chart. Even as some experience peace and prosperity others are also experiencing pain and conflict, and while some experience pain and conflict others experience prosperity, and failing to recognize and help those in struggle sows the seeds for our own future undoing, delusionally believing that the problems of others need not concern us even when we are in fact active participants in those conflicts such as is currently occurring in Sudan, Palestine, Israel, Ukraine, Russia, etc. The key to making the future a better place is not to control it but simply to show up for opportunity and do our best, not to disengage from life and problems just because we can, or to pretend bad things and bad people don't exist just because we are afraid, and engaging in control and authoritarianism is distinctly *not* our best because this assumes control of fate and the free will of others over which we truly have no power, while withdrawing from conflict oppositely denies all free-will of which we are still possessed. Fundamentally, the same lessons of personal Neptune transits apply to these cycles as discussed in the upcoming chapter on Neptune and spirituality and successful outcomes of Neptune challenges and transits. Taking care of the needs of all and not only our own is the proper course to take regardless of transits, influences, or other world trends.

Before his death in 2019 Andre Barbault also wrote in an article of the Astrological Journal that "It may well be that we are seriously threatened by a new pandemic in 2020-2021." One of the most other common and fateful cycles is the relationship of Pluto with pandemics, which occurs because Pluto is the planet of power and control and it is during pandemics that we are most at the mercy of nature and reminded of our true station on this planet as a humble animal species which does not have control over such things as fate, biology, human nature, or the nature of microbes. Several aspects with Pluto can be accompanied by pandemics, most notably both Saturn and Jupiter. The former demands that we adopt structured and disciplined behavior in order to survive pandemics, such as social distancing, self-care, masks, vaccination, washing of hands, contact trancing, education and awareness, etc., with those refusing to accept our vulnerability to disease often succumbing to it (a common human coping mechanism being denial). But with Saturn it also risks the authoritarian implementation of those requirements which can and does then do as much harm as the pandemic itself since authoritarianism is always led by fearful, insecure, and stupid humans.

When pandemics are accompanied by Jupiter it often results in a more uncontrolled pandemic that explodes out of nowhere (an instance of the growth effects of Jupiter not always being auspicious as is so commonly and errantly attributed Jupiter in most astrology practice). The 2020 coronavirus pandemic indeed accompanied the 2020 conjunctions of Saturn and Jupiter to each other and thus *both* also to Pluto. This trend similarly also occurred at the 1981-1982 beginning of the HIV/AIDS pandemic first with the Jupiter conjunction with Pluto followed closely by the conjunction with Saturn. The 1918 Spanish Flu pandemic also occurred with Jupiter conjunct Pluto, as did the 1968 Hong Kong Flu (one of the deadliest in history though we almost never hear about it). The Sixth Cholera Pandemic which lasted for decades first began after Neptune was conjunct Pluto, which makes a lot of sense because Neptune rules not only the illusory but also the unknown, and although cholera had been identified in 1854 by Dr. John Snow as a water-borne pathogen which required water treatment to prevent its spread his discovery was denied and obfuscated by politicians and public health officials and it took many, many decades before countries around the world finally moved to adopt clean water sanitation infrastructure and practices. The Fifth Cholera Pandemic likewise began in 1881 at the approaching conjunction with both Jupiter followed again shortly by Saturn, as was the Third Cholera Pandemic in 1846 following Pluto conjunct Jupiter and approaching conjunct to Uranus, before that the First Cholera Pandemic occurring around the 1819 Saturn conjunction with Pluto. In fact, the 19th Century was pretty much the cholera century, which likely began with the approaching reset of the Neptune-Uranus cycle from their conjunction in 1821 which was also squared to the Saturn, Pluto conjunction and as such was squared to Pluto for several years.

It also appears that more world-wide pandemics occur at the major aspects such as conjunction, but that smaller, more localized pandemics occur at the more minor aspects such as the 1957-58 flu pandemic largely restricted to China during Pluto square Saturn. In fact, we are still currently in the Seventh Cholera Pandemic which began in 1961 at the approach of Pluto conjunct Uranus which was also opposite Saturn, which is why it is not so known and widespread as major pandemics as the 2020 coronavirus but was likely "surprising" in the conjunction with Uranus because nobody expected cholera to be such a problem in our contemporary world. Such was also the smaller Fourth Cholera Pandemic that began at Jupiter opposite Pluto in 1863.

Uranus cycles are also quite interesting because Uranus being the planet of unpredictability, revolution, and technology (because technology is inherently revolutionary), always comes with surprises and instability, yet whether they are benefic or destructive depends on the quality of the alignment with other various planets. For instance the Uranus cycle with Saturn is a dichotomic lesson in opposites, Saturn being the planet of order and Uranus being more or less the planet of chaos, it's conjunctions such was what previously occurred in 1988 is a surprise but orderly revolution of systems and institutions such as the deconstruction of the Soviet Union away from a planned economy to mixed, as well as the beginning of new surprise challenges to systems and order such as the overdue recognition by institutions of oncoming climate change which at the moment are not yet highly consequential due to the harmonic nature of the conjunction, but will be in later, tense aspects. At the following 2000 Uranus, Saturn square began the implosion of the "dot-com" bubble, Hurricane Katrina rolled ashore at the 2005 inconjunct to devastate New Orleans, the 2008-2010 opposition was accompanied by the sub-prime mortgage financial crisis, and during the 2021 closing square was the pandemic-associated financial crisis (with recoveries then occurring in the sextile, trine, and semi-sextile influences). We currently find ourselves in the closing sextile of this current cycle as the literal government of the United States is being dismantled by a right-wing populist, and while sextiles are a benefic influence it might not *appear* to many that this is anything of a benefic nature, but we have in fact been living under a very toxic, harmful, liberal economy that has taken much prosperity and wealth and transferred it to the few, topmost people, and the karmic function of such events is to tear down or reform structures and institutions which are outdated, stagnant, or destructive, and the United States itself has had a long history of violent, merciless exploitation and oppression of humanity that many of its citizens prefer not to acknowledge as we go about buying shit and participating in destructive capitalism and the rape, theft, and murder of other countries and peoples, and the events of this time are in fact a direct karmic conse-quence of the refusal to acknowledge such harm or to change evil systems in favor of, you know, not killing other people and taking their resources. So while the destruction of the United States as we have known it is stressful for many within the United States it is actually a karmic development foretold in the stars (but also something anyone not stupid could see coming anyway) which will likely lead now to a new and more moral governance as the younger generations finally supplant the older, traumatized authors of the past.

Uranus also concerns the people (as opposed to institutions and the elite), and the 1988 conjunction with Saturn occurred in the sign of Scorpio which is also the sign of power, sex, control, taboos, and shared resources (since sex and sharing always inherently involve power dynamics), and since 1988 much of this cycle has also concerned people like myself whom are LGBTQIA+ (or as the adorable epithet the right-wing likes to call us, *the Alphabet Mafia*) which further highlights this ongoing theme through conflict with institution and rules often destructive to human life and wellbeing, especially of minorities (Uranus) whom are not in majority (Saturn). In fact, the first coming out day was held in 1988, after which there is also a marked acceleration of anti-discrimination legislation protecting the rights of LGBTQI people, because of the involvement of Saturn which governs structure and rules (which is literally what legislation is), and simi-larly antagonistic influences such as at the square in 2000 when major, national organizations and candidates for high office were super bigoted and campaigned on platforms opposed to the rights of minorities, and the 2008 opposition when bigoted legislation even in California's prop 8 in spite of electing a "leftist" (he was really a centrist) Democrat and first black President of the United States or, after having lost the war against gay people, the later shift to persecuting our transgen-der family that reached a fevered pitch at the 2021 square before people finally began to realize at the 2025 sextile how absolutely hateful and disgusting their persecution of such a small minority group really was (and since the opposition in fact won control of government and no longer cared about those issues because it was always just a distraction from their pursuit of money and power). Because of this dynamic of the Uranus, Saturn cycle, clues to the themes of the next cycle can be observed at the upcoming 2032 conjunction (every 45 years) in the sign of Gemini which, being the sign of knowledge, may concern such trends as the newer generations having both lost the ability to learn, being so reliant on technology and entertainment, or the introduction of much manufactured misinformation to destabilize and destroy entire systems and institutions on a scale never before seen, but which will lead to greater progress in communication and knowledge at the benefic aspects.

Also, as nothing in existence is exempt from the laws which govern reality there can also be observed at such intersections of history the nature of their unfolding and progression by yet other concurrent transits, for instance their activation by the Sun as it passes over them, or aspects to Mars, Uranus, or Neptune which simultaneously describe action, surprise, or delusion, respectively. Many astrologers have already established long ago these kinds of broad cycles specifically dealing with the bigger, slower planets, but there is also a highly errant practice of trying to divine the outcomes of such events as elections and their victors by birth chart interpretation of the candidates involved. This is laughably archaic and extremely unspecific because a person's birth chart contains the information for their *entire* life and personality, and it is *impossible* to divine a specific outcome of some event like an election based on such extremely generic data, as a person's chart could indicate leadership destiny and even perhaps election to a higher office but doesn't mean it will specifically be *the exact* election at issue. During the 2024 United States Presidential election I instead looked at the concurrent transits of the two candidates and it was very difficult based on their comparative transits before, during, and after the election to believe the supposed leftist candidate (they were really a centrist, anti-immigration, anti-trans supporter of genocide and so naturally lost) would come out the victor, because they had difficult transits like *Saturn opposition natal Uranus, Mars square natal Mercury, Pluto square natal Mercury,* and *Saturn opposite natal Pluto,* which are all frankly terrible transits to have while running for President. The right-wing candidate had far, far more favorable transits such as *Uranus sextile natal Saturn* and *Uranus conjunct natal MC* which is the most auspicious for something like this which indicated a change in fortune and increased power and opportunity. While it doesn't mean winning can't happen with such bad transits it would require a tremendous amount of effort, presence, luck, wisdom, and destiny that was unlikely considering the other favorable transits of the other candidate. Most astrologers predicted the supposed leftist candidate as winner, justifying their choice by birth chart, but in reality it was because they preferred her to the other, and I was not at all surprised in the least when the right-wing candidate won, especially because I was just fucking paying attention to the situation, even without the astrology.

The most elusive of great cycles is that of Neptune with Pluto which concerns the unconscious function of the entire world, Neptune being that which is intangible and Pluto the ruler of control and power and occurs every 495 years since each planet moves to excruciatingly slow. But one of the problems which plagues astrology and hampers practical usefulness of its information is excessive concern with the esoterica of astrological influences, divining fate and connections in the birth charts of disparate actors or overly dissecting the step-by-step astrological hand-holding of world events and trends to effectively remove any semblance of free will from the players concerned, and while the Neptune and Pluto cycle is quite important to the quality of time for the entire world this book is already overly long and I think I've very clearly made my point about these influences and their mark upon the course of history and will simply mention the cycle's existence.

While it is a great party trick to be able to predict the general course of the future using such cycles as predictive models they also demonstrate how the course of history and reality is a function of immutable metaphysical laws which construct metaphysical reality just as the physical laws construct physical reality, in which cause and effect (causality) are conjoined from the beginning of time to the present day and will continue long into the future, which is not the domain of humans but of the Universe and its masters and what we experience to be God. All we can do is our best, and trying more only results in frustration or unintended consequences. For instance, had Andre Barbault been able to apply his prediction of a global pandemic in 2021 to systems or institutions it still would have occurred, and since we also cannot control the behaviors of people there was in fact no way to prevent the pandemic from occurring anyway. Indeed, however, control systems were erected to do just that, to mandate the wearing of masks, social distancing, and even vaccination, and the more mandates decreed by government the more certain populations of people rebelled against them, even when it meant their own demise, because it is human nature to spite and rebel against real or perceived encroachment on our freedoms and independence. But we also risk erring in our efforts to control outcomes, even when our intention may ostensibly be good (in truth our desire to control pandemics is usually a selfish one disguised as otherwise), for instance the many people who shamed users of antiparasitic medications during the coronavirus pandemic because it was a virus, not a parasite, but many people's immune systems are in fact debilitated by parasitism (pretty much anyone with existing metabolic illness like obesity, diabetes, etc.,), and the use of antiparasitics, while misguided, did in fact help people survive by helping to restore at least some immune function. This does not mean all the other approaches to surviving COVID were legitimate, as many of them were outright insane or entirely duplicitous, and the point is that believing in our personal wisdom as absolute is the complete opposite of it. Instead the most useful application of this knowledge is simply to increase our effectiveness, through more informed decision making, and to understand why life works the way it does, which is never to stagnate, or to do as we desire, but instead in repeating cycles which can be depended upon to repeat again and again throughout not only our lives but also those of nations and history, in the fractal cycles which make up the metaphysical nature of reality.

22. The Spirituality of Neptune

Just as the behavior of others is not within our control, the existential matter of our existence is also not something that falls within the scope of our power and responsibility but is instead reflected in the nature and influence of mysterious and elusive *Neptune*.

First discovered in 1846, some astrologers delusionally believe Neptune was not an influence in astrology before this date (which is an ironic delusion to have of the planet of delusion), but this is simply a willful control mechanism meant to cover for insecurities about inconsistencies in astrology, which is silly because inconsistencies in science does not invalidate it, they are simply corrected as we gain new data and better understanding, and of course we would not have discovered Neptune until we possessed the tools to do so, and such is the very nature of Neptune also to be itself illusive! As indeed it was not even discovered directly but instead was suspected to exist through indirect observations of other planetary bodies very similar to how I eventually discovered cystic fibrosis as the root of my health problems, which is why I almost died in my earlier thirties, during a Neptune transit of my sixth house finally arriving there from indirect evidence and symptoms including cancer and diabetes, the risk of both which are greatly increased by cystic fibrosis. There are likewise many things in life which both exist and we have no idea about but which nonetheless have influence over us.

I hate when astrologers compare the mythological namesake of an astrological influence like Neptune to the nature of the planet, because really they have absolutely no fucking relationship to each other, and the planet Neptune has nothing to do with the Roman God of the Sea, and such esoterica serves mostly to alienate the general population which are not usually interested in or cannot relate to the myopic minutiae of historical, astrological tradition and esoterica and instead desire practical astrology which can assist their quest to live and succeed. Instead, Neptune demonstrates governance over that which is spiritual, intuitive, ethereal, idealistic, dreamy, wishful, illusory, delusional, and all that is unknown and unseen, and naturally rules the 12 house of the Zodiac and the correlate sign of *Pisces*. Because many people do not recognize the difference between spirituality and religion, Neptune is often mistaken as being in charge of religion, which is instead the domain of institution ruled by Saturn and the tenth house of society and philosophy ruled by Jupiter and the ninth house of the mind, as religion is an intellectual, institutional, philosophical construct meant to manage, control, codify, and exploit spirituality but which is entirely separate from it. Faith itself is Neptunian, however, because faith is a delusional coping mechanism used do handle parts of life we don't or won't understand, while religion which encourages faith (delusion) is of Saturn and institution, while the mind which forms the philosophy of religion of Jupiter and a function of expansion and growth but which, like in cancer, is not always good.

Neptune also rules the unconscious mind, demonstrating how spirituality is not something that can truly be described or intellectualized, or sometimes even identified, and the difference between spirituality, illusion, delusion, and the unseen are also often considered difficult to perceive but in reality they are quite easy to distinguish and is only difficult because most of us entirely mistake religion and dogma for spirituality, when in fact those are simply tools which attempt to control spirituality and is itself entirely separate from them. But understanding the difference between the spiritual and that which is delusion is whether it serves our ego, or subverts it. Most people use belief in God or religion entirely to serve their ego, their fears, hated, to harass others, condemn, or exalt their own personal ambitions and desires, especially in the pursuit of materialism, which is all delusion in service of the ego, not spirituality, since the primary function of the ego is for the survival of the body from potential threats to our wellbeing, while spirituality instead turns those concerns over to a higher power such as the Universe. Truly the most delusional and invalidating function of religion is to claim God loves some more than others, betraying God with the most hateful and vile of all human behaviors in complete violation of love and balance, as God does not play favorites among any creation and it is humans, not God, which have a long history of claiming the contrary in order to justify our actions based in fear, hatred, and insecurity. Beliefs, ideals, tenets, morals, mythology, dogma, standards, tradition, and ritual are all control behaviors, while spirituality is an existential connection to the greater will of the Universe which, unlike those control mechanisms, is not a construct but something that literally exists in reality because spirituality is literally the energy which makes us a living being, a deeply and singularly personal experience entirely separate from other people, institutions, philosophies, and certainly not anything which can be written down (you can write about a spiritual experience but the experience itself cannot be recorded for others to experience). Anything which serves the ego is delusion. Anything which frustrates it is spirituality.

People may read the phrase "the greater will of the Universe" and balk, both religionists because it does not specify a deity, and atheists because it admits existence of a higher power, but both are an identical control mechanism which seeks to define the indefinable for human convenience and comfort, even as religionists have never met nor seen God and atheists clearly walk on the Earth put here by powers beyond our comprehension. Because religion functions as a control mechanism it *cannot* actually facilitate spirituality because the very function of religion is to control life, to support the ego, to excuse control behaviors and construct rules by which we can expect specific outcomes in life, none which are actually possible. Spiritual experiences had by followers of religion occur *in spite* of their practices, not because of them, during rare moments when behavior *incidentally* aligns with the laws of reality. Neither side has a monopoly on drug abuse but for many people both within and without religion the chemical alteration of the physical and psychological state is the only way they can achieve moments of peace and introspection, when both stress and desires for control are unrelenting due to lack of more effective life skills which could achieve those ends through healthy and useful means.

As spirituality is derived by surrendering the ego, spirituality is derived from *surrender*, but not to other people, institutions, events, ideas, or philosophy since none of those things actually determines the composition of reality but to the conditions of reality itself, to find acceptance for the limits of our existence and humble mortal station. Because control mechanisms are by definition the *opposite* of surrender they thereby oppositely impair spirituality. Therefore practicing surrender by giving up control behaviors is also the practice of spirituality, as demonstrated by the act of doing inventory therapy even when we would rather do literally anything else but which is why it facilitates spirituality, or of meditation or prayer which requires taking time out of our day to stop doing things we want to do and instead surrender to a moment of quiet and sit in peace, without concern for our ego and its demands, and instead of talking at the Universe or God simply listen.

A practical example of this difference between ego and spirituality is that which occurs in the context of platonic and erotic relationships—When engaged with friends or family members we often shut down at moments of silence because without anything to fill the gap we become painfully aware of our emotional vulnerability and risk of becoming close to people who then could have power to hurt us, and our ego spontaneously and instinctually erects defensive barriers and we exit those encounters with pithy excuses like the kids are coming home from school or we have some work we really should get done. This is the ego at work, defending us against the vulnerability of spiritually connecting with others manifested in its effect of self-sabotage and self-defeating behavior in search of control and security. Oppositely in moments of erotic passion we instead ignore this defensiveness, driven by raging hormones and a desire to engage in sexual encounters barrel straight through to intimacy, remaining present and interested even after there is nothing left to fill the silence and persist willingly into that state of vulnerable discomfort which then has the effect of enchaining us to another and incidentally in the process also facilitating spirituality. Whether it is erotic or not all intimacy cultivates spirituality because it subverts the ego and connects us to a soul that is not our own, purposefully discarding control and turning that over to higher forces such as fate, God, and the Universe instead. Sex can be and usually is a very spiritual experience, even during casual sex or spontaneous encounters with strangers, because for most people the very nature of sex demands surrender in order to first have the experience. Of course some of us use force or sex as a control behavior in which case we never experience its spiritual function due to the pure expression of ego using the vulnerability of others for our own selfish control desires which are in truth based in fear. But this example makes it easy to distinguish spiritual influences such as are represented by Neptune from those which are delusional, simply by recognizing whether our ego is gratified or subverted.

Unlike Pluto transits during which we have no choice about the nature of its influence, Neptune transits do come with a choice between delusion and spirituality. All Neptune transits in fact are meant to center spirituality but because many of us do not know what spirituality is or are driven by unresolved trauma and control mechanisms will instead experience Neptune in a state of delusion, naivety, or ignorance, while study, introspection, self-care, and service instead take us toward the spiritual direction. Neptune's relationship to reality is one of demonstrating what spirituality is, practically, how important it is for our lives, how real it is, how unprofitable delusion can be, and how to actualize spirituality if we pay attention and avoid delusion. Even when Neptune's lessons seem harsh they are always ultimately for the benefit and wellbeing of our spirituality and higher purpose, and keeping that in mind can make any transit of Neptune more productive and fulfilling.

Neptune transits are very, very long, and whenever Neptune transits a house, sign, or aspect it often starts either with delusion or ignorance and later progresses to spirituality if we are paying attention and not totally obstinate in our behavior. The ego is an incredibly powerful force in all of us because it is meant to promote our literal survival as an evolutionary human animal, and Neptune transits are especially difficult for a species like ourselves with the power of language and the trauma of culture which can provide incorrect and delusional characterization of reality to which we would more easily submit were we more like other animals who do not have such complexities. The entirety of a Neptune transit can be delusion if we resist its lessons and continue trying to control life and people in it, but the consequences are especially harmful and can even descend into literal mental psychosis or death if the level of delusion is the most extreme a human mind can conjure. As an example, Neptune's transit of Libra's zodiac sixth house which began in 2011 when my health problems became impossible to ignore, and initially I tried to control them, as would anyone, by increasing my diet and exercise, but these efforts not only didn't pay off they directly exacerbated my condition and I was forced by reality to confront my misconceptions about life and biology, listen to my intuition, and begin study and understanding of reality in opposition to what I had been told was true. Persisting in study, research, and investigation is not delusion, but if I had any expectations of controlling biology and arriving at solutions to my health problems just because I was putting in effort that would be an example of delusion, where instead my continued study and efforts were simply the only option I had in dealing with my health, hoping that I would *eventually* find useful answers if the Universe willed it so. Many other Libras during this time similarly experienced significant health problems, even in the tragic case of the beloved Michelle Trachtenberg who had untreated diabetes and died, while others had problems derived of self-destructive behaviors like eating disorders or excessive exercise and abuse or neglect of the body because such behaviors are delusional of the reality that such behaviors always come with consequences as what happened when my health began to shut down in 2012 after years of starving myself and pushing my body too hard.

Likewise, Neptune transits which happen in youth are very often exactly those which, like Pluto, also bring trauma to our identities, often too because we are raised by people who do not themselves possess effective skills to teach us, because they in turn were raised by parents without those skills, and so on and so on. Previously when Neptune transited my zodiac fifth house, the house of creation which encompasses art, sex, and children, I was entering and growing through young adulthood from eighteen to thirty-one and worked hard to use my considerable artistic skills to earn a living. Initially I had grandiose and unrealistic ambitions to work at established video game companies, having had an internship as a teenager at one small, independent studio. The dashing of my delusions during this time from daily, mostly unrewarding work as a graphic artist supplanted my fine art skills and at the end of the transit entirely lost my considerable illustration, painting, and sculpting skills from lack of practice. This is a prime example of how fighting Neptune transits can be harsh, fighting against its influence in my life cost me not only the ability to draw well but the desire to even do it, so beaten down was I by working in business for predatory employers instead of pursuing my primary interests and life goals, due to fears of financial insecurity. If instead I had lived a more monastic existence (a modest apartment and lower bills), and pursued my art for my own satisfaction and not anchored my hopes and dreams in professional success I probably would not have lost it, and ironically found greater success in the future because of a willingness to submit to reality rather than control it.

Neptune is also the planet of dreams and wishes, and whichever house it is transiting will also be the themes of our greatest fantasies and desires. For instance when Neptune transited my zodiac fourth house of home I dreamed of having a family that loved me and a place I could feel safe and accepted. When it transitioned to the fifth house of creation I dreamed of being an accomplished comic book artist or filmmaker or video game artist and having a family and children, and when it transited my sixth house of labors I fantasized about all the companies I would make and how they could transform the world, and of being a healthy specimen of human with no health problems to get in the way of those dreams. No doubt its transit of my seventh house of relationships I will also dream of the perfect man and a relationship in which to be totally fulfilled the way it is in my favorite films and stories. While it is great to dream, unfortunately this is also often a reaction or coping mechanism to demoralizing limitations also brought by this transit wherein we cope with our inability to act by instead living out those unrealized dreams in our mind. There is nothing really so wrong with doing this but excessively retreating into the tempting coping mechanism of fantasy and detachment can seriously prevent us from living in the present and being effective within the bounds of our real limitations and meeting our real life obligations, and finding a healthy space in which to dream but also to be present and reasonable can transform this experience from one which is impairing to one which is beautiful, romantic, and soothing. Using inventory especially to inventory disappointments and frustrations can help illuminate the divide between what is unreasonable and debilitating and what is compassionate and healthy.

Escaping reality because we cannot live up to our dreams also indicates severely unrealistic expectations of ourselves, often due to unresolved childhood trauma in which our parents or other authority figures in turn also placed unreasonable expectations on us and thus modeling that behavior. For instance my suicidal depression, trauma, and young age was always going to prevent me from effectively entering the film industry because I could not comprehend myself capable of even simple steps which would have achieved that goal, and living in my fantasy was more a way to cope with the inability to accept that my real limitations would never allow me to realize those dreams at that time. It was also not my job as one person to save the world and liberate people from the limits of society, greed, and exploitation the way I had been indoctrinated in the religion of my family, as a child urged to convert others to our religion at any possible chance (which in fact is a harmful tool of indoctrination meant not to find new converts but to further trap young minds). Instead of having unrealistic expectations of ourselves, practicing inventory can help us understand what we actually can or cannot do, to help define our focus by weeding out the unimportant and unrealistic, to then care for our own needs and simply do the best we can whenever we can.

Worse, fantasies which accompany Neptune transits can sometimes make us feel so disillusioned with life that we hold others responsible for our personal disappointments, fears, and unrealistic expectations. As someone who was not just a single, solitary person but also autistic and severely traumatized I very often found myself the target of persons who sought to make me responsible for their failure to achieve their own dreams, fantasies, and expectations, and let me tell you that entertaining delusions regarding our responsibility which this transit brings can risk the loss of very valuable relationships without us even knowing what is really happening because our resentment and selfish disregard for the wellbeing of others (borne from unresolved fear) will fully and delusionally convince us that others truly are the problem even when the fault is fully our own and even when they are actively engaged in our support. Over a period of four years I cooked nearly every single meal for my partner nearly every single day we lived together except when we went out to restaurants or when I was hungover, and took care of the lawn and garden and often did the laundry and cleaning. He often went without eating for hours and hours and didn't think of making food until he was ravaged by hunger. Feeling sorry for him I instinctually picked up the job of making our meals and making sure he stayed fed, since I liked cooking anyway, and he was in turn better able to focus on his work (which I also frequently helped with too), not only having to do my own work but also take care of his needs and the maintenance of our home. Then one day after a business trip he told me it had been nice to be somewhere he was actually appreciated in complete delusion to my constant admiration and praise of his talent and care for him in our relationship and putting up his frequent abuse and histrionic, entitled behavior. Months after we separated and he was forced to take care of himself he finally messaged me one day saying he hadn't realized how much I had taken care of him. This is not uncommon, and people often interface with completely invented versions of people they have in their lives which are based in unresolved resentments, trauma, and personal insecurity, even living with entirely invented problems that don't actually exist such as one person I coached whom lived in fear of their partner chiding them for spending money. I was confused when they told me about this behavior because they didn't spend unreasonably—simply keeping the house stocked with food and occasionally buying furniture, clothing, and other supplies as needed in a completely reasonable behavior. Inquiring deeper into the problem it turned out their partner had *never* actually gotten even remotely upset with them for spending money, and the insecurity instead originated from unresolved childhood trauma surrounding money, having the whole time, however, resented their partner for this imagined behavior they were not even doing which had burdened the relationship for literally years.

While Neptune is compassionate it is as unyielding in its lessons as Pluto, because the very nature of delusion is a harmful desire to control reality, something we can in fact never do, though most of us are taught to believe in our dreams and pursue reality as if we also possess the power to will them into being or have even an adequate grasp on what reality actually is. Reality is in fact so difficult for many of us to accept that we risk just as much loss during Neptune transits as Pluto transits if we obstinately refuse to heed its lessons *because* we also will not even understand that we are mistaken or living in a completely imagined facsimile of reality. Having not tools such as inventory therapy which would help us surrender and connect to a greater purpose and serve our fellow humanity, our principles, dreams, fantasies, and desires during Neptune transits are very likely to be delusional, which risks leading us down destructive or ineffective roads fighting imaginary monsters, and it is imperative to resolve trauma and directly address fears and insecurities in order to first properly understand the reality in which we are actually living and after that to understand Neptune's influence, and center our focus on spirituality, service, and healthy dreaming in subversion of the ego.

Neptune transits do not mean giving up in the face of adversity—quite the opposite my journey is nothing if not an example of persistence, of showing up for opportunity and doing our best. However, most of us also wish for our fears to be resolved by experiencing the *opposite* of what we fear, for instance acceptance in a relationship instead of rejection, or wealth instead of poverty, or health instead of disease, or youth instead of aging, but life does not work like that, and we will all at some point in our lives experience those things we would rather not, and in such cases Neptune is even more insistent that we submit to life and accept its terms or risk being unable to successfully navigate difficult experiences. If we are listening to Neptune the transit progresses to resolution as we near the end, with a more enlightened sense of purpose, humility, spirituality, and understanding of our real place in the Universe, especially when these transits happen in adulthood and we have an increased capacity to learn from those experiences rather than be traumatized by them. As Neptune transits are very long each of us will only experience some of Neptune's influences during our lifetimes and it is most useful to identify where Neptune currently resides in relationship to our Sun sign, then trace its progression through our life in correlation with past experiences, frustrations, disappointments, delusions, and hopes and dreams in anticipation of our current and future experiences to understand the conflict between delusion and spirituality as reflected in our life path and personal challenges. As all people older than ourselves or the Sun sign which precedes ours will have already lived the transits we currently experience it can also be helpful to discuss those experiences with them, if possible, to gain greater insight.

Specifically, Neptune transiting the first house is a disturbing awakening to who we are as a person and our purpose in life, since the self and our identity is the foundation for everything we are and do. First house transits are often characterized by entertainment astrology as easy or fulfilling because it is the original house of the self, and conjunction transits are classically considered benefic, but confrontations with the self are often disturbing or disappointing, especially if they are informed by unresolved trauma, fear, or delusion, so this transit is not inherently easy or difficult but instead relevant to our level of delusion, unresolved trauma, and severity of coping and control mechanisms, where those who resist this transit are most of all destined to be crazy as delusions borne in the first house transit then directly affect and inform all the following houses, so understanding this first house transit and not deluding ourselves is extremely important, otherwise its following transit through the following houses will be far more difficult. It is very important to remember that *all* solutions to Neptune transits are pursuit of the spiritual—We are not what we do, our past, what we have, or how others think of us. Inside each of us is an inherent self separate from the life around us, and fear of letting go of trauma or superficial trappings to our identity prevents us from understanding who we really are, then we spend the rest of our lives in pursuit of validation which usually means the sacrifice of morals, relationships, sometimes even our very humanity as we live in service to the ego. In the nuance of Neptune transits we have a choice between that which is spiritual or that which is delusional, so when a Neptune first house transit is not spent in self-care and service then delusion is the only remaining alternative, which is a very poor foundation indeed on which to build an identity. It is important especially during this transit to remember that religion is *not* spirituality—religion is merely an institution whose sole purpose is to serve the ego, exploit spirituality, and excuse control and coping mechanisms and is not useful for understanding this transit which, if the case, will only serve to empower delusion and subvert spirituality rather than fulfill it. Delusion serves the ego, spirituality confounds it, and identity of self as concerned by the first house can be found not in service of our personal insecurities, fears, and hopes and dreams, but of others.

Since money is required to afford food, clothing, shelter, fun, and even often romance and relationships, Neptune transiting the second house can be very distressing for those who do not fare well in the first house transit. But the second house transit is most perilous if we are very attached to materialism, considering even the absence of financial security as a threat to our survival when it absolutely is not, and even when everything is, in fact, perfectly fine. There are many things of value in life and when our material priority is based on something as intangible and fake as money those with a Neptune transit of their second house ironically find money and resources fleeting and insubstantial but blame others for their dissatisfaction and leave behind relationships and life goals as they seek simply to neutralize their all-consuming fear of not being rich as a means to control life and in service of one of the most acrimonious ego control mechanisms. As such this transit often forces us to acknowledge other things of value as health, relationships, energy, integrity, talent, opportunity, or even time until we learn to let go of superficial fears of want and trust in life and God to care for our wellbeing even as it demonstrates such care (it just maybe isn't exactly what we want). Fighting this transit can even take the form of thieving as occurred to me not once but several times from those with this transit because of delusions we can commit such behaviors as theft without consequences. Freedom from debt is a hard lesson many learn during this transit as we also take on loans for things we don't need or instead of living within our means, or pressure other people in our lives to subsume what is in reality our personal responsibility. There can also be problems like being victim to scams or losing money in schemes, because our delusional nature is to also disregard warning signs and be willfully ignorant in service of our delusional dreams and desires. This is all easily avoided if we simply direct our energies to self-care behaviors and service of others rather than our own personal enrichment, and there is much joy which can be found in this transit if effort is made to find the spiritual consideration of resources, especially in sharing and helping alleviate material suffering in others in which this has its greatest productivity. Surrendering to the unreliable nature of reality, trusting in God and life to provide as it always does, and finding value in things that aren't money can produce instead the comfort that life gives far more than it takes, and that we do not need to rely only on our power for our material wellbeing, that life will provide if we but show up for opportunity and do our best.

As Neptune transits are very long its influence on our lives are similarly grand and singular themes that each of us seem to deal with through long stretches of our lives, and when Neptune transits the third house it will likewise correspond with delusions about knowledge and communication—specifically from where and how we acquire information, how to use that information in our lives, and then also how we share it with others. During the writing of this book those experiencing Neptune in their third house appeared to behave like the metaphorical ostrich with their head in the sand, trying to purposefully ignore information presented directly to them, to run away from uncomfortable realities, blame others for their failures, literally avoid communication, or in turn refuse to share it also with others. Many desired careers or success in media and publishing but were delusional about the consequences of their actions which prevented that success, or about the process of getting such opportunity in the first place (such as that it doesn't just fall in our lap—it has to be worked at and committed to). A prominent right-wing bigot with this transit was expressly denied opportunity to write on television shows and films as was their dream because of the consequences of the heinous, vile, racist, homophobic, and transphobic things they wrote and did, showing how delusions about consequences are often a major factor of Neptune transits. Yet others also were deluded about what information was truly useful, often conflicted between wether something was true or useful or simply a lie or unhelpful because the lies and untruths promised things they wanted such as effortless liberation from oppressive life circumstances, which again is rather easily understood in the context of whether it serves our ego or not—for instance if a solution to health problems is advertised as just having to take a pill or buy a product it is in service of our egotistical desire to be liberated from requirements to instead care for our health every day and learn about our biology because continuing education is inconvenient to our ego, but doing so subverts the ego and finds spirituality in submission to realities we would prefer not to accept.

Because the third house rules not only communication but also siblings and spiritual siblings the consequences of fighting this transit is to engender isolation, loneliness, and helplessness, because none of us can on our own solve our problems or meet our needs and we are required to communicate with others, especially our peers, not only for solutions to our problems but simply to exist and have a healthy social life and fulfilling relationships. Without serving others we become pariahs and cyphers, imprisoned in cells of our own selfish making, the key to our release not any moment of success, promotion, financial windfall, or stroke of luck but surrender of the ego. Part of the delusion to this transit is to think our siblings or childhood friends are part of the problem, when in fact they are as much a product of their own intergenerational trauma as are we, that everyone is trying their best and it is our own personal trauma and biases which are deluding us to the nature of our relationships and dependency on others. Remembering that the ego blocks spirituality while surrender engenders it, surrendering to the vulnerability and intimacy required in relationships such as those with siblings can cultivate the true spiritual nature of this transit in deepening our relationship with them, and admitting our powerlessness over life can deepen understanding of it through the acquisition, use, and distribution of knowledge in service of others (this does *not* mean putting up with abuse—on the contrary, learning how to set healthy boundaries can also be a function of communication as this transit concerns). Showing up for opportunity and doing our best during this transit can also result in very fulfilling experiences of creativity in fields of communication as writing, film, acting, books, and other forms of communication, though because Neptune transits are long these might not happen on our desired timeline, as they saying goes instead, *in God's time*.

Neptune passed through Libra's zodiac fourth house of home and family nearly the entirety of my childhood in which I never had one single place to call home nor even parents or siblings on whom to rely, and before the age of seventeen in 1998 when the transit ended we had moved fourteen times, no more than about a year on average in any single house. Being too young to understand how life actually works this was so devastating a traumatic experience I would shortly after attempt to kill myself, not so much the frequent moving that was traumatic but the complete absence of real family as my parents were as much a threat to my wellbeing as our physical transience. As mentioned before, the fourth house is one of those at a square aspect to our Sun sign, so transits through here are often wrought with tension and stress as we are forced to directly relate with people of disparate dispositions, worldviews, and individual traumas, biases, and prejudices. A more mature perspective relevant to Neptune's transit of the fourth house is that no relationships last forever, and while it is wonderful to have good parents and family on whom we can rely we are all of us individual mortal animals who also live in a mortal world and as such everything is inherently fleeting, impermanent, and unreliable, and the value of relationships and love comes not from their persistence but that they ever occurred at all, and are made no less important by their ending or scarcity. While impermanence can sound ominous it can also be a good thing, as I was finally freed from my abusive family when coming of age and forced out of the house and able to begin my own life, which would have been a better experience if I had understood my own inherent worth as an individual such as what can be found in inventory. This transit can also concern the literal place of domicile, though, and if we think that homes too are permanent (such as when we are children) this transit can be especially unsettling because they are as transient as anything in mortality. If my parents had been reliable the moving would not have been so traumatizing, and being a child and having no reference from which to understand life it was always going to be impossible to understand how to prosper in this transit, while adults could instead seek to understand our personal delusions, unhelpful behavior, and unrealistic expectations, to instead explore more enlightened concepts of home, seek fulfillment of those needs through a spiritual orientation, having compassion for our ourselves, to surrender to the reality of our need for security and belonging, without trying to manifest it through force of will. Better yet, while family is important and has very real effects on our lives we can also be our own parents and care for ourselves as what we expect from them, which they are not able to give because they also do not possess those skills (otherwise they would!), and part of the delusion of this house is that home can in fact be *wherever* we are and not dependent on any one place, and family can be whomever we choose (which similarly can be more successful when empowered by inventory and not delusion and control behaviors).

The fifth house is that of creation, which means it encompasses creativity, art, new love, casual sex, the making of children, etc., so problems that arise during Neptune's transit of the fifth house can be delusion or illusion in terms of all these acts of creation. Some people might have trouble conceiving and not know why, or often stymied when searching for romance, or will feel creatively blocked or ineffective when pursuing goals of creating businesses and projects as was the experience of Libras from when I was 17 to 31, the very end of which I began a relationship with someone whose nature I deluded myself as what concerns this transit. Self delusion in terms of romance is a primary peril of this transit, and indeed every one of my romances during this time were based on willful delusion, and as a young man I was always mystified as to why no other boys I liked seemed to like me back, when in reality I was often interested in people who were not emotionally available (due to unresolved trauma), and fell into many delusional relationships with men who were charismatic, unstable, or abusive, mistaking the intensity of their obsessions for love having never experienced it before and not knowing what it looked like (the unknown of Neptune). Spiritual unity with another is the other possible course for this transit if we resolve control behaviors and unresolved trauma. Initially I also wanted to be a comic book artist or make video games or film, but the demands of life and adulthood quickly got in the way and I settled right into working as a graphic artist for the stability and safety and stopped practicing my art. This transit need not be stressful at all, and that experience is instead a direct function of significant unresolved trauma and unrealistic expectations of ourselves such as having to make people love us, especially what occur when we are still young and inexperienced—for instance there was no way to have gotten a job in video games without an education or experience which would facilitate that. I did not need to make people love me, and simply did not yet know it is enough to simply be, and if instead had much compassion for myself and my life experiences would perhaps have found a man sensitive to those needs instead of antagonistic to them, for how can we expect others to love us when we don't even love ourselves? Remembering from my previous works that love is *not* a feeling, but *action*, having love for ourselves which is required for resolution of trauma comes not from mindset but from action—practicing self-care, doing inventory, and meeting our needs as required (as opposed to expecting others to do it for us).

Other perils of the fifth house transit are people using us in relationships, such as for financial gain, citizenship, popularity, money, etc., but deluding ourselves into believing they are interested in us for ourselves. While using people in that way is detestable it is, just like our behavior, borne of fear, insecurity, and other Neptune transits they are experiencing, and they are no more responsible for the relationship than we whom delude ourselves to what's really occurring. Or it could also be unrealistic expectations of potential partners, searching for marriage when others are only interested in superficial sex and fun, or the reverse, instead of being honest about our intentions. The fifth house does not rule partnership nor power and control, which are themes of the seventh and eighth house, but love and sex as an act of creation, so it frequently handles the beginning of love affairs, casual sex, and the creation of children, and those later themes of established partnerships occur in subsequent upcoming transits. To fare decently this transit requires keeping an open mind, not taking things too seriously, and finding joy and fulfillment in what we want to do and can do rather than what we should do or think is expected, and always have compassion for our mistakes, shortcomings, weaknesses, and failures because we are doing our best.

At the time this writing Neptune is nearing the end of Libra's zodiac sixth house transit, the long years of fateful health struggles in the house of labors as what lead to these works since it was the collapse of my health and career ambitions which catalyzed this journey of healing and discovery as what resulted in these books, and it was no coincidence that my severe health problems began after 2011 at the start of this transit when I was thirty-one, abusing my body with destructive dieting and exercise habits motivated by unrealistic expectations and yet unresolved trauma, and as my health problems progressed it was extremely difficult to understand what was actually happening to me as doctor after doctor failed to diagnose what I would later find after much work a full decade later to be cystic fibrosis. Not even when I went to an emergency room after coughing non-stop for weeks and completely unable to sleep did they think to do anything but x-ray my chest (which is *not* how cystic fibrosis or many other conditions are diagnosed). This situation was particularly interesting because cystic fibrosis in men prevents fertility, so being homosexual also never had any occasion to fail at impregnating anyone which would have provided occasion for discovery. As most humans do I first tried to handle my illness by doing my best and even ignoring a lot of it, because the prospect of dying was so abstract I could hardly comprehend what was actually happening to me. Joining Alcoholics Anonymous in 2015 because I could no longer delusionally pretend my drinking wasn't a problem was entirely an act of submission where I would learn the tools of inventory therapy which empowered me to resolve the trauma which otherwise prevented my acceptance of life's terms, biology, and my shortcomings as a writer (can you imagine it being worse than this? Because it was). During this same transit I witnessed other Libras handle things like eating disorders which are nothing but exercises in self-delusion, both in terms of our self-worth as well as how eating and dieting negatively affect our health and sexual attraction. Especially when we have unresolved fears of disease, helplessness, dependence, death, worthiness, etc., this transit can trigger existential fears which then exacerbate harmful behaviors that actually worsen the problems.

This transit is especially broad because it *also* covers other daily obligations like work since daily, mundane activities of responsibility take up the majority of our time and attention, and it is also said by many good astrologers that when Neptune transits the sixth house work our opportunities and ambitions are confounded unless they they are rooted in the good of others or service to others. My ambitions to progress as a professional 3D artist and animator or make video games were always stymied during this transit, but as I made accomplishments in health and research and shared them with others I found an engaged and interested audience whom actively supported me in turn (though I made far less money). But even when I tried to help people during this transit I was also often treated with contempt and abuse by controlling people who view the receiving of help as powerlessness, or the gatekeeping Karpman 'helper' type whom strive to maintain discord and dependence in service of their ego which further underscored the delusional side of this transit that it is not possible to help anyone unless they want it. It may be tempting to concoct some supposedly selfless scheme to find success during this time but such schemes are anyways in secret service to our ego, and any motivation for personal gain during this transit will always be met by disappointment because we also are not those whom decide what is service and what is selfish, and must be a anchored in a genuine concern with the wellbeing of others, which also includes leaving alone those who wish it even should they suffer incredibly. Many Libras I know personally are mental health therapists, and while I have no small amount of contempt the profession as mostly populated by Karpman 'helper' types they too saw some success if the nature of their work was service to others, but not personal gain if they for instance attempted to translate it into social media fame. Eventually I decided even not to have a paywall that prevented people from obtaining my work, according to their income levels, that changing the world was more important than my personal profit and financial success, and had to do much inventory concerning entitled thieves taking my life's work without offering so much as a cent for a decade of labor and research at great personal peril.

This transit could be plainly seen in prominent Libras of this time (and others in their similar upcoming transit of this house), many of whom struggled with frustrating, persistent health issues or professional frustration, some even becoming literal antagonists to humanity in violence and hatred. Other prominent Libras instead whom sought to serve humanity such as Alexandria Ocasio-Cortez who famously entered politics during this transit from a desire to help her community and toppled an established and powerful asshole found immense success, while other Libra politicians which instead served the ego (such as in sustaining exploitative, liberal capitalism) instead met failure after failure in their quest for power and control. The unknown can be very perilous in the house of health and work, and delusion means trying to live our life as if we do not have problems or consequences, while the spiritual path in this transit is to both admit and have compassion for our mortal limitations, to reorient our selfish and superficial desires for health and profession toward healing, service, and humility, and there will be plenty of opportunity in the future when this transit is done for more individually fulfilling pursuits.

Neptune transiting the seventh house is entirely oriented toward partnerships, usually those which are romantic but does absolutely include close business partnerships, friendships, or political partnerships. As all Libras are up next for this transit I haven't a great deal of personal insight into it, but seeing Virgos struggling previously with delusional expectations of their romantic partners here in the middle of the astrological gauntlet (the sixth, seventh, and eighth houses) imagine the problems of a seventh house Neptune transit as the delusional idealization of others, expecting too much of them or wanting them to be different than they actually are (as was similarly committed during its fifth house transit). One peril of this transit is thinking partnerships can solve problems that we must actually solve on our own, and one person with this transit I knew briefly during this time was frankly not very attractive in personality or appearance (they were rude, contemptuous, and annoying), and they were in a relationship, inexplicably, with an extremely cute, kind, and talented person and, from what I heard, the sex was really great but they purposefully broke up with that far lovelier partner to pursue shallow relationships. I don't know how this person's romantic life turned out but there is a good chance they never found someone as good as the one they let go and were deluded in the opposite direction typical of this transit, which is that their partner was a lot better than they perceived and certainly better than they deserved. In fact, the person might have felt that way themselves, insecure about having landed such a catch and self-destructively exited the relationship in order to stop feeling constantly insecure, but in the process being deluded to their own self worth, which comes from nothing but our very existence, and that our actions are what make us admirable not our looks or money or other superficial and transient estimations, which instead only serve as magnets for other peoples' insecurities, and that to be worth of a relationship all that is required of us is to show up for opportunity and do our best.

Sometimes we purposefully ignore problems with partnerships due to fears we aren't capable of solving conflicts, or of being alone or lonely as a result of conflict, or that of being financially unstable or from personal insecurities about self-worth, but in so doing jeopardize the partnership or even our own wellbeing in both the present and future due to our refusal to address them. When I met the ex which would later ask me to marry him but then abandoned me when I got sick there were no planets in my seventh house but Neptune (in the fifth house) was opposite my natal Venus, which rules love and the sign of Libra that is associated with the seventh house, also indicating willful delusion since opposite is a challenging aspect, and Uranus, the planet of surprise and unpredictability, entered the seventh house about a year into our relationship to reveal the consequences of Neptune's position at the time we met. Such transits do not always mean delusion, however, as opposite or square could also mean spirituality which is achieved through difficult conditions or hard work, and seemingly ominous Neptune transits like through the seventh house can equally mean deep spiritual experiences but for which it is required to surrender to vulnerability, intimacy, and spirituality, as astrology can never be used to predict events but only to understand them and the general nature of our experiences, and instead of experiencing delusion we instead experience great heights of spiritual enlightenment as what can only be experienced through others, or finding someone who is actually reliable and dependable because we did not chose them to serve as a bulwark for our fear (of which it is only our responsibility to address), or it could take the form of getting to know someone for who they really are rather than imposing our expectations and idealization onto them. This is difficult or even impossible when burdened by unresolved trauma and control and coping mechanisms, and requires uncompromising willingness to subvert the ego and listen to what life is trying to tell us, for which practice of inventory therapy in terms of all relationship experiences past and present can be very informative and illuminating, to help cultivate an open and humble approach toward relationships and their issues, remembering that it is equally possible to be deluded or spiritually fulfilled through any partnership and not only those which are romantic.

Transiting the eighth house, Neptune is likely at its most illusory since the eighth house deals with power and powerlessness, and the entire point of the conflict between delusion and surrender is in also inherently about power and powerlessness. But this position is not very intense since the themes are agreeable, unless the person is naturally inclined to significant control behaviors. Often this transit simply confuses us as to what we can and cannot control, and control mechanisms are rather directed at everything and anything which tends to then dilute their severity even for people who are not very cognizant of what they're doing. At the writing of this book many Leos with this transit seemed a bit lost and confused as to what they should be focusing on, or how to handle control behaviors coming from others such as parents, friends, bosses, work, government, etc. As sex is entirely an act of power or surrender many with this transit probably have unpleasant, embarrassing, or frustrating experiences with sex and sex partners who used sex as a way to control or shame the person with the transit, to which we then respond also with control behaviors either by avoiding or alternatively trying to assert even more control, the former leading to loneliness and the latter leading to stupendous confrontation and recrimination. Sometimes it can be hard to differentiate between what is control and what is not control—to be very specific an example for instance of control such as by sex shaming or pressure surrounding sex, the act of control by us would be to do it also in turn to them, engaging in arguments or shame or derision in recrimination as what is received, while not-control is compassionate, healthy boundaries and equal give and take, making amends whenever required and never seeking to make someone do what we want or punishing them for behavior we don't like. We are not excused of our control behaviors simply because someone does it to us, but neglect of our needs is also not the opposite of control and we all have the right to care for ourselves, set healthy boundaries, and communicate our needs, allowing others to meet them if they will or can, and if not then making wise, compassionate changes as required. A controlling response would be something like *'you're so demanding,'* to someone whom is demanding or hurtful instead of *'I can't give you what you want,'* the former an attack on a person's character meant to disarm and manipulate an opponent, but the latter is a frank recognition of the conflict at hand and both persons' right to desire but without ceding any ground on our needs and self-care nor compromising our integrity.

Shared resources are also a major theme of the eighth house, so investments, taxes, royalties, shared bank accounts, etc., could be sources of uncertainty, delusion, or even secrets or the unknown. Taking loans we cannot actually handle or which are given by duplicitous and unreliable sources could be a pitfall. As with sex, the major peril of this transit is a sense of entitlement, since those who are entitled feel they are owed or can take from others without consequence, and this delusion is likely to be sorely rebuffed during this transit. Since most people are not like this it is more likely to be rather muted, but even self-delusions such as being excessively self-sacrificing or letting a partner abuse our shared resources are still control mechanisms and the point of this transit is to teach the difference between delusion and surrender in terms of control, so nobody is likely to experience this transit without some overt effects, but they do not need to be very severe or consequential so long as we let go of control behaviors, trust life, and work to resolve trauma which motivate it, turning over control to those higher forces actually in charge.

In the ninth house of the mind Neptune takes the form of professor and attempts to teach us which of our ideas and beliefs are delusions and which are grounded in reality, or even how to transcend reality if we are particularly tuned in. Neptune is not intellectual, but the ninth house is, so this transit is more muted and particularly benefits from logical, rational thought and time spent in education, meditation, and self-reflection. Many of us have been brought up in communities and religions which sustain mythological and delusional coping mechanisms, and during this transit these ideologies become a profound liability and risk becoming *even more* deluded which can have especially disastrous consequences later. Failing to understand reality and spirituality this transit can result even in cult behavior and support of duplicitous or delusional figures who tell us exactly what we want to hear. A shortcut to navigating this transit is that no person in any reality is deserving of worship or idealization—All humans exist in the same space of mortal limitation and dependency and it is impossible to court the divine or entreat fate, and those who proclaim to do that are either traumatized and delusional or purposefully duplicitous and exploitative, and allowing people to prey on your own personal wishes and desires for money, salvation, or excitement is our own control behavior which will lead into entirely delusional ideologies and even voluntary loss of money, resources, and relationships given in service of false prophets and false idols. As with everything in life we are all individually capable of experiencing spirituality *without* intervention or dependence on others, and the Universe has truths to teach us but we have *nothing* to teach the

Universe, so humbling ourselves even when strongly held beliefs seem threatened is God attempting to teach us how reality really works, not how we think or want it to be, and resisting such lessons is exactly opposite of the humility often ironically required of religious beliefs anyway.

Sometimes the transit through the ninth house is marked by a desire or opportunity to travel and experience other places and people, to escape the intellectual stress which comes from our useless knowledge and rituals, but sometimes this journey can occur even very close to home as we are exposed to different ways of life and experiences not familiar to our own lives. Or it can also encompass learning in an institution such as school, with the peril that we delusionally think an institution is always right simply because it is an institution, and oppositely experience awe and understanding if we relent control, subvert the ego, and allow the Universe to teach us through experience. This transit is much easier to resist than others since consequences of doing so are not always obvious, but doing so is still perilous because there are real truths about how the Universe works which can enlighten us to the nature of God, who we are as a person, and how life actually works which will benefit us in years to come and more difficult transits, which can never be known if we do not surrender and willingly shed those delusions borne of fear and trauma and the desire to tell life what it is rather than let life tell us who we are. Oppositely if we let go of our ego and remain open to learning it is possible to experience revolutionary new spiritual insight and philosophy during this time, but which requires absolute dedication to shedding service of the ego and instead its subversion and service to others.

Neptune transiting the tenth house is a more tangible experience since it directly and conspicuously encompasses career, reputation, and public image. Many institutions do not actually have our best interests in mind, a reality which can be rather upsetting during this transit through consequences of membership or belonging, and often continues directly from themes in the ninth house since the institutions of the tenth are informed by ideas from the ninth. Knowing many Geminis experiencing this exact transit during the writing of this book it became clear that clinging to institutions when they do not serve our best interests comes with a profound cost in the public's estimation of us. Like all Neptune transits it is entirely possible to cling to delusion throughout the entirety of this transit but the price of doing so will be especially steep during tenth house transits since it also squares our Sun sign. Fundamentally, the idea that our public, outward image is defined by us and not those whom possess the opinion is the delusion at the heart of this transit, and it is not possible to define or control what others think, and doing so during such a transit as this will always reap unintended consequences, perhaps even being deluded into believing we are actually accomplishing our ends but which instead have effects we are not even aware of which later come to cause us much loss and misfortune.

As with all problematic transits we only behave unproductively because of fear, and learning to accept and overcome our fears can give us the strength to surrender to the lessons life is trying to show us, such as that what matters is our own integrity and not the opinions of others. For instance, an example entry under the 'better way' column in a fear inventory of public opinion of us would be *'who the fuck cares what people think?'* and then to honor your own personal integrity rather than what you feel is expected, politic, or opportunistic. Right now many people with a Neptune zodiac tenth house transit as it transitions into the sign of Aries are beginning careers and projects based on grifting and opportunism, and like was the experience of those previously with this transit will have to pay the bill later as the natural consequences of acting without integrity, because there are *always* consequences to our behavior whether they are now or in the future. Such opportunism comes from suddenly realizing our best interests are not a priority for institutions with which we have long identified, and thinking instead our advancement can be achieved through subversion, dishonesty, and appealing to people's fear, hatred, or anger. But delusional behavior during this transit wastes our energy, skills, talents, relationships, money, and health where oppositely listening to Neptune in this transit can produce very rewarding spiritual culminations of previous life experiences and real progress in our lives, in spite of what other people think. A productive tenth house Neptune transit very frequently results in abandoning one career for another, leaving exploitative institutions, or finding ways to be of service of others through our public face and career. Spirituality is an entirely personal experience while membership in an institution is not, which is a reason why spirituality often seems elusive when people search for it without when in fact it comes from within. Pride most of all motivates resistance to this influence because negative opinions of others can be especially embarrassing due to evolutionary desires to be accepted by the group. Those expectations can be subverted by humbly making amends for wrongs and fixing past mistakes, with no intent for our benefit, and otherwise not acting on concerns for reputation or public perception but nor by avoiding them either. Giving in to fear (fear of others, of criticism, etc.) is the delusional option in this transit, while surrendering the ego and trusting in life to take us where it will is the spiritual path.

In the eleventh house of fortune Neptune exposes our delusions in regard to friendships, groups, wealth, and luck, or facilitates spiritual experiences in those areas. Many of us have friendships or association with others borne of convenience and superficiality, by association with institutions of the tenth house, or our wealth, status, connections, or even our own personal control mechanisms. Because we evolutionarily require other human beings for our own survival, fulfillment, and wellbeing, rejection from the group is a fundamental motivation for associations which merely serve to nullify this fear rather than being based on real and substantial connections, love, and spirituality, and when the latter is not the case those friendships and fortunes will likely be lost during this transit as the superficiality of our associations is made painfully clear, and many people may feel lost or alone even when we have plenty of people in our lives. Much like my experiences with Neptune in the fifth house of creation we can also be deluded to the nature of others, mistaking charisma and intensity for interest and reliability and may be rudely confronted by the real nature of those we thought were otherwise reliable and relatable, or we may place unrealistic expectations on friends and become disappointed when they do not live up to those expectations in which case the fault lies with us and our lack of compassion for others (which originates from the absence of compassion for ourselves). The recognition that we also only serve this purpose for others is particularly upsetting because it also triggers the very fear of abandonment inherent to our instinctual biology. But not all relationships are so shallow, and developing a strong sense of self and refining our beliefs and interests and resolving unresolved trauma can help us find and cultivate meaningful and fulfilling associations.

Because this transit also encompasses other types of fortune such as wealth some of us may find that our investments or savings does not provide as many benefits as we though it does, or it may abandon us in a time of need and we find our luck does not always hold out, and like other delusions are not consistencies on which to rely. There is also the distinct ability of making investments into insubstantial, duplicitous, or false schemes and losing our wealth because of our delusional choices or undisciplined or unreasonable hopes and insecurity. The instability of life is the only constant on which we can rely, so such experiences are meant to help us find meaning in things which are not so transient, insubstantial, and material, and not entertaining control mechanisms and coping behaviors just because they make us feel safe, that we do not need money or riches to be fulfilled and happy and while such experiences may be stressful they will ultimately strengthen our connection to things of a spiritual nature.

The twelfth house is the natural home for Neptune and is marked by profound experiences in solitude, reflection, meditation, and the culmination of lessons learned throughout our lives. As the twelfth house is thus also that of the unseen, the unknown, endings, and spirituality we may find this occurring literally, with less participation in public and outward life, especially if we are not young, forced through circumstances or by feelings of inadequacy or confusion to experience spirituality. This is ironic considering that spirituality is not often obligatory or forced upon us, but there are spiritual cycles in life which are in fact obligatory which includes loss, endings, and rebirth, frustrated ambitions, etc., and before any rebirth can occur such as starting a new cycle in the first house a loss must come first, or least the letting go of things that no longer serve us. This is not always very traumatic but instead the death of the ego, of expectations, of delusions, or that which we previously mistook for spirituality. The delusion which occurs here is that we can find satisfaction in life without true spirituality, and resisting this transit can bring profound feelings of emptiness and dissatisfaction from events and pursuits in our lives, especially if we expected satisfaction and fulfillment from them in the first place. We can persist in disappointment too if we refuse to surrender and turn inward, which is exactly what is required of this time.

Many Aires during Neptune's transit of the twelfth house were thus less active or visible, some saw literal conclusions to life pursuits or probably felt a sense of emptiness even though they had achieved success or the realization of life goals and pursuits, wondering why these things did not bring the sense of satisfaction we perhaps thought would come. Others were challenged to discard their old or incorrect ideas of spirituality but might have instead clung to them doggedly in the insistence that we, not the Universe, control life. This transit does not mean we must abandon the outside world during Neptune's twelfth house transit—spirituality can and should be found *while also* existing in the world, not apart from it. Attempting to escape life is a form of control, by resisting, controlling life by avoiding it rather than finding the bravery to submit ourselves to fate, fear, and trust in and heeding the forces of the Universe. Other Aries oppositely met the challenge of finding spiritual surrender to the realities of life and were thusly privileged to peer into the vast expanse of reality and find more joy and fulfillment than before, as spirituality is not separate from the world or from life it is the very energy which makes life possible, so it can be found everywhere (especially in nature) and we do not need to remove ourselves from life to find it, and these truths are merely found inward, through reflection, meditation, quiet, inventory therapy, letting go of control, etc.

Refusing to subvert our will during twelfth house transits we can, during this time, become also the source of loss for others, reflecting our own fear onto the world around us and avoiding experiences we need to have will produce unfortunate consequences for friends, relatives, business partners, etc., as we seek to hold onto things which in truth we must let go. We are never in charge of life. That power belongs only to God. This truth learned during this transit can bring a deep sense of fulfillment as well as spiritually satisfying experiences and healing conclusions to certain themes of our lives, or risk significant self-deception and loss if we refuse to let go of fear and control. As with all the zodiac the twelfth house is not itself an end but merely precedes the first house in a ceaseless cycle, and learning humility, understanding spirituality, being in nature, and accepting the unknown will establish a healthy foundation from which to build the sense of self in the upcoming first house transit. Without learning surrender during Neptune's twelfth house transit delusion will make personal identity and purpose in the first house confusing and frustrating, where instead surrender and spirituality will make identity and purpose in the upcoming first house clear, fruitful, and awesome.

Transits to our natal Neptune also contain these influences and effects, depending on the angle and quality of the aspect. For instance whenever the Sun transits trine to our natal Neptune we are usually called upon to offer service to others without any regard for our own benefit. One quite profound experience I had several years ago (they will not all be so obvious) was after a 12-step meeting in which by chance afterward in the parking lot of a grocery store when doing my shopping I happened upon an older man from the meeting. Saying hello and being friendly he soon confided in me that he was having a very hard time, and I listened to his problems and then gave what I thought to be helpful advice, that it was *okay* to have a hard time and that we are only one person doing our best, and to have compassion for himself in what he was going through. I don't know if it helped but that wasn't really the point, which was instead to do something that had no immediate benefit for me at all, as will everyone experience on a regular basis (this one twice a year, to be exact). Difficult aspects to natal Neptune like square or opposition similarly might bring periods of delusion or required spirituality through conflict or tension, and other planets such as Venus, Mars, Uranus, Pluto, Jupiter, Saturn, Chiron, the Moon, etc., will likewise bring their natures to aspects with natal Neptune and affect our spirituality (or delusion nature) according to the quality of the aspect in the domain of their rulership such as love, ambition, spontaneity, control, growth, structure, healing, and emotions, respectively.

Neptune is a lovely planet which embodies all the mystery of existence, whose transits of the Zodiac directly correlate with our inner, spiritual, and delusional natures, and clues to what we do not know, is secret, illusive, illusory, delusional, or spiritual lies in its transit across the heavens and relation to our birth chart.

Understanding that we are mortal humans incapable of knowing all things brings empowerment during Neptune transits, making sure to always serve the spiritual function if we are to avoid delusional potential, that while expecting any definite outcome from a Neptune transit such as believing we are in fact being spiritual and thus in control is exactly the kind of delusion it brings. Frequent practice of both meditation and inventory therapy is the best way to experience spirituality and all influences of Neptune, to better understand its purpose not through rationalization and intellectualization but through simply being, existing, and connecting to the spiritual undercurrent that flows through all of life. Instead of focusing on outcomes we instead shift our attention to our behavior which, if in service to the ego (such as by refusing to make amends as required) is delusion, and subversion of the ego, especially in service to others without concern for our benefit (such as making of amends as required), is the path to spirituality.

23. Money and Means

One of the most enduring uses of astrology is to concern ourselves with wealth and riches. especially in entertainment astrology, which often feature also advertisements for predatory psychics and other exploitative bullshit meant to prey on people's desire for the security they believe money can bring or shelter in fantasy and illusion, and there is a fixation on prosperity and money which is as obscene as it is incorrect.

We desire wealth and riches when we fear life, fate, do not trust in the Universe (God), and rely only on our own sorry talents which deep down we know are not sufficient to subvert things like disaster and loss. While astrology can portend periods of wealth or its likely loss, that knowledge *cannot* be used to actually subvert fate and make ourselves rich if that is not already in the plans for our personal journey in the first place, and anyone using astrology either to anticipate wealth or alter the course of our lives to obtain it will be sorely disappointed, because, again, we do not possess the ability to control reality, even when armed with such knowledge.

But the primary reason for our troubles concerning money is that money is not actually real, and wealth and prosperity is in truth every bit as fleeting and cyclical as any other part of life, not something which is directly determined by our will or desire but merely the scenery through which our path on this Earth progresses. Money is the curse of humanity, but not because money is entirely evil but in fact *because* it can also be good and necessary, and therefore people pursue wealth and riches thinking we are in fact caring for our needs or doing things which are productive and useful when in fact we are usually acting on fear which at best creates ineffective or at worst destructive, harmful behaviors.

While fear is a survival tool that helps to keep animals alive it can and does also make us act completely irrationally and ineffectively. For instance, the fact that rich people fear being in the very social class in which I and many literal *billions* of others live is completely psychotic, getting along not only fine without lots of money but in many ways live a life far less complicated and far more fulfilling, productive, and wonderful than people with multiple homes and jet-setting lifestyles can ever experience, as the rich mostly spend their time and energy in service of that money and the systems which allow them to hoard such wealth, which is otherwise uninteresting to normal people, giving over their entire lives to institutions and systems which do not care about them in the least and do nothing for anyone, constantly in conflict with yet others who desire wealth in lawsuits, recrimination, and even heinous and morally abhorrent behavior that costs far more than money could ever pay, ironically hardly ever truly enjoying their lives, constantly fixated on status, chasing wealth, and abuse of drugs or superfluous experiences to numb them from the empty void such behavior creates, unable even to enjoy common, everyday luxuries like grocery shopping or taking a stroll through the city. Because we mistake money as something inherently good, because it gives us the delusion of control, we then make ineffective or even misguided decisions based on the delusion that all pursuit of money is likewise good, when in fact money is only ever good in the way it is used and not inherent to what is, in reality nothing more than an artificial construct.

As discussed in the chapter on rational astrology there exists together both fate and free will, where we are the ones who decide our destiny but that destiny is also a function of the metaphysical laws of reality in which we will always make the choice we are going to make, due to our life history, environment, human biology, the society in which we live, and all the other factors which have more influence over our life course than our own personal will. Free-will cannot exist without fate which firstly gives us opportunity to exercise it, just as fate cannot exist without free-will which is the agent which brings fate into being. Having experiences wherein we learn new information such as in the reading of this book and the resulting tools it may give us to be more effective in our lives is itself not a life-hack we stumbled upon accidentally and can now cheat life to suit our needs but instead something that was *always* going to happen to you in the course of this life. We must show up for our life and do our best, but what we end up doing is also fate and destiny which cannot be altered, least of all by the amount of riches we receive.

Yet there are also useful laws of astrology which can be used to better understand our position and the course of our lives, and most relevant to our finances and wealth are transits, positions, and related influences to the second *House of Means,* eleventh *House of Fortune,* and *eighth House of Power.* Where the first house rules the self, which encompasses our birth, having then resources that we may continue to live and grow is the next most immediate step and requirement for the progression of life and thus why the second house rules means and resources. The eleventh house similarly rules wealth, separate from means is more about resources *after* we have gotten them and what we do with them (and why it also includes friends). The eighth house does often concern finances but in terms of the power dynamics which result from them, either in sharing of resources with spouses or business partners or owing taxes to the government.

It is not possible to jump ahead over progression to other life themes like making friends or starting a family before we have the means to even live, so the karmic progression of the zodiac thusly demonstrates the requirement for means as the second step or stage as represented by the second house in the logical progression through the Zodiac. Of course, the second house is equivalent to the sign of *Taurus*, as Taurus also rules the Earth beneath our feet and things like minerals because those are all resources. Money is also such a banal representation of more fundamental principles of life and existence which in my own experience were much more substantial than merely how much money was in my bank account. Both the land in which we grow food and the food which comes from the land is also a resource, and having energy to work the land (or show up for any job) is also a resource. In fact, the second house appears to *literally* concern means, not money, in the sense of *opportunity* to get money or other resources and not only that we get them. So while some transits concerning the second house did bring changes or challenges to my income they also highlighted other things of value which were not money, especially the *opportunity* earn money or income, which is itself means and more relevant to the influence of the second house.

The true purpose of money also is not for wealth but instead the ability to *participate* in life, to both care for our own needs through the gathering of resources and thus relieving us of dependence on others, in the process also directly contributing to society through our exchanging of things of value with others and working to produce value and contribute. After all, in its best conception money is nothing more than a contract between us and others without whom money is then entirely meaningless, and in fact the sign of Taurus and the second house also embodies principles of *sharing*, which is why Taurus are so often generous (if not debilitated by unresolved trauma), as they especially love *participation* in life, not hoarding wealth (because they do share so readily it does risk the 'buying' of relationships, however).

Therefore, the second house is not only about mundane financial status but the ability to take care of our needs and participate in society through the production, acquisition, and distribution of resources and the means by which we achieve this, which yes does in fact include money but which is not limited only to it, and many people who neglect their body often find insufficient energy or health to meet the demands of second house transits, although material means and resources are often the primary focus. From the end of 2012 to 2015 Saturn began its transit of the zodiac second house for all Libras, and while there had been some moderate financial stress in my life before this time it also coincided with the beginning years of Neptune's transit through the sixth house of labors (work and health), showing how all transits function together and not in distilled isolation, and many like myself likely experienced the burgeoning chaos that was to be our lives for some time, first in the demand of Saturn that our means comply with structure and discipline all the way till it joined Neptune in the sixth house years later. Almost immediately at the outset of Saturn's transit in the second house arguments with my partner over our income became more frequent, even to the point he demanded I also work as a barista at a coffee chain even though I made five times that rate of pay doing my normal career as a motion graphics artist and animator. This happened because where Libras had recently had Saturn in the second house Aquarians had Neptune instead in their second and Saturn in their tenth, the house of career and public image, and where Saturn demands structure and discipline Neptune inflicts delusion and illusion (or alternatively blesses us with spirituality if we apply our efforts to service of others instead of our ego). Because our relationship had been in a downward spiral already for some time it ended when I got really sick, after literally nursing him back to heath from an especially traumatic case of appendicitis, and before the end of Saturn's transit I relocated back to Los Angeles during my recovery from cancer and alcoholism, sleeping on friends' couches while putting my life back together and working to finally achieve financial stability, as mentioned before, when Saturn finally gifted me with a well-paying job at the end of its transit of Libra's second house of means.

Indeed if at the time of reading this Saturn is transiting your second zodiac house it is something to look forward to even if at the present moment financial affairs are very stressful—because at the end there will come a wonderful gift not only of relief but likely generosity if you are paying attention and doing your best. Indeed though the Saturn transit of the second house can be stressful for many people this is in truth not a function of money and finances but of our perception of our own self-worth and the purpose of life and resources, believing wrongly that our life is only up to us and that nobody else can save us from the pain and despair of loss and loneliness (which is indeed an onerous burden to bear). As my experience demonstrated when my life fell apart I found opportunity nearly immediately, even unsolicitedly offered a job at the company I was freelancing with many nice coworkers I got to see on a daily basis and given money in turn for my contribution to their workplace because the Universe (God) not only has a path for us but, contrary to what we may fear, is also constantly trying to help us, which does not mean we get what we want but that all we are *required* to do is show up for opportunity and do our best.

As discussed before, transits through the zodiac appear to play far more the role of fate in our lives where the ascendant Houses are instead more relevant to our freewill, and as Saturn transited my second ascendant house immediately after its zodiac second house transit (because of the position of my birth chart angles), was when I simultaneously began pursuing this current work of writing, research, and producing literature and science for consumption as a means for future livelihood to replace the unfulfilling corporate career, even if I didn't at the time realize that was what I was doing, and even though I would make *far* less money because it was far more fulfilling, but still demonstrating the concept of means as opposed to simply money, that it is what we *do* for money which mostly concerns the second house.

But the location of birth planets in the chart can also greatly affect our money and wealth nature and life path. If for instance Saturn is located in the natal second house we likely have a shrewd and responsible approach to earning and managing money and resources (perhaps even to the point of being miserly?), and will likely achieve a stable financial status throughout life due to this structured and responsible approach to finances. Or if Jupiter is in this place may have a life lucky with an abundance of resources. Mars might indicate a life spend in pursuit of material success, while Mercury would mean success through some kind of communication or learning, Venus probably through love or partnership, Uranus unpredictability and instability, and Pluto's placement there likely control issues surrounding money or the merciless or ruthless pursuit of it (which, because of the dual nature of Pluto, could also be transposed to more spiritual application of resources though life-changing events). As previously mentioned, Neptune is in my second ascendant House and so my life path was marked by delusion and illusion surrounding means, such as initially believing I could make a living as a comic book artist or video game artist but could not find a way into that industry due to limited opportunities, psychological illness, and lack of education. But when I was around just five or six years of age a friend and I accompanying his mother to the grocery store did not yet understand the concept of money and pocketed two candy bars which nobody noticed until we got into her car and began joyfully unwrapping them. The horror and embarrassment which came afterward as she scolded us and marched us back and shamed us in front of the store for stealing made me horrified at the thought of doing that ever again for the rest of my life. Indeed, at the age of thirteen I made an acquaintance in school who one day showed me his trick of slipping candy and other un-purchased items up the sleeves of his shirt, and afterward his shoebox stash of illicitly-gotten candy, and then never hung out with him again after (I should have just encouraged him to be honest?). As wealth and income have always been a challenging part of my life, for years I thought that having this placement doomed me always to failure even when I tried my hardest, since delusion is a fundamental property of Neptune, and while money is not very important to me there are still things like a garden and a family I greatly desire which require it (especially in this time in which we find ourselves), and believing that money is not required in life is itself too a delusion. Yet during the course of my research into astrology and practice of inventory I finally realized this placement also has the same positive potential as what is suggested often in Neptune transits, which is that delusion and illusion are only one side of an influence which is or can be equally spiritual and enlightened, and that it perhaps has allowed me to recognize there is far more in life which is valuable than money, such as our time, integrity, respect, relationships, intelligence, wisdom, health, etc., and as is often given as advice for Neptune transits the service of others is one of the most appropriate and actionable ways which we can actually be effective during this influence, realizing then that instead of valuing money I have come to work toward the uplifting of others, in building knowledge and science and the production of books, knowledge, resources, material, and reputation which will serve as resources in the long term rather than simply working have comfort now, and in that have instead found the purpose and success which was otherwise denied when pursuing other, more self-centered, myopic career goals. Such also can Neptune's transit through the second house be handled, by redirecting the energy spent in selfish insecurity, anxiety, and fear into the selfless service of others and more enlightened use of our time and energy.

Because transiting Neptune is quite slow and his transits quite long his lessons when moving through the second or eleventh houses can often be demoralizing, sometimes seemingly unbearable since we are never relieved of them in the way we want or anything less than far too long. After splitting up, my old partner with Neptune transiting his zodiac second house took all our furniture, money, our dogs, and even kept almost every birthday and Christmas gift he had ever given me though we left a home in which we had lived rent free for two years because of the work I did in its remodel, or when I paid his rent for a few months early in our relationship so he could pursue his own creative ambitions, so angry and resentful was he that I did not earn six-figures. Previous to that transit of Neptune it was in the second house of means for Capricorn, and *both* family relations whom were Capricorn also stole from me during that time, such is delusional nature of Neptune in the house of means, not only meaning delusion about the consequences of such behavior but that even such action is necessary in the first place! Which is founded on yet the other delusion that we alone are responsible for our wellbeing when in fact there is not only destiny, fate, the Universe, and God but also job opportunities, career advancement, time, wisdom, other people, friends, family, social services and government programs to help people in need, etc.

You will remember from earlier in this book that a choice must be made during all Neptune transits between delusion or spirituality, either to surrender our ego and seek the spiritual lessons it gives or else persist in delusion and illusion as we challenge the laws of the Universe and tempt fate such as through behaviors like theft, neglect, willful ignorance, etc., as if there will be no consequences to such behavior. In truth, as a very mild-mannered person I did not even retaliate against any of the people who stole from me (I did start to once but then decided to let it go). But this does not mean there weren't consequences to their behavior, as not only did they immediately lose my respect and trust, they *also* removed *themselves* from my life because my presence reminded them every time of what they had done to me, not only continuing to hurt me but also robbing themselves of a relationship which otherwise would have been of value to their lives. The law of cause and consequence is *immutable*, and can never be transgressed, even if we avoid immediate or material consequences for our behavior there are *always* consequences, and those which are immaterial are even worse because they affect our very concept of ourselves and our ability to find fulfillment! Desperation which accompanies the fear that we have not enough means we believe are required (or that we are entitled) is the consequence of not choosing the spiritual path, which is a direct consequence of not trusting in life and God to have a path for us in the first place (considering the religious indoctrination of my family their thieving dishonesty is either ironic or exactly what we would expect), where the Universe is the power which brings us means and other resources, not ourselves, where our only responsibility in life is not to determine the outcome of any situation but merely to show up for opportunity and do our best. Earning money at a job is only an illusion of control over such things as material resources, because those can be taken from us at any time for any reason, or we can be paid so little as occurs in capitalist economies that not even one job is enough to survive, because in reality we can control nothing but our own choices and decisions and the rest is up to fate. Compared to bending the course of fate, which we cannot do anyway, the responsibility of simply doing our best is indeed most simple, but if we believe the opposite is required of us, to control the very nature of reality and force others to do what we want we then resort to desperate behaviors like stealing from others in order to accomplish the impossible task we have set for ourselves.

The delusions of Neptune are of course not a foregone inevitability, and most people of course do not steal during Neptune transits, only those whom feel most afraid that we will not meet our high or impossible expectations, because of unresolved trauma and coping mechanisms and the pursuit of wealth as a replacement for self-worth. One person I know was raised in an even more religious home than myself, and the poverty of their parents was used as a mechanism to abuse the children and draw attention to themselves, always impressing upon the children how very near destitution they were and making a show of their poverty in veiled solicitations of pity and charity from their religious community which in turn imbued the children with desperate fear of the need for financial security as adults that they did such things as steal or marry rich people they didn't actually like just so they could be cared for, in one of the sorriest consequences of not believing in our own inherent self worth that we waste our entire lives with people we don't even like just to treat such fears and insecurities. Practicing inventory therapy such as for fears of financial insecurity, fear of loss, fear of being poor, of being alone, of our incompetence, resentment of the poor or resentment of being poor, or against family, friends, or business partners concerning money and finances owed, taken, or absent, taxes and government, thieves and theft, capitalism and money itself, economies, landlords, etc., can help us reorient the delusional fear surrounding money to humble spirituality, a less severe worldview, and finding of joy and value in things which are not money such as relationships, integrity, joy, peace, opportunity, respect, life skills, wisdom, nature, health, rest, stability, etc., and to also use our wealth in service of others rather than to selfishly hoard and keep that which we take, for indeed there is no rich person on all the Earth whom has not taken more than they gave.

In my book on psychology I relate a story about when my mother came into my room one day and told me she was afraid I wouldn't be able to provide for a family when I grew up. She delivered this scathing estimation of me at the same time that I was co-captain of my swimming team, on the honor roll and having the best grades of anyone in my family, working on my Eagle Scout badge (which I got later), having an internship at a computer graphics company, a position on student council, studying for the ACT test, and working for my father on weekends and other "free" time. It was an absurd, abusive thing thing to say, likely triggered by my obvious disinterest in materialism which otherwise preoccupied her entire existence. I was also exhausted from taking on too many responsibilities, obviously not aware in the least of also being autistic and having cystic fibrosis, and resisted working for my Dad (not *with*) because of similar abuse which occurred every time we were hauled to his job sites and made to sweep sawdust or lay sod. Never had my parents actually taught us how to manage finances, let alone how even to mentally handle money (because we cannot teach skills we do not ourselves possess), and instead teaching through their own poor behavior to fear financial insecurity and do everything possible to live in the upscale, snooty neighborhoods, even though we were frequently poor, like build houses we couldn't afford and live in them until they sold before moving to the next one.

Pluto entered my second ascendant House in 1995 when I was fourteen years old, and would be there until about 2009 when at the age of twenty-eight it entered my third house, which correlated directly with these experiences of trauma as a child and then my resultant dysfunctional life as an adult, since Pluto transits are so long.

Because of this abuse and neglect, my only skill as an adult to handle finances was to *ignore* them and, just like my parents, I then often found myself without enough money to make ends meet, living paycheck to paycheck, *even* when I was in fact earning more than enough to live comfortably. Like people often do I tried to handle my financial instability at times, making balance sheets or using software to stay on track and avoid frequently over-drafting my bank account. But the problem was not a lack of discipline or structure but of unresolved trauma surrounding money, finances, and my ability to provide for myself which had been thoroughly dashed by this abuse, because even while earning plenty as a sought-after 3D artist and animator I still carried this undeserved self-doubt and poor self-estimation instilled, and no matter how many spreadsheets or financial plans I made all my financial choices originated from the fear of being unable to provide for my own needs or what others expected of me, so instead of making responsible financial decisions such as after crashing my car drunk into a concrete wall at the age of twenty-seven I instead leased an even *more* expensive car and bought presents for people I wanted to be my friend but whom didn't care about me in the least just to prove I wasn't a loser lacking effective life skills. After the pandemic I was making about $100,000 less a year than I did as a graphic artist, but never once in eight years of living in technical poverty did I overdraft my bank account, have any unpaid bills, or get into any debt, because of the way inventory therapy helped me learn new life skills like the ability to even handle money in the first place, like recognizing what money really is (a meaningless construct) and how unimportant it is to our quality of life and self-image as compared to other, more important and real things like confidence, friendship, peace of mind, time, health, good food, and knowledge (especially knowledge). The transit of Pluto through my second ascendant House thus correlated with the desire to control life and reality by making my finances do what I wanted, which is in express conflict with how reality works, and because it was in conflict with laws of causality I was never effective in solving the problem, which finally occurred when finally learning the skills required which have nothing to do with investing, saving, documenting, etc., but in self-esteem and understanding the constraints of causality.

Money and means are not always so glum and worrying, and Jupiter's transit through the means, wealth, and power houses is often am opportunity to also improve our financial standing, although these are often misrepresented by untalented astrologers as being windfalls instead are often the result of past actions which help provide for our means, opportunity to balance shared resources, or to grow or secure our wealth. Personally, the transit of Jupiter through the second zodiac house which occurred for Libras in 2016 and 2017 was the time when I first started accepting financial support for my blog before then turning it into a book, and still required the production of content such as articles and information for people to support, not to simply sit back and watch the money pour in without effort. Jupiter transits are nearly always auspicious, just not as magically and wondrously benefic as is so often peddled in exploitative and entertainment astrology meant to prey on people's dreams and fears, which are in turn simply based in unrealistic fears of our own incompetence. Jupiter always serves to expand and grow, so there will certainly be opportunities for increasing our means during its transit of the second, eighth, and eleventh houses, and even its transits to the Ascendant or Midheaven can also sometimes come with financial results, but because Jupiter transits only last a relatively short time it is important to take advantage of the opportunities which come and not expect them to last forever. If for some reason you miss a Jupiter transit it will be back there again in another short 12 years, and in the meantime will touch many, many other parts of life as it continues without end around the Zodiac.

Uranus transits of the financial houses are something also to be very wary because Uranus being the planet of *surprise* and *unpredictability* is not harmonious with the concept of financial stability, and the transits here will certainly present with unexpected bills, expenses, and other financial themes. This can also occur in the eighth house too, which is the house not only of power and control but also for shared resources, which can then bring unexpected tax bills or problems with a spouse or business partner (or partners) which have joint custody over investments, savings, and income. When Uranus transited the eighth house I received two surprise tax reassessments due to government incompetence, but I *also* received unexpected reprieves from them later, because Uranus does not only bring unpleasant surprises but also those which are pleasant. In the house of means this can and does often take the form of job instability, bouncing around from one place of work to another, or feeling an inconsistent ability to be effective at work or however we bring in money, and both the loss of employment opportunities and gaining new opportunities can *both* be expected during this transit. The lessons of Uranus often seem completely unnecessary, but they do still exist and we must contend with them, so if you currently have Uranus transiting your second house or that transit upcoming simply try to stabilize your finances as best you can, live within your means, pay off all your debts and obligations, and behave as if sudden and unexpected interruptions might come at any time, for any reason, during the entirety of its transits, not out of fear and anxiety that danger lurks around every corner, which is still itself an unhelpful control mechanism based in the delusional fear that we are alone in the Universe, but instead the acceptance that we can and will meet any challenge that comes our way and that we are not, in fact, alone in our journey through life.

Uranus does not only deliver debilitating surprises, it also delivers some which are also auspicious, and the nature of its transits are to bring a mix of both good and bad things into our lives which should both be expected. Uranus so happens also to be located in my zodiac second house, and if you also have such natal placements you might unreliable, unpredictable, or prefer self employment in terms of means (I have worked as a freelancer or self-employed nearly the entirety of my adult life), and while this can be unsettling for some and also limit the degree of income we win it also feels somewhat natural and expected, being part of my personality, even feeling frustrated and inhibited in a long-term, steady job anyway, limited by myopic employers in the scope of what I can otherwise accomplish. There is nothing wrong with such Uranus placements and transits but it can be very helpful and responsible to also pare down your life so things which do require income like housing and food can be easy to provide, to then better facilitate the freedom, passion, and versatility demanded of our nature.

Transits of Uranus such as through the houses related to finances can be understandably stressful, and it certainly helps to learn a worldview of money which rises above the mundanity and facade of financial security that is normally motivated by fear, to recognize that all life is nothing *but* constant change, that we are not the only resource which can provide for our needs, and that we can simply ask for help when needed. In fact, that is one of the lessons of Uranus—to be prepared, flexible, take ourselves less seriously, and not rely on unrealistic expectations and ego which are so easily subverted. If we delude ourselves into believing life is constant and fortunes unchanging, or if we require financial stability to enjoy life or find fulfillment, we will then never be fulfilled and can expect to be disappointed quite often. If instead we understand that nothing in life is permanent anyway not only can we never truly be surprised, there is so much else in life we can find fulfilling.

There are, however, worse transits of the finance houses, Pluto being the most dire and perilous and potentially the most consequential depending on how we respond to its influence, and if care is not taken to submit to the power of Pluto much harm, chaos, and loss *will* result, not only to our finances but many other things of value, including potentially our very freedom. This is not because Pluto is cruel or especially difficult, however, and in fact Pluto transits need not be difficult *at all* which instead is only a result of our own control behaviors, and Pluto is simply *absolute* and allows no room whatsoever for usurping his domain of power and control and cause and consequence. During a Pluto transit we are, unfortunately, more likely to do just this and to learn the hard way. Remember, the second house rules means, not only money, so it in truth governs *how* we earn money, so our control and power impulses during a Pluto transit of the second house are then directed toward what we *do* for a living and the means whereby we acquire our resources and money, which is not so perilous as long as it is done with integrity and without amoral or manipulative behavior, but when it is will be very destructive both to others and then our own bottom line as we later reap the consequences for those actions, while instead remaining moral and reasonable results in a very productive transit. Sagittarius recently had a Pluto second house transit, and Taylor Swift is a great example of this dynamic when, at the outset of the transit predatory businessmen took ownership of her music catalog against her wishes—itself a power and control behavior by others, due to Pluto transiting other areas of their charts—and was rightly angry at being treated as such. After attempting to get them back without success she instead re-recorded her entire catalog so that she, and not the people trying to control her catalog, actually had control over her art and income, which demonstrates initiative, not control and manipulation, which they instead did to her by trying to take over *her* work. Because she behaved exemplary in this instance she rightly saw success in the outcome, but many people do oppositely assert control in ways which are entirely inappropriate or even amoral and thus end up losing means and other resources of value, even to commit crimes and hurt people and then lose their very freedom. So the Pluto transit of the second house is best handled by doing our best, being resourceful, showing up for opportunity, but *never* using manipulation, control, resentment, etc., to achieve our goals and instead submitting the control of life and outcomes to the Universe instead.

One very karmic and important aspect to the Zodiac which can and will directly affect our experience in transits which affect our finances is the North and South Node of the Moon as discussed in the upcoming chapter on Karma which govern destiny and fate, and whenever the North Node transits a house it accompanies fateful developments in our life path. When the North Node transited the zodiac second house for Libras in 2013 and 2014 were the years in which conflict with my partner caused me to lose everything I had worked to build, upending my life and later caused me to change course and pursue a different life direction than in the past. The North Node moves backward (retrograde) through the Zodiac, however, as if we are swimming against its current, and when it transited my first house after the second was my time in recovery during which I first learned inventory and the cure to alcoholism and addiction which started my newfound work and purpose which lead to new opportunities for productivity and advancement I could not see at the time. Not all North Node transits are so dramatic—the nodes move quite quickly and so we experience many transits through each house during a lifetime, and my experience was likely a result of other parts of my life path coming together at that time, but these transits will always concern those themes which they touch. Most importantly, because the North Node and South Node are always connected, the South Node is always opposite the North and thus points toward the complimentary themes of the past, as opposed to the future, with which we must contend, and the house opposite the second is in fact the eighth, which is why finances are so often a point of control and power dynamics within relationships because means and resources sits directly opposite power and control, and this transit of the North Node in the house of means especially concerns the abandonment of control behaviors as expressed through finances, often through opportunities to do so as through conflict with others, which may not seem like opportunities when they occur but are very much opportunities as far as the Universe is concerned.

As discussed in the chapter on birth, signs, and destiny the placement of the angles is also a useful tool to understand the long term, eventual fate of our material success and life path, when placed in signs associated with finances are typically destined to be centered around materialism, wealth, or control of them, for better or worse, and remembering at all times that we can waste our lives when misinformed to the nature of money, wealth, and means these placements are not often the boons they might at first appear but instead burdens of delusion which deceive us into a false sense of security and purpose which are best transmuted to higher spiritual functions of resources and means such as using them effectively to improve life for all.

While money is a primary theme in all lives it is nothing more than a contract between people for the exchange of goods, and should not be treated as a source of self-esteem or to attract love, companionship, or wield power because those will in turn be as false as money itself. We are each inherently possessed of value simply by being alive, let alone all the immense qualities and talents each of us may possess (even if you don't yet know what those might be), which is more worthy of our time and investment than simply making numbers get bigger. Take joy in the things directly that we wrongly expect money to buy—love, joy, good food, effective behaviors, relationships, quality homes and communities, access to opportunity, and collaboration with other people, because in truth it is the unification of humanity through which money has any power in the first place, which can be had entirely without it.

24. Karma and Healing

Planets and alignments in classical astrology were given the quality of 'dignified' or 'debility,' which meant whether a planet was benefic or harmful, good or bad. But this is in fact a delusion of the human psyche and not a real characteristic of *any* of the planets, borne simply from a desire to control life and destiny through the seeking of fortune and anticipation of misfortune. The entire practice of astrology itself emerged as a mechanism of control in service of the human ego and, in truth, remains that today, even this book, since learning, knowledge, and understanding are all ways in which we try to control our experience, and the more we know the less we fear or can act more effectively. Predatory and stupid astrologers will no doubt use this book and my work to justify their exploitation of hopeful, ignorant victims whose vulnerability is also caused by the same desires for control as their offenders, which can be avoided by teaching ourselves, doing inventory, and relying only on reputable astrologers whom demonstrate integrity and wisdom.

But ancient peoples were beset far more frequently by misfortune that we are today, surrounded by high rates of mortality, loss of many children, inexplicable plagues and disease, and almost endless geopolitical wars and other conflict, and astrology was a way in which people tried to anticipate misfortune with the implication that bad things can be avoided if we know about them beforehand. Astrology does not at all work like that, and we can never, ever subvert fate even if we know full well ahead of time what is going to happen. As mentioned earlier, fearful and cynical people will respond to this concept with cynical positions like 'why try at all, then?' and contemplate giving up, but giving up is also a type of control, still upset by the fact that we really have no power over fate, continue attempting to control it by avoiding it (or what we understand as avoiding) which we just cannot do either. Instead of thinking of astrology as a system to subvert misfortune and chase fortune, for which planets are deemed 'dignified' or 'benefic,' astrology is simply a tool to better understand the nature of reality which can inform our choices and insight that we might be more *effective* in our lives.

The truth is that no planet is good or bad, in fact, and the entire anthropomorphisation of planets as 'dignified' or otherwise fails us in our understanding of the nature of the cosmos and contaminates reality with the shortcomings of the human psyche. It is understandable that we want to avoid misfortune and suffering but if, for instance, a person believes that Pluto is an evil or bad planet may fear its transits and try desperately to avoid its negative influence and will through their own behavior end up causing the very evil they wanted to avoid. The negativity in fact more often comes from ourselves and our insistence on control, not the planet at issue, where Pluto's ultimate role is to take from us not our possessions, relationships, or life but the onerous and suffocating burden of the responsibility of control itself. As individual, mortal humans we cannot alone carry the burdens of our own lives let alone those of others or the very nature of reality and cause and consequence. Indeed since we are not even capable of control in the first place Pluto is a benefic planet even when its transits are quite stressful or upsetting. Those of us who strive desperately to herd human lives, subvert disease, or amass fortunes are entirely miserable, paranoid, fearful, and often unfulfilled, and transits of Pluto, Saturn, and other planets and influences previously characterized as bad or harmful are in fact liberators and guides which help show us more effective ways to live and consider the Universe. Pluto's desire is to take our burdens and shoulder them for us, in reality a savior which comes, though forcefully, to make our lives better. Also for example, Jupiter is described as the most benefic planet, but throughout my life almost no Jupiter transits have brought me luck or fortune but instead benefic consequences of past choices and behavior. But it also can very easily bring expansion of unwanted things like delusion when in stressful aspect to Neptune, or control mechanisms when in aspect to Pluto. As discussed in the earlier chapter, Saturn has been traditionally regarded as a debilitating planet but it has throughout my entire life always guided me back to my true life path and helped me overcome obstacles. The lessons may have been tough but Saturn even leaves a gift when it departs signs, and calling Saturn anything less than a wondrous, benefit planet is simply misguided.

One way in which these terms are properly used in astrology is how planets are either *exalted* or *debilitated* in the natal chart, for instance Jupiter's muted influence on my life is likely due to its natal position in the sign of Virgo or its close proximity to my natal Saturn, and this is a natural dynamic of the planets in which they influence each other, but not because any is specifically good or bad, and believing a planet has harmful influences instead reveals how we fear life and distrust God. Of course there is much suffering, and loss can be devastating, but at the same time death is a step of life which happens to every single living creature and is necessary for us to progress to the next stage of reality, whatever that may be. Being ignorant to what happens after death and fearing that unknown is not the same as death being a bad thing, we are just afraid of the unknown or what we cannot perceive, which is a survival instinct befitting mortal animals that they may better persist as a species and not anything inherent to the nature of death or the unknown. As I have oft mentioned in my work, if death did not exist we would not exist either. My cynical self might exclaim 'maybe that would be better!' but as a mortal human I cannot know the full workings of the Universe and why we must be here and experience mortality, and the salve for that uncertainty is not finding certainty in mythology and religion but simply accepting the conditions of reality and turning over power to those forces which do in fact govern fate and reality, which has constructed all which came before us, including our ancestors, back to the dawn of Earth and the stars from which we are made which had to die for us to exist.

Karma is a word that we have appropriated from Hindu religious tradition, but primarily because it does appear to be a natural law of the Universe, specifically related to metaphysical causality in which the consequences of our actions are *never* optional. Yet Karma is not about vengeance or comeuppance and instead it is about *causality* from which can learn wisdom and empathy. There have been times in my life for instance when I refused compassion to people who required it, such as my little sister when she wrote me that email that was in reality an attempt to connect with me that I did not recognize as such (having been taught to spurn emotional vulnerability), and she never wrote me again and when I later in turn required compassion from her she similarly withheld it as I had done. Of course, I had my own struggles, and it had not been very long since I had tried to take my own life, but as we reach the latter part of our teenage years our choices do still become our own, passing into the autonomy of adulthood we must take responsibility for our behavior, and as we grew older still my sister fell deeper and deeper into the confines of her religious institution and further and further from emotional relationships with me and other family members, finding belonging and purpose among doctrine and dogma as a replacement for the warm embrace of kin mostly unknown to her (and most of my family, for that matter). Although I tried to be there for her in later years after finally recognizing the gulf between us it never bore fruit because the damage had already been done.

In astrology there are several aspects which govern things like karma, healing, and destiny but they are not in fact what we might first expect. For instance, the asteroid Chiron rules healing and is often concerned with experiences as I just described. In fact, I distinctly remember seeing a transit for my sister at this time in which the astrologer advised to be alert to the opportunity for showing compassion to someone when it is not expressly requested but instead implied indirectly, to avoid a "painful sting" if they instead showed coldness, and remembered how much my past action toward her was in fact characterized by that exact same framing and likely occurred during that time. Chiron often serves to help heal unhealed wounds but does so often by ripping them open again, which occurs because we cannot heal wounds we are not aware. This can often be an uncomfortable process, especially in the earlier parts of the transits, and I will admit always being relieved when a difficult Chiron transit passes, sometimes being more foreboding than even those of Pluto, since it is human nature to avoid pain and ignore things we find inconvenient and uncomfortable. Because of this nature of Chiron and wishing never again to cause harm to another person if I can possibly help it, I always look out for Chiron transits (by Chiron or to natal Chiron) so I can be more cognizant of how my behavior affects not only others but also myself, as behaving in the way I did was not only harmful to my sister but also to myself and my own personal wellbeing, in part by shunning a relationship which should have been very important to me, further isolating myself from the family which had rejected me, and in turn harboring guilt for my cold behavior which added to the burden of self-hate and insecurity that weighed me constantly.

During another fateful Chiron transit (opposite to my natal Sun), was portended to be a dispute in which I should similarly be gentle with others or risk losing those relationships and provoking a "blow below the belt." Worried about what that meant it finally became apparent when, after a mild and completely in-character conflict with my own Mother, which had been occurring for the last twenty-five years, she one day lobbed a scathing, ill-conceived text message in a fit of rage after being told I did not want to see her unless she was coming over to apologize for supporting homophobic and bigoted institutions and politicians. Unlike what I might have said when I was younger, it was not in any way inappropriate, with no manipulation or attempt to control her, but simply to state my problem (for the umpteenth time) which is that her behavior directly and negatively impacted my life. Specifically my mother would avoid eye contact with me and treat me with the impatience and condescension we often have when interacting with someone we don't like or deeply resent, and in addition to her offensive support of racist, homophobic, vile persons, which I could have overlooked, it was literally impossible to have a relationship with her by her own choice of refusing to actually engage with me, hanging up the phone after only a few minutes talking or when at her home excusing herself to go clean the already clean home, or watch a national news broadcast even though I hadn't seen her in person for three years, and my presence in her life served nothing more than service to her ego.

Our family is a classic narcissistic family structure which usually sacrifices one or more of the children as a scapegoat (me) which holds the narcissistic structure together, and as such also prejudiced my siblings and their children toward me which then prevented me from ever being able to have any real relationships with any of them, and the control I had been demonstrating was in trying to have a relationship with people who clearly were not invested in having one in the first place. But her response to my disinvitation was a rambling and lengthy tirade in which she referred to me, her own son, as "you people" and characterized me as a an actual geopolitical adversary rather than her child asking for acceptance by his own family, dumping all of her political anger, frustration, and stinging resentment built up during years consuming political anger entertainment and reactionary politics as if this dispute was one of a political nature rather than a deeply personal, family dysfunction. It seems rather serendipitous, or rather inextricable that it should correspond demonstrably and succinctly to a transit described perfectly by a talented astrologer in a prewritten, automated entry on their website. But as is often the case with astrology was presented as if we can do anything to prevent the transit from occurring or, by extension, that it is somehow

our fault, because I had nothing to do with her consumption of political entertainment nor the formation of her attitude toward me, which in truth is also a
function of human evolutionary survival psychology and not even a machination of
her own. I supposed I could have not said anything at all, but I was as demoralized
and tired of my relationship with her as she was with me, and had previously asked
her not to contact me for the same reasons anyway, and maybe could have handled
the situation differently except that I just really didn't want to have anything to do
with them anymore, which her response thereafter made extremely easy to do.

Obviously, this sounds the opposite of healing but this is exactly how Chiron
transits work, as what occurred after this reopening of a wound was the realization that I had been hanging on to my relationship with my parents because of
an indoctrinated fear that should anything bad happen to me in my life that they
would be the only ones on whom I could call for help. This is in fact a common
abuse dynamic which occurs in narcissistic family structures to children in which
parents intentionally or unintentionally engender a sense of helplessness by not
only withholding necessary life skills (like a healthy relationship with money) so
that children are in turn dependent on them for life, but also brainwash us into
believing we are incompetent and vulnerable to peril. While having my parent's
love and companionship would be lovely I did not in fact actually need them to
be happy, fulfilled, and productive in my own life, which I realized after sitting
down to do inventory on this episode several months after it occurred. Not only
that, but all that I expected my parents to provide my life (i.e. parenting) I was in
fact entirely capable of doing myself, and could and should be my own parent now
that I am grown, and do the things for myself that I would expect from a healthy
parental relationship which they are clearly incapable of giving, not because they
are bad people or bad parents but because they simply lack those skills, as did
their parents, and theirs before them, and so on, since we cannot give to others
that which we do not ourselves possess. Chiron transits are nearly always like this,
where vulnerabilities are often revealed which, while sometimes unpleasant, results
ultimately in greater growth and finally effective healing (and not necessarily the
kind of healing we want but what is probably better for our long term growth,
development, and experience).

All Chiron transits are very specific too, for instance a particularly wonderful
Chiron transit I wish could last the entirety of my life was Chiron trine Neptune
which was described as being sometimes overcome with feelings of bliss and in
harmony with nature. I was not prepared for just how painful it would be to be
treated by my darling nieces and nephews the way my parents and siblings had,
which was an entirely different type of misery I could not endure, and after ending
relationships with my hateful family moved to a new apartment away from them
but which was near a winding river, living fully alone and recovering more of my
health not only due to my research and discoveries but from taking regular strolls
along the river to get much needed Sun exposure, take in the funny ducks, geese,
and the breathtaking mountains capped with snow, where there were even wild
mulberries growing over the trail, and I was in awe of nature and the dichotomy
between needless human conflict and the beauty and simplicity which exists all
around us, feeling the fresh breeze on my skin, the energy to walk a great distance
(which I did not even have in my mid-thirties), stronger and freer than at any time

in the last decade, and in such sights and solitude would be overcome with feelings of contentment and reverence for nature and this chance at real happiness I not only never thought would come but did from such truly simple things.

Chiron was also in Libra's sixth house of labors (work and health) with Neptune from 2011 to 2018, and indeed I was able to achieve a great deal of healing as told in my first book in part because of the influence of Chiron here, though it has since transited my house of relationships and I haven't yet been able to see its effects there except maybe in the fact that more than ever I have grown beyond the need for a relationship to find love and self-esteem. But this also came with an uncomfortable Chiron transit opposite my natal Pluto which through an unfortunate experience online forced me to recognize and admit my suspected undiagnosed autism which I had been in denial ever since I was younger when someone unsolicitedly called me a 'savant,' even though it was meant as a compliment, and I nearly bit their head off because I had been so afraid all my life at being treated differently. Even during these difficult angle transits Chiron ultimately brings healing, and its transits always end up having wonderful effects on our lives even if they might be achieved by reopening old wounds. But there are also frequent, smaller Chiron transits such as the (lunar) monthly transit of the Moon to natal Chiron which, for me whenever it occurs, often takes the form of dancing around my apartment to really good music, and the path of Chiron through our chart very specifically concerns the healing journeys we take in response to the trauma and pain of our past.

More fateful for the long arc of our life and what also fully concerns laws of Karma are the *Nodes of the Moon*, an aspect of astrology quite unfamiliar to most people as it is not commonly a subject of entertainment astrology, but is still of *very* consequential importance and worth understanding in order to have a broader view of our ultimate destiny and purpose in this life. The Nodes are derived from the intersection of the orbital plane of the Moon with that of the Earth's orbiting the Sun, and because of this there are two points, the *North Node* and *South Node*, and because they are equidistant they also always oppose the other directly in the Zodiac, much like an orbiting Midheaven, Nadir axis.

There is much information on the Nodes of the Moon available from astrology sources, but often they are similarly characterized as if they play a role in the personality, but instead are explicitly concerned with our ultimate fate in this life as a natural progression of Karma. I am not entirely convinced of the legitimacy of that concept, and there is a real danger of charlatan purveyors of fake 'past-life' therapy (no person can do such things, as that would invalidate the very purpose of life in the first place), but the practical application of the Nodes to our personal experiences in this life does in fact appear to be not only real but most importantly also useful, and functions much like the Midheaven and Nadir except concerning our Karmic destiny in this life rather than the material and inner life paths. While this may not be very clear, the difference is that the Midheaven is ultimately our professional, material, and public success and achievement but the Karmic destiny of the North Node is how we grow as a soul, journeying from our familiar and comfortable nature indicated by the sign placement in the South Node to our growth destiny in the placement of the North Node. This occurs because opposition is always a position of conflict (remember that not all conflict is bad!), so the nodes of the Moon indicate or describe the conflict between our familiar, innate self opposite to our ultimate, unfamiliar destination into uncomfortable or unknown destiny which occurs by the end of our lives. The South Node thus represents where we feel most safe, familiar, and in control, while the North Node is where we oppositely feel vulnerable, incompetent, exposed, and not in control but where we must in fact venture through the course of our life reach our destiny.

As an example of this dynamic my Node placement also happens to be the same as my Midheaven and Nadir placement, the North Node in Leo, the sign of individuality and self expression, and my South Node in Aquarius, the sign of the group and group consciousness. Normally only the North Node is drawn on charts, and the South Node is inferred, but I think it would be good for charts to indicate the South Node with a small notation because people will not easily intuit that it should be there. But this therefore means I feel most comfortable in the group, anonymized by humanity or family and buoyed by common purpose and support, while my destiny instead is fated for publicity and individuality where I feel most uncomfortable, against my very strong desire not to and in spite of my efforts to avoid this kind of life. Very often it is said that those with a Leo North Node become famous, or at least publicly visible (the Midheaven here would confer some prominence anyway), but are not naturally comfortable in that position such as Princess Diana, Barack Obama, and Ghandi. It is also said that this position obliges some significant contribution to humanity, having some kind of debt to humanity as indicated by the South Node placement, and that success in this life is not granted until this is achieved. This implies that people like me with this placement might have been a horrible person in the past life, which obviously my ego does not want to admit or entertain, but such concepts also risk the encouragement of blaming people for their hardships and misfortune in life as is already such a common human behavior, the irony being it is the very behavior which stupidly incurs such karmic debt in future lives, if they do indeed exist, as we are not excused from our bad behavior just because others are also bad, but also represents our own personal insecurity in which we have no compassion for *ourselves* and thus fear misfortune, stupidly believing we can avoid misfortune by shaming others for theirs, which is not at all how reality works, and doom ourselves to a life of petty recrimination and voluntary resentment instead of enlightened contentment.

As such grandiose proclamations demonstrate, this aspect is not something that occurs on a small time scale but across the entirety of a life, from birth and our origins to our ultimate fateful purpose in adulthood. Interestingly, the Nodes move always in retrograde motion through the chart (clockwise, as opposed to counter-clockwise as most planets do), and this can lead to confusion when trying to understand its influence in our lives since its thematic progression is exactly opposite and counterintuitive to most other transits, much like swimming against the current. The nodes do transit somewhat quickly, however, spending only about 1.5 years in each sign, and it seems that their transit through the signs and our zodiac houses concern fateful developments along our life path which concern the ultimate goal of our long narrative arc. For instance, the last time the transiting North Node was conjunct my natal North Node, indicating the start of a new cycle related to my karmic destiny, was December 2017 exactly when I was putting ten or more hours a day into writing *Fuck Portion Control* and then offered it up for preorder. Most recently the North Node reached my zodiac sixth house of labors when finishing up *this* book, requiring long, long days of writing and editing as well as some of the last bits of research concerning FPC content. So every time the North Node visits a house (especially the zodiac houses) it indicates events that concern the fulfillment of our karmic destiny, and if we are paying attention may become more effective in its fulfillment.

It takes the nodes about eighteen and a half years to fully transit the zodiac, so the phases of return to their birth position are thus also the very stages and progression of personal development and growth from child to adult to elderly that are the major books that contain the content of our lives, with each sign transit of the Nodes being the chapters within those books. During the entire first transit of the nodes through the zodiac in our young lives it is an introduction to those fears and insecurities to which we develop our coping and control mechanisms. For instance, the North Node was transiting my third zodiac house of communication when I was in sixth grade during which I had formative experiences like reading the word *bologna* aloud for the first time ever and, being laughed at by my classmates and having already an abundance of trauma and autism mistook it as a lesson to keep silent in order to avoid ridicule rather than the joy and fun of being a silly human. If I had instead responded by not taking myself seriously (which was never going to happen considering we unconsciously learn these behaviors from our parents) my classmates would instead have been endeared to me and my popularity would have risen. I don't mean to imply that popularity is important, because it isn't at all, but acceptance can greatly affect a child's trajectory in life, and that small difference doomed my life to a dearth of social support rather than its bounty, but which also later taught me the true value inherent of friendship that was apparently so crucial to my karmic evolution. It was also the time I first tried writing, because we had been gifted a small typewriter, but the embarrassment of my parents reading my first attempt was so painful I never again shared my writing with anyone until I was twenty-eight and the North Node was in opposition to its natal position (conjunct the natal South Node) which did not go great but also wasn't totally bad either.

Such is the nature of the Nodes, and when the North Node transits come around again for a second visit in our adulthood those are opportunities for growth and resolution of the related negative experiences we had as children, if we are not entirely obtuse and petulant, which concern our karmic life path. Then it also happened that the North Node visited my first zodiac house again at the same time as when my life fell apart in 2014-2015 just as it had eighteen years earlier when my family kicked me out of the family for being gay, and I was forced again to entirely reevaluate my life path and make a major course correction, not just for professional or aspirational reasons but to literally save my life. At the time of this chaos my partner's North Node was transiting his *ninth* zodiac house and conjunction with his natal Pluto, demonstrating his own life path developments concerning both control, worldview, and personal philosophy. I've mentioned before how the Ascendant and Midheaven concern fateful life experiences and it could be that the North Node triggers such fateful developments in our lives (not only misfortune but fortunate ones too!), to resolve and restart every time the

Node returns to its natal position, and in the future ultimately become the very points where our supposed destiny is in fact fulfilled.

Wherever the Nodes appear in the birth chart is thus one of the greatest astrological influences of anyone's life, as it encompasses the primary arcs of trauma, growth, control, consequences, and acceptance and surrender from a person's birth to their death, and the purpose of the Node positions and transits in our chart are specifically to experience things of which we are afraid and which make us uncomfortable, to discover new and unfamiliar horizons and experiences to evolve beyond what we have already known and reach for greater and more important purpose.

The placement thus of the Nodes in the signs begin with Aires, being the sign of independence, decisiveness, and action these are all things that frighten and unmoor those with their North Node here. Across the chart from Aires is Libra, the sign of partnership, dependence, and diplomacy, and those with this alignment feel most comfortable in the swaddling arms of relationships, but at the same time fear conflict and are often simultaneously codependent, since a consequence of conflict can also mean losing the relationships where we feel most comfortable. Relying on others is not bad but the alignment of the Moon's nodes here show we are *overdependent* on others and uncomfortable with our own power, and during the life of one with the North Node in Aires many experiences will be had which challenge our codependence and force us to trust our own power, take responsibility for ourselves, and learn how to be an independent, driven, and motivated person.

Taurus is the sign of resources and materialism, so those with the North Node in Taurus might fear their own earning power and independence, where the South Node is in Scorpio, the sign of power, control, sex, and taboos and so may be preoccupied with sex and power dynamics with others as a way to avoid confronting our own self worth and ability to care for our own needs. Because we might fear taking care of ourselves we may spend all our time, money, and attention taking care of others. Ultimately we cannot care for others without first caring for ourselves, so the consequences of this placement may become demoralizing and profound until we understand that we must take care of ourselves, which includes associating with others that also care for us and not only us for them.

Gemini is a sign of communication and collaboration so those with the North Node here are often egotistical and convinced their own positions are correct simply because it's the one they have, even without any supporting evidence or research. Sagittarius is the sign of independence, travel, worldliness, and freedom and while those qualities are wonderful we are also human and rely on others for our survival and needs, so those with the South Node here are often at odds with others, self-righteous, dogmatic, and careless, as their fear of facts, truth, dependence, and loss of freedom motivate many self-destructive behaviors that harm relationships and even our own health and wellness, believing that enlightenment will take away the supposed freedom we find in ignorance. Being wrong is not a moral liability but being obtuse and petulant is, and other people can often impart great value to our lives if we can sit and listen to them, because knowledge and enlightenment is empowering, but it only comes from exchange with others, not from within.

Because Cancers love to cook, reliance on processed food or eating at restaurants is a funny but relevant effect with the North Node in Cancer since our contemporary, industrial food systems so easily enable those to avoid cooking which the sign of Capricorn where the South Node is located does not love to do. Two of my own siblings have this position and I don't think I've ever seen them once prepare a real meal for themselves or anyone else in the entire time I've known them. Cancer is also the sign of home and family but hardly can any human avoid having a house or family since both are essentially required as a result of other functions like having sex and partnering. Instead, the South Node is in Capricorn which is the sign of profession, rules, and institution, and many with this alignment find refuge from the fear of domesticity and emotions in career and stoic detachment even while building a family. This is unfortunate because such a life is also very lonely and loneliness often brings fear and self-doubt, and we can and do also reinforce our loneliness by refusing to ask for help or cultivate real, intimate relationships with others, mistaking the consequences of our own actions for the way we think life is. Greater emotional stability can be found by trusting others, not that they won't cause hurt or disappointment but that humans are not responsible for mortality and its limitations, that no relationships last forever but that also does not mean they aren't valuable. That other people want relationships with you makes this, in practice, a relatively easy task, and all you have to do is let them in.

Leo embodies the individual while Aquarius embodies the group, and for anyone like me with the South Node in Aquarius hiding amongst the crowd brings a great sense of safety and comfort. Because my Midheaven and Nadir are also in this alignment it meant important shelters like my family and friends were later ripped away from me and I was forced into independence and the uncomfortable spotlight as the condition for success in this life. Other people with this node alignment might compromise their own wellbeing or standards for inclusion in groups, because sticking out brings fear and discomfort. Throughout our life we will have opportunities to learn that our individuality is important and enough to stand on and we should not fear being seen—plenty of people experience publicity in life and do wonderful things and lead wonderful lives, and it is only our inherent insecurity which prevents us from pursuing this destiny.

Virgo is the sign of habit, structure, and responsibilities and embodies things we must do on a daily basis such as work, caring for our health, or even that of pets. So those with this position tend to be especially scattered and undisciplined due to the South Node in Pisces, the house of spirituality and illusion or delusion, seeking to avoid facing the real world and its causes and consequences. Our lives demand structure in balance and the consequences of avoiding routine and respon- sibility may catch up to those with this position before they are able to act in spite of their fears and adopt better self-care skills and take responsibility.

Being the sign of relationships, having the North Node in Libra means fear and distrust of relationships, which is very difficult since most human needs are met through others in partnership. No matter how much we desire it we cannot actually do everything on our own, and because of this a person with this aspect likely feels contempt for their reliance on others as well as those very people on which they feel dependent, having the South Node in Aries, which is quite unfair and superbly self destructive. Defensiveness and blaming others might be a hallmark of this position, but we are every bit as mortal and fallible as anyone else, and belief to the contrary is merely a deluded coping mechanism meant to help us cope with mortal limitations and is not truly an objective analysis of reality. It's okay to rely on others, to take breaks and relax, to just enjoy life. Many people strangely misunderstand balance to mean doing everything by ourselves and being absolutely healthy, productive, and successful, which is not even remotely what balance means. The more we can understand that relationships with others and needing others is not only good, it's also necessary for our own wellbeing, the less frustrating and difficult life will be.

The North Node in Scorpio finds comfort in material, superficial trappings of life, having the South Node in Taurus, and is quite frightened of their own powers of passion and intimacy, of conquering or being conquered, of transformation and surrender. Real intimacy is one of the richest and most intense experiences a person can have, and fearing it we can construct many distractions with money, health, or other resources which empower our desire to avoid the vulnerability of intimacy. But because we are also human we all instinctually crave intimacy and so this can be an especially challenging conundrum. Destructive, emotional, histrionic behavior and drama also subvert intimacy and are used as tools by people to avoid what they truly fear. After all, who wants to get close to someone who is volatile and unreliable? Yes, intimacy comes with risk but that's the entire point. It would not be intimacy without risk. Risk is what makes it intimate, and it's okay to let go and experience intimacy. It is far more rewarding than anything material.

With the North Node in Sagittarius and the house of the mind, travel, freedom, and worldliness we greatly fear instability and find comfort in what we mistake as cold logic and reason, for if our mind is always full and distracted and we are always talking to people and engaged in conversation or business nothing could possible creep in during some unexpected silence and force us to find deeper meaning in life. Control of uncertainty through obsessive behavior is a risk of this position, as well as excessive use of communication technology like phones or the internet because they are a safe distance from anything real and substantial where we can observe the world without risking participation in it. But we cannot control life no matter how much we desire to, and may make poor decisions and ruin opportunity entertaining our fear of letting go and trusting in life. It is not meant for man to be in control of life, so let go and let God and let fun and spontaneity abound, stop worrying so much, for life cannot be lived only on rationality and trying too hard not to make mistakes or too make life go the way we want absolutely will result in mistakes and our life getting entirely out of hand anyway.

Capricorn being opposite to Cancer means the North Node here causes overdependence on home and family and great fear of personal and professional responsibility. To those with this position everything is always someone else's fault, because we fear an inability to effectively meet personal challenges, criticism, and obligations, and we may even pathologically avoid formal employment or repairing mistakes we've made. To make this worse, many life experiences may seem to reinforce this false belief which then perpetuates and hardens it, but in reality we experience those consequences not because we cannot take responsibility but because we *do not*, which we very much can and should. Sometimes this requires the literal learning of skills like how to make a proper amends for mistakes we have made (which is taught step by step in *The Perfect Child* in the chapter on amends). Personal empowerment through the learning of skills we lack can bring more confidence to take on personal responsibility, and no person is beyond healing of past pain and trauma, which only requires practicing self-care and introspection, such as inventory therapy.

Opposite those with the node in Leo, those with the North Node in Aquarius feel unsure of the group and most comfortable being an individual. This can bring also great distrust and uneasiness in relationships which can bring immense internal and external conflict since we can still desire companionship, sex, marriage, and other aspects of relationships but not bring ourselves to relax and fully participate in them. Refusal to respect or accept the balance required in relationships can also lead to obsessive, dissociative behavior like always cleaning or finding other ways to preoccupy our time than connecting with the people in our lives where we feel most vulnerable. Exposing ourselves to love and dependence that is required in relationships can be very scary—loss, heartache, and disappointment are the reason many people avoid relationships in the first place. But with great risk also comes great reward, and there is strength in bonds and connections that no other resource on Earth can match, but for this to happen you have to let people be who they are. The notion we can't lose something we never had is especially not true in the case of this alignment, for missing out on a primary point of life such as intimacy with others is truly one of the greatest losses of all.

Pisces being the last house of the zodiac is a culmination of all the previous signs, and as such is often too full which can prove challenging to any astrological positions. So it is with the North Node in Pisces that a vagueness of general fear in those with this alignment. Pisces is also the house of spirituality and many born with the node here will, like the placement of the South Node in rules loving Virgo, cling desperately to the standards and expectations of religion in hopes that obedience and compliance will bring control and quell the existential tempest within. But one of the very points of living is uncertainty and unpredictability, that we may come to terms with mortality and humble ourselves before the Universe and let go of control and fear. Unfortunately this alignment may cost important relationships as we choose institutions over people and relationships, but if we learn to accept those things of which we are afraid and fully embrace real spirituality will eventually we will find peace in uncertainty.

The Nodes are some of the most interesting aspects of astrology, but one which is often overlooked that can otherwise help us understand the path on which we tread and more effectively embrace our destiny rather than avoiding or fighting it. If there is such a thing as reincarnation, which makes more sense since why would we live one, short life in the billions of years that the Universe exists, we will probably have to repeat this life path in the next one if we don't succeed in meeting this challenge now. Because our fate is already directing us in this direction, all that is then required is to show up for opportunity and to do our best.

The Lost Scout

In a dark and lonely forest a young boy scout had wandered away from his troop and was suddenly lost. For hours he wandered and called for help, and was beginning to feel hopeless, when suddenly out from the dense understory leapt a great, grey wolf.

"Ahh!" shouted the boy, so afraid his knees began to shake. He looked around and saw a nearby rock, and picked it up quickly, ready to bash the wolf if it came close.

"I heard you calling for help—" said the wolf, "I'm not here to eat you."

"But you are a wolf!" said the scout. "You eat things like me that are smaller than you."

The wolf shook his head. "I have never eaten a boy in my life, and if I was going to eat you I would have already."

"I *am* lost, it's true," said the boy, "But you are a wolf and if I let down my guard it will be the end of me!"

The wolf snickered and stepped a few paces back. "You've been told too many fairytales," he said. "Your troop is directly behind where you stand, about a mile from here, and if you turn around and head straight in that direction you will find them."

The scout scowled. "How do I know you aren't lying? That you will lead me into a den of more wolves where I will be devoured piece by piece as their dinner?"

The wolf sat back on his haunches, growing amused by the trenchant little boy. "I suppose you don't," he said. "But what choice is there? You're lost in the woods, and the only help you have is from me whom, if I wished, could eat you up right now. The truth is I know the woods, because this is my home, and have a keen sense of smell and can smell your other kind just over there."

"No!" shouted the boy, still shaking. "Get away and don't come back!"

The wolf shook his head, "Alright," and turned around and walked back into the woods.

Alone once again, the boy looked around at the forest which surrounded him, which now seemed more quiet than ever. Sitting back against a tree, he kept both hands on the rock just in case the wolf decided to return from the forest to eat him. Many hours passed and the sky began to grow dark, the air still and cold. Though the boy knew how to start a fire, since he was a scout, he imagined the great wolf using it as an opportunity to catch him off guard, so instead stayed where he was, shivering and shaking with cold and fright.

Very soon the light disappeared altogether and night fell upon the forest, and though he listened for any sound of rescuers the boy heard nothing but creaking boughs and the occasional call of an owl. Soon the air grew so cold the boy had no choice but to let go of the rock and hug himself to keep warm, and before long his eyes grew too heavy to keep awake and in spite of his great fear, fell asleep.

In the morning the boy woke with a start, but was relieved to find himself uneaten and no sign of the wolf. But still there was no sign either of any coming rescue. Tired, sore, cold, and hungry the boy finally gathered the courage to rise and continue searching, and when the sun began to pierce through the forest he felt a renewed sense of wellbeing and hope.

But the woods were vast, and seemed to constantly change and shift around him, as if the forest was conspiring to keep him. Soon his stomach began to grumble and groan, but never having lived in the woods the boy had no idea what he could eat. A small stream gave him a drink, which felt refreshing and invigorating, but only made him more aware of just how hungry he was.

Before long the boy came across a large fallen tree, overgrown with mushrooms and moss had left a large hole in the canopy above, and the boy climbed onto the log and let the sun warm his cold body.

"I am not going to find my way out," the boy said to himself. "And unless the people looking for me find me here this is where I meet my end. In fact there were many people searching for him, but they were on the other side of the woods far away, unbeknownst to the boy. He then remembered the wolf and how it had, in fact, offered to help him. It was true the wolf could have eaten him if it wanted, but instead had sat there, kindly, and also departed when asked. Suddenly the boy realized he had been wrong about the wolf, and that if he had accepted the offer for help he might already be back home, eating a warm meal and sleeping in a warm bed. Feeling sorry for himself, the boy called out, "wolf? Wolf!" But there was no answer, for he was still frightened that something even worse than a wolf might hear him. But soon fear of the wolf was replaced by fear of never getting out of the forest, and the body got down from the log and called out with all his might, "Wolf! Come back! I need help. Please, help!"

The boy set off again in the direction he thought the wolf had said to go, continuing to call out, "Wolf! Wolf?" Suddenly, before he had walked even ten minutes the enormous form of the great grey wolf came trotting through the forest directly at him, then stopped a few feet away.

"I'm sorry, I refused your help when you offered it," said the boy. "Will you still help me?"

The wolf nodded, "of course."

With that the wolf turned and set off through the woods, "come on," he said, and the boy followed willingly, though careful not to get too close. Yet shortly they came to an area of the forest which looked familiar, and the boy grew happy. Without warning, the wolf stopped suddenly and his ears pricked up, the hair on the back stood on end. Then the boy heard the faint voices of the people searching for him, "we found them!" joyful and relieved. "Oh, thank you, thank you, wolf!"

Then boy ran toward the voices, "I'm here!" he shouted. Reaching the edge of the forest the boy turned around to wave goodbye to the wolf, but it had already disappeared. For rest of his life the boy remembered every time he was afraid he could always ask for help and would find it, even from those least expected.

25. Forgiveness

Anyone reading my work will be well familiar with my life story, coming from a family possessed of hateful religion and hateful politics whom kicked out their own son as a sacrificial scapegoat as so commonly required of narcissistic family structures, which caused a deep depression and attempt at suicide before finally realizing that rather than living for others I could live for myself.

Yet over the many years after that experience trying to stop rumination and regret over their rejection and the pain and heartache it caused continued wrestling with depression, anxiety, and trauma which motivated alcoholism and other self-destructive behaviors until finally finding the opportunity to learn new and effective life skills in inventory therapy and discovering the cures to depression and alcoholism which finally relieved me of those problems.

In retrospect, part of my struggle was very much that no matter how often and sincerely I chose to forgive them for what they had done it not only failed to change anything about my situation, with them continuing to treat me as an outsider and regarding me with much contempt, resentment, and animosity, it also did nothing to quell the internal tumult that was a direct result of both past and continued abuse. This is because the act of forgiveness does not also resolve the effects of those things which are requiring of forgiveness. For instance one of the effects of abuse is to destroy a person's self-esteem and instill insecurity and ineffectiveness, and simply forgiving someone does not also enlighten a person to new life skills which can empower them to resolve those deficiencies. For many years this disparity between my experience and what were the otherwise promised effects of forgiveness was a demoralizing exercise in futility and confusion, not knowing how to proceed in life nor even reach goals such as starting a family or pursuing the right kind of career, so overwhelmed I was with the negative effects of the trauma suffered at their hands which refused to depart my life no matter how much I strove to be rid of it. Enlightenment finally came in the form of the recovery programs and Alcoholics Anonymous, introduced to the practice of inventory realized its true therapy in the rehabilitation of the human psyche, which not even those who made and run the program are aware whom, through the process of trial and error, stumbled upon the very process by which the human mind learns, exploiting that pathway for the teaching of new life skills and thus a new spring of hope. The many years spent afterward in introspective practice to then resolve my unresolved trauma and teach myself new skills for handling life not only brought resolution to many things formerly out of reach but also great insight into the human psyche itself, my own mind now unburdened from so much care which had before monopolized so much time and energy, recognizing the patterns and coping behaviors of human evolutionary biological psychology in my friends, family, acquaintances, others in recovery, institutions, and later in those I started coaching.

Most horrifying in my experiences helping others was the revelation from several women of a certain age they had been raped by their own fathers, a heinous sin which though I knew occurred was shocked to find horrifyingly common. As what occurred in my own life, many years, even decades, had been spent by them attempting to forgive the men who were supposed to be their protectors of such ultimate betrayal, their own psyches forever shattered by the cruel reality of life, sometimes beyond any repair, and just how futile the act of forgiveness actually was in rehabilitating the heart and mind, not only from the worst of human atrocities but even those which were relatively less traumatic.

After several years of pondering this frustrating disparity someone one day admonishing me on a social media post of mine bringing attention to the vile assault by religious people of transgender children in defiance of all that is good and right and making a mockery of God whom loves all his children that I should choose instead to *forgive them,* not to offer solidarity or help even though in that moment the ongoing assault against children by those claiming ownership of God raged with unchecked hatred and opportunism, to provide cover for the vilest of deeds in destroying the innocence of children, hurting families, interfering in private medical issues, bullying and exploiting vulnerable people, and threatening even the confiscation of children, to forcibly raise them in the religious homes of strangers. All my childhood people like this unkind and insecure person had similarly excused the persecution of gay children and adults which led to my attempt at suicide as a young man, admonishing us to accommodate their bigotry as religious people, refusing to intervene for our protection even when they electrocuted us in the name of therapy or beat us even to death in our homes and communities. Because of my work and research I suddenly recognized in this person the Karpman *helper* type (as discussed in my book on psychology) which finds purpose in conflict and duplicitous intent, enabling and sustaining the abuse and conflict in which they find purpose such as the demand that victims forgive their transgressors even in the midst of the transgression. It is objectively true that forgiving the people who at even this very moment exploit transgender people for their own political gain would not only fail to stop their hateful persecution nor help those targeted by them but emboldens attackers with promised impunity, and his behavior was nothing more than empowerment of those who do evil in the sacrifice of the innocent for the benefit of the *helper* to feel a sense of purpose, security, and usefulness, failing as they did not only to do anything to stop what is happening and thus aid those in distress but to outright ally with and excuse wrongdoers under the guise of a negotiator.

Recognizing that forgiveness does not itself stop heinous behavior or hold transgressors to account, forgiveness as a concept that we are oft instructed is in fact nothing more than a control mechanism invented by those very wrongdoers, narcissists, power-hungry abusers and their supporters and enablers which functions in reality to relieve themselves of the responsibility and consequences of their behavior, in convincing the victims and wounded that the very state of being victimized and wounded is the problem but not those people or behaviors which cause it, and serves only to support and encourage conflict in order to sustain systems of control, oppression, and exploitation which advantages two of the Karpman types at the expense of the third. As another example of this most heinous human behavior, one past love interest of mine whom was sexually abused by his uncle went to his mother finally for help, and rather than coming to his rescue and protecting her own son she scolded him for talking about it! Admonishing *him* never to tell anyone, in such a stark demonstration of this monstrous behavior which shields pedophiles and excuses those with power in service of control, oppression, and violence, even at the very expense of children and those most vulnerable to abuse and exploitation. There is only one conclusion to such behavior, which is that such people actually do not have a problem with child sexual abuse, but instead support it and demand of victims their silence in support of systems of control and exploitation for their own personal security at expense even of their own children.

Unsurprisingly, many abuse victims consciously or unconsciously harbor shame in our abuse, having been told by both our abusers and their supporters that not only is it our responsibility to forgive them but that we even played a role in our abuse, through our naivety or vulnerability as if we had control over such things instead of being the natural state of childhood and children which is exploited by knowing adults, conspiring to sacrifice us in service of their desires for control and power, the admonition to unsolicited forgiveness one of the very tools whereby abusers and exploiters accomplish their abuse and exploitation, by severing whatever small amount of empowerment a victim might retain and thus diminishing the likelihood of consequences for both perpetrators and those whom enable them.

All those who experience trauma (and be certain—not only the Karpman *victims* do) find forgiveness ineffective in the resolution of pain and trauma because the act of conferring forgiveness does *not* also correct and repair the harm and damage which has been done, which otherwise must be accomplished to heal and move forward. If someone stabs us it does not matter how fervent our forgiveness—we will continue to *bleed* unless we receive or apply medical attention, and our emotional bodies are no less vulnerable to trauma than the physical yet many of us wish to transcend victimhood by denying we can be victimized such as by choosing to forgive those whom have wronged us as if that very act stops a bleeding wound, to pretend we have not been stabbed even while bleeding out, to ignore and neglect the open spiritual wound and the spiritual blood we continuously lose because we lack the skills required to actually repair the damage and facilitate healing. In fact, just as the unhelpful commenter demonstrated, one of the ways by which unsolicited forgiveness empowers abusers at our expense is that it also serves to *distract* us from the wounds and injury which need attention, and I was only able to finally heal from my experiences as a child and young man not because I chose to forgive my family but because I finally learned those skills which are required to fill those wounds and patch myself up, which was achieved through behaviors of self-care and self-compassion like the practice of inventory therapy and caring for my body instead of trying to dominate it. None of that involved the act of forgiveness, not one of them asked for it anyway, and it was through acts of self-care and empowerment that I achieved the peace, tranquility, confidence, compassion, and love which are otherwise and duplicitously promised of forgiveness, because they do not in fact come from forgiveness at all which is nothing but a distraction to prevent them in the first place, but instead through actionable behaviors of self-care we can do for ourselves without relying on anyone else, least of all our offenders, who almost never offer amends anyway since they similarly lack skills required to resolve mistakes, say sorry, amend wrongdoing, and ask forgiveness.

Forgiveness is instead the doorway to the making of amends by which the offender, not the offended, must walk, to request forgiveness in an appropriate manner, and those who are offended have the choice of whether they will open it or not, but for which must be asked and not given unsolicited. The act of offering unsolicited forgiveness is a (understandable) control mechanism which desires to be rid of the effects of abuse and trauma, and by offering forgiveness before it is requested we instead behave in the Karpman *helper* model which is to liberate offenders of the consequences of their actions which in turn serves only to empower them and others like them at our own peril and that of others they will victimize. This obviously does not mean we continue to resent those whom have done us wrong but continued resentment implies yet unresolved trauma such as what is resolved through inventory practice, which is the act of self-compassion that helps achieve healing, where instead forgiveness without request for it is a similar, delusional (but understandable) coping mechanism characteristic of a limited tool-set similarly at the root of our otherwise ineffectiveness.

Forgiveness too concerns the offenders, by centering the dynamic around them and not ourselves or our wellbeing, except in that we desire relief from our pain and suffering, and being a victim does not also liberate us from the nature of reality, the laws of cause and consequence, and the responsibility for our own wellbeing, but practicing inventory therapy helps us operate within the bounds of reality, to tease apart our responsibilities from those of others, to recenter our personal narrative on our own wellbeing and the things we can control as separate from those we cannot, turning over such things as the behavior of others to God which then liberates us from that burden. One of the saddest experiences I had helping others was in the counseling a woman who had been raped by her own father many times in her youth whom refused to practice inventory and preferred instead to believe she was in control of her fate which then unfortunately included shame and remorse in believing (even if unconsciously) that she somehow played a role in the abuse, because our great fear of helplessness such as the experience of being a child is an overwhelming terror for many, desiring never again to feel so helpless and thus taking on responsibilities no human being actually has, such as bending fate and taking control of reality. The truth is that even as adults we are no less at the mercy of nature and fate than when we were children, it is only that we have an increased ability to make autonomous decisions, and many an atrocity still occurs at every moment of the day around the world as we naively play our video games or fill up our embarrassingly oversized SUVs with fossil fuel in complete disregard for the future of our children and this lovely planet on our daily shopping trips and commute to a meaningless job, and we do these things and many other coping mechanisms to distract us from that inconvenient reality in which we find ourselves, even to the sacrificing of our own humanity as we bid others to submit to their rape, their death, or the ruin of their children so that we may comfortably exist in our manufactured delusion that serves only our ego, even if that also means holding on to responsibility for the abuse done to us as a child rather than recognizing there are things in life we simply do not control.

Another person I know had been similarly admonished by confidants to forgive her childhood rapist (in this case a cousin) and move on even though the rapist himself had never asked for it, and I spent some time teaching her my opinion on this matter as what had arisen from my own experiences—that forgiveness should be earned by offenders, and our efforts should be instead on resolving the effects of that trauma we experienced by practicing self-care such as inventory and nurture of the body, which seemed to better assist her own road to recovery, no longer obligated to liberate her abuser as a condition of her own liberation as those around her errantly insisted, and reported back to me later that she was much relieved of her dilemma (and I think also practiced inventory on the matter).

As was my realization much too late in life (though thankfully did occur since to some it never does), what we think must come from others (such as healing from trauma) are very often things we can actually do for ourselves—I could get to know myself, have fun with myself, take an interest in my interests and show myself the kind of care that normally comes from parents, friends, and partners. Maybe this was a problem of being autistic that I did not learn it until midlife, or maybe because my own parents did not know this either and could therefore never have taught it (and a reason why they rejected me, because they also do not like themselves). But it is the key to the healing we normally attribute to forgiveness, which is *providing* for ourselves that which we normally expect of others, or believe to come only from others, such as care, concern, interest, compassion, time, attention, support, etc., and are instead far more capable of doing for ourselves than we have ever imagined, simply because we have heretofore been ignorant to it, where unsolicited forgiveness is done in service of the offender that we may be liberated of their influence on our lives but in reality we can heal ourselves without that by learning new life skills and resolving psychological conflict entirely unrelated to or dependent upon them or others. Why, of course, should we expect others to show us compassion when we do not show it to ourselves? How can we expect others to care for us when we refuse to? How can others love us when we do not? As so also patience, kindness, sex, etc? Yet many believe love is a feeling when in fact (as discussed in my book on psychology) love is *action*, and in the act of taking care of our needs we show ourselves love, and in the act of refusing to care for our needs reveal the absence of love for ourselves. When we are offended by others it need only last the duration that they do, and beyond that our life need not concern them, where relief can then be found by seeing to our needs and taking care of that which requires care, and if later our offenders request forgiveness (in the steps appropriate of amends as discussed in my book on psychology), then we may, or may not, give it according to our own discretion. But healing—that is something we can achieve at any time and any place so long as we put in the effort to learn the skills required for it. Unsolicited forgiveness is otherwise a distraction from that.

We can, however, ask for forgiveness of those by which we have done wrong. Probably the most herculean task any person can undertake is going to someone we have offended, admit our wrongdoing, and ask for their merciful forgiveness and doing all that is required to fix what we have wronged, including the commitment to *never* to do it again. This is difficult for several reasons, least of all that we desire control in reserving the right to still offend others as we deem necessary for our own selfish purposes, which is why most of us fail to ever reconcile our ideals with our actual behavior, not fully understanding in fact why it is that we behave in such self-destructive ways as to refuse to repair damage we cause even to those we supposedly love or to commit to treating them with respect and love even as we hypocritically demand those very things from them.

Rejection is one of the most painful experiences we can endure, sometimes even more painful than other forms of loss, because our human species is uniquely impotent in terms of evolutionary survival tools in which our evolution discarded sharp teeth, quadrupedal speed, or even a fur coat in favor of social cooperation and dependence on the group which obligates us to others for our own survival and, fearing for our survival. Being exiled is not a fear possessed of bears, sharks, or snakes, because they can and do survive just fine on their own, blessed with tools by which to gather food, endure inclement weather, and winning opportunities to mate, where each and every one of us human animals is wholly dependent on the assistance of others in our species, thereby one of our primary motivational instincts is the horror of rejection, even to the point of outright denial, to refuse anyone the opportunity for our expulsion such as admitting that we have made a mistake and hurt someone. Admitting we have made a mistake is explicitly then an action of handing power to someone else, an act that many of us would rather die that do, but as we have taken control from them which is what caused offense in the first place we must give it back as the price for peace of mind and comfort in resolution. Or, not even so dramatic, we would rather hurt people we supposedly love and care about than give them the power to do the same to us, which is an ironic antithesis to love, for if we love someone should we not give them the power to rend our heart at their whim? Putting trust in someone does not mean trying to control their behavior or even expecting them to not make mistakes. Trust is giving people power to do with it as they will, part of the act that is love, without which it is not love but want, for love is action and refusing action cheapens our desires to nothing more than a selfishness exercise in narcissistic delusion.

Realizing at the age of thirty-four I cared more about sparing myself further heartache than loving another person brought half a lifetime's worth of regret crashing down upon me in an instant, especially because so many years had already been wasted. We fear giving others power over us in the making of an amends but that is the price of gaining forgiveness and the bonds of peace and love which follow. It's also not actually that serious, just get over yourself and say sorry, that you will never do it again, and do what you can do to fix what you've broken. In gaining forgiveness it is always the offended party's choice whether to give it, or anything for that matter, and trying to win forgiveness so that we may retain a relationship, benefit from association, or be relieved of our guilt is antithetical to the very principle of forgiveness, which is the resolution and restitution of what we have done wrong, *not* further service to our ego or simply only saying sorry.

This can be especially difficult for those who have not done inventory, continuing to lack skills we need to effectively repair the damage we have cause continue also to be lead by our unconscious fear and control behaviors are therefore ineffective, so doing inventory *first* can be greatly profitable and productive by first liberating ourselves of the fears, insecurities, and control and coping mechanisms which motivate self-destructive behavior and thereafter learning new life skills that are not based in fear and insecurity (since we have removed those). Just as most things in life are not within our control, it is not our responsibility whether others choose to forgive us when we have requested it and done the necessary steps to repair that damage, or even whether we are allowed to continue a relationship with them, and part of the parameter of amends is to accept whatever boundaries the other person or persons expect and request, otherwise we are still at risk of being a control freak that might cause yet further harm and therefore be ineffective in our efforts to declutter our lives, and whether people forgive us for our mistakes is the responsibility of the Universe and God, where ours is only to ask for it and take the steps required to repair the damage we did.

One of the reasons it can be so difficult to actually say sorry and ask for forgiveness is that we *know* what other humans do when they have power, because we do it ourselves, and we are at great risk of persons exploiting our humility should we offer it, so we don't, so nothing ever gets better and we lose more and more of life, relationships, integrity, good will, etc. A common saying is that resentment is like drinking poison and hoping the other person dies, and refusing to ask for forgiveness is the same—it is not other people who are at peril when we refuse this (they have already suffered), but ourselves, in the loss of our own integrity, peace, and the respect, admiration, intimacy, resources, opportunity, fulfillment, and support of others, and many old people who all their lives have refused to make right the wrongs they committed end up alone and unsupported, even in their decrepit old age refuse to discard their ego and instead persist in completely unnecessary, self-inflicted suffering and ruin. I have related before how we as humans withdraw from relationships in which we have offended other people, having this occurred to me from people I loved whom hurt me and then withdrew when my very presence reminded them of what wrong they had done, thus wounding me more than once. Several times in my youth I similarly made mistakes when trying to make friends, and not knowing how to apologize and care for relationships (since I had never had opportunity to since my parents also did not possess those skills) instead also withdrew from them in embarrassment, when all I had to do to win friends and built relationships was say sorry and fix my mistakes. Of course, not hurting people in the first place is ideal, but we are human and it is inevitable that we will hurt others even when we don't intend to, and the lack of skills for the making of amends and to create the conditions requisite for requesting forgiveness imperils our position and robs us of close relationships and the resources of companionship, support, and togetherness which is required for a successful human life, and it is a delusion to think that we somehow improve our position by refusing to acknowledge the humanity in others wounded by our misdeeds or oversight which, in the end, will cost us more dearly than even what we did to them.

We can also both solicit and grant forgiveness of *ourselves*, although only when that concerns *only* ourselves and not our behavior towards others, for instance if we have offended another person the act of forgiving ourselves for doing that without also making amends and asking for forgiveness is as delusional a behavior as possible meant only to spare us the discomfort and ego subversion required of amends. But many of us wrongly believe that many things we do are wrong which are in fact simply difficult or we might not be good at them, such as failing to clean our homes sufficiently often or well, or suffer from addiction to drugs or alcohol, or have a difficult time managing money, holding down a job, not making a large salary, or even being a sociopath or narcissist in the first place. Life is hard, and every person on the planet is in fact doing their best, even if that best is not very good at all, and if our best means it's hard for us to clean our home, or gather courage to go outside, or meet a new romantic partner, or to feel feelings we think we should feel we are in fact doing quite well and should give ourselves a break! Nobody who is sociopathic or narcissistic chose to be that way or did anything that caused it, which is different than the actions we take which, if harmful, are certainly our responsibility, but such conditions are instead a function of hormones and neurotransmitters (which can be addressed as discussed in *Fuck Portion Control*). Forgive yourself for not being good at life, for trying and failing, for doing your best even when it was not enough, for being mortal, for failing to meet the insane and unreasonable demands of our parents, siblings, and partners, for saying something stupid or embarrassing, or not getting someone to like us back, for not being rich or sexy, for not being young or healthy. Unlike forgiveness of others, forgiveness for ourselves (so long as it truly is of ourselves and not offenses we've committed against others) is entirely within our power to do so, and we only fail to do that because we do not love ourselves, because we have not practiced self-care such as in inventory therapy, to find out what about us is worth loving.

There is no happiness in hatred. Yet the path from that dark place is not in self-will or greater control over life, but enlightenment to our shared human condition and empowerment to more effectively meet life on life's terms. Many of us refuse to do this kindness to ourselves because of control behaviors like self-pity, which is a common and understandable control mechanism which serves to provide a sense of self-worth when none is forthcoming from our environment, and by the measure of self-pity we estimate our ability and try then to operate within those boundaries in order that we may yet avoid more and future pain, disappointment, and heartache. After all, how can we fail at something we know we cannot do and so do not try? Why try to clean our house knowing we can never succeed in keeping it clean? Why date people who actually treat us kindly when we know we could never get someone like that? Why start a business when we know it's impossible and we will fail? Why reach out and establish relationships when we know we are unloveable? Why sit down and do inventory when we know it is impossible? All these kinds of behaviors are nothing more than control mechanisms meant to preempt our ruin by first anticipating it, divining the parameters of life in fear of overstepping them. But we do not set the parameters of life, and we should instead meet our ruin and forgive ourselves for failing rather than never try at all! Risk love, friendship, and success, for even if we fail and others hate us for trying there is nothing we can really lose and at least we will have done our best. Those who forgive themselves know that we can fail over and over and over again and it does not mean we are the less for it. In fact, failure makes for greater wisdom and knowledge, without which we are ignorant. I did not write these books and do this work because I was good at *anything,* but because I was bad at life and failed at nearly everything, and the more often we fail it increases the odds of our success through experience and wisdom if we are paying attention and not simply obstinate in our arrogance.

So no, of course you're not fucking great at life—so what? Who gives a flying fuck? Forgive yourself. Stop taking things so seriously. Have some fun. Care for your wellbeing. Resolve your trauma, fear, insecurity, and control mechanisms. Say sorry to those you've hurt. Fix the past and secure your relationships. Cut off those who refuse to treat others with kindness and respect. Even if you're alone there are plenty of movies to watch, games to play, books to read, projects to start, food to make, pets which need rescuing, people to meet, things to see, things to grow, and things to be done. Each of us is a lot more interesting than we give ourselves credit, and life is far more wonderful than not. So go easy on yourself for being mortal and having a hard time. You're doing your best, which is the most that can be expected of anyone.

26. Uranus and Fortune

The exciting eleventh house and the planet Uranus govern *fortune*, and while, yes, this word does encompass material wealth, material things are not the only resource which in abundance constitutes wealth, and as is always the case in astrology it concerns far broader concepts of reality than our insular, fearful, myopic, insecure human minds typically comprehend, since money is the defacto resource we chase in order to feel a sense of control over the world around us. As a concept, fortune could be said to be the fruits of our labors—all those things we receive as a result of all the experiences and transits that have come before, the trauma and heartache, the learning and growing, the effort and the persistence, what we save and retain in order to not only work and strive but also enjoy and relax.

One of the primary resources besides money which Uranus and the house of fortune concern is that of *friendship,* and transits to the eleventh house or concerning Uranus can and do also bring with it friends and the dynamics of friendships. It is arguable that friendships are more valuable than money because friends can help us in times of need and provide companionship, joy, and help us to better define our sense of identity, where money in fact isn't even real and merely serves to delude us to the security of our position on this planet.

In 2015 after my life fell apart, unbeknownst to me Uranus moved into a trine aspect with my natal Neptune, portending surprise friendships and spiritual support, and indeed that was what happened when I went through the 12-step recovery program, guided by many wonderful and interesting people which had similar experiences to mine. For a time this had a singularly awesome effect in my life which I had never before experienced in which people were genuinely not only accessible but took an active and vested interest in my wellbeing—constantly checking in with how I was doing, listening when I needed to talk to someone, and provided nonjudgemental insight, camaraderie, and sharing of personal experiences when I required guidance. It was truly was a different emotional and mental experience, as for the first time in my life felt part of a group of people rather than a temporary visitor, and it had literal effects on my hormones and neurotransmitters like I had never known was possible and was truly wonderful to experience. Indeed I believe this experience is probably how most other human lives simply are, with access to family members, friends, and society in general which is not prejudiced by religious hatred and bigotry, not having to think twice about whether your very safety is in danger. In my book *Fuck Portion Control* I discuss research (in animals) demonstrating that addiction and alcoholism occurred when access to social groups was impaired such as when rats were contained to individual cages and denied the opportunity to play, mate, and socialize with other rats, but did not occur at all even when given free access to drugs like LSD when instead rats were housed together and given many enrichment opportunities.

This is an oversimplification of the role of social access in addiction and alcoholism, however, because social support does not fully heal damage caused by isolation and trauma once it has occurred, because there also occurs neuro-logical damage from prolonged psychological stress which further perpetuates hyperawareness, obsessive control behaviors, and which causes even more stress in response to believing we are alone and, disempowered by a lack of effective life skills, also then constantly fear for our own wellbeing since we intuitively know we are not good at life nor able to effectively handle its challenges such as rejection, financial obligations, disease, political turmoil, death, etc., and my experienc-ing healing in the 12-step program was amplified and more complete due to the simultaneous treatment of my cancer which incidentally treated my learned hopelessness using the tools discussed in my other books which in turn facilitated healing of my nervous system in addition to the generous, unmatched emotional and social support of my peers in that program and being first exposed to the prac-tice of inventory.

Wherever Uranus travels through the zodiac he brings with him not only surprise and unpredictability but also themes of friendship and fortune, just as Neptune brings spirituality or delusion, Jupiter growth and progress, and Saturn structure and discipline. When I was a young teenager with an unreliable and unpredictable family and home life in which we moved on average once every eighteen months both Neptune and Uranus were transiting Libra's fourth house of home and family. In my older teenage years, however, Uranus moved into Libra's fifth zodiac house of creation which was a time I unexpectedly excelled in my art and design and found refuge from my tumultuous world in art and creativity and camaraderie with other artistic rejects in my art classes. Then as Uranus transited into the sixth house began to occur hints of the health problems I would later have when Neptune barreled there later, but also made friends in my pursuit of work and purpose. When Uranus more recently visited my eighth house of power and control, which very often concerns shared resources such as taxes, I was unpleasantly surprised by a tax assessment (in spite of earning near the poverty line), on the taxes I had already paid. Luckily my life for the last ten years since my recovery has been entirely financially stable, for the first time learning through the practice of inventory how to live within my means and no longer have an insecure compulsion to *feel* financially stable and was able to easily absorb that challenge. But throughout the transit Uranus also sent me friends in terms of power, wherein a few relationships began because people approached me as an authority on health problems they were challenged (and I certainly could have abused that position if I had not understood these concepts). Uranus also kept dropping financial surprises such as some rental relief as several free months of rent when I moved apartments, or unexpected sales spikes and attention toward my work. For anyone whom harbors trepidation and anxiety around money the transit of Uranus through the 2nd, 8th, and 11th houses can indeed be aggravating since Uranus brings chaotic, unreliable, and unpredictable developments as likely to lose finances as gain them, but once control behaviors, fear, and insecurity are resolved and the realization is made deep within the psyche that money isn't even real the financial challenges of life simply become part of the background noise and colour of life rather than real concerns.

Uranus transits to Jupiter also directly concern economic cycles as well, especially those which are disruptive or based in technology, since Uranus also rules technology which is inherently a revolutionary karma. The very recent resetting of the 14-year Uranus, Jupiter cycle on April 20, 2024 occurred uncannily at the exact same time of the last *bitcoin* halving on April 19, which was not a set date but instead a function which occurs every 210,000 added bitcoins, and there will also as such occur related events at the major transiting aspects to unfold over the course of this next cycle. While cryptocurrency enthusiasts may welcome such connectedness to astrology its association with Uranus is also the reason for its unstable, volatile, unreliable, and unexpected behavior. This is not to say that digital currency could not be stabilized in the future, only that for now it is a revolutionary function and as such governed by Uranus and given to all its properties.

Likewise, Neptune transits to Uranus or the eleventh house often bring delusion to wealth and friendship, in which we might overestimate how much fortune we really have, take friends for granted, or lose our wealth in risky investment schemes. Similarly, many people with an abundance of wealth delusionally misunderstand it as a bulwark against misfortune, believing that because they are rich they are also immune to laws of mortality and causality which in turn often becomes their very undoing. The problem is that wealth shields people from the consequences of their actions, surrounding themselves in turn with other fearful and insecure people whom want access to that wealth, equating those plainly ulterior motives to their quality as a person and deluding themselves of their own nature, when in fact none of those people would want anything to do with them if they weren't wealthy, because any person who hoards money while others starve and languish in poverty is indeed a loathsome individual with few, if any, admirable qualities that would genuinely attract others. Consequences to our behavior is often the very thing that builds character, the making of mistakes and their consequences what inform us to what is okay in life and how to hone an honorable and admirable worldview, ideology, and person, but when people let us get away with bad behavior such as occurs when we are wealthy we fail to learn such valuable lessons then spend our whole life as a tolerated pariah in an invisible cage of our own making.

Even worse is the part of Neptune which rules the unseen, unknown, and secrets, which are things that we often do not even know are occurring in relation to our wealth, such as schemers and swindlers or just simple ignorance and arrogance. Many people have lots of friends but none of whom would stay by their side should misfortune come calling. Some even encourage self-destructive behavior that costs us dearly simply because it gives them greater power over us and supports their own control behaviors and coping mechanisms at our expense. One famously ignorant and obscenely wealthy person from history whom was destroyed by his wealth was King Henry VIII of England, whom literally had women murdered when he wanted to have an affair, sacrificing his very humanity because his wealth kept him from facing direct consequences to his actions. Henry ate a diet of Kings high in protein and low in healthy fruits and vegetables, because he could afford the most luxurious of diets, and nobody could tell him to eat better. But this kind of diet promotes ammonia producing parasites which made him shockingly obese and metabolically ill, suffering from gout and covered with painful boils before dying a very early death at the age of 55 in spite of having every resource at his command. Less fortunate people with limited means who could only afford a high plant-food diet would ironically live longer and be healthier even if they didn't appreciate such a diet simply because they could not afford to be so slovenly. Just as Jupiter does not determine the quality of friends and wealth, we find in the eleventh house that neither does Neptune, and having what we want is often a lesson in disguise which can cause us much suffering as any more obvious misfortune because what we want is usually grounded in fear and insecurity, not wisdom. Nobody was going to tell King Henry 'no' because they were so afraid of him, and his wealth gave him power to never hear no, and so solutions to his immense problems never even reached his ears and eyes, which he could have discovered himself if he'd had any self-awareness, and he died a psycho, diseased, murderous loser that not one person in the entire world liked. The fable of King Midas with his golden touch was supposed to be an allegory about this very thing, but having to infer such lessons through allegory is beyond most people, and plainly stated the eleventh house does pertain to fortune but fortune is not always good.

Uranus's transit during my early recovery was also in Libra's zodiac seventh house and was a reason my partner became unpredictable and unreliable—these words by the way are also interpreted with bias, as being unpredictable has equal potential to be fun and thrilling as destructive, so the influence of Uranus indeed also seems foreboding if we continue harboring fear and desires for control and anxiety for the future, but for those who don't mind surprises or even cherish unexpected thrills that surprise can bring Uranus can and is also a planet of fun and whimsey. Earlier I made the analogy of ten people in a room and one of them calling us fat which can cause us, depending on our trauma and control mechanisms, to incorrectly believe the whole world is against when in fact nine of the ten people did no such thing, with only 10% of people in that scenario being unkind and inconsiderate. The same is true of all life, and Uranus transits are very often just a handful of experiences which may be unexpected, not the entire time, which is otherwise simply normal life proceeding as usual with a peppering of excitement or disappointment. Of course in situations such as I found myself the instability of my partner during Uranus's seventh house transit was quite catastrophic, but afterward I moved on and rather enjoyed the rest of the transit, having a lot of sex with new, fun people, making new discoveries, new relationships, and achieving things I had never thought I would, and rather than my warden Uranus ended up being my liberator.

Uranus entered Aries nearly the exact same time as Neptune entered Pisces, so looking back over the last fifteen years it may be difficult to differentiate between the effects of Uranus and those of Neptune exactly, and Uranus was also in Pisces immediately before Neptune which I only just realized when writing this sentence, meaning that the same effects of Neptune over the last fifteen years were also occurring before that due to Uranus's transit there immediately before it, but where Neptune is one constant and uninterrupted influence Uranus only occasionally sparks some chaos or bestows some benefit, like a manic game of Russian Roulette.

During Uranus transits its nature and events (like any planet) can also be triggered by other transiting planets such as Mercury, Mars, and especially the transiting Sun. Earlier I related a story about having a bag of flour explode during transit in the mail and basically getting most it for free (that shit is expensive) on a day when the Sun transited opposite to Uranus in my zodiac eighth house (concerning shared resources). One of my friends at the time was a Sagittarius with Uranus transiting their sixth house of labors and suddenly came down with severe stomach problems and reached out to me for help, demonstrating how the Uranian energy of friendship was channeled through that experience. Also, as my life fell apart and Uranus went through my seventh house of relationships several friends let me sleep on their couches, and one even had an apartment he was trying to sublease which I took over for several months, fully furnished, and similarly through friends found assistance that was so sorely needed. One of my friends with cancer had Uranus transiting their ninth house of philosophy in which I also attempted to expose them to the practice of inventory and astrology, to help them better handle cancer and the uncertainty inherent to life. They mostly ignored me, however, and preferred instead to cling to their self-pity and control mechanisms as life never forces us to do anything against our will.

Indeed if friendship was valued as much as money by everyone we would not even need money as means for living, because each of us would participate in society simply to help our friends and be around people. This already happens in many ways due to simple human behavior, for instance that it has been well established most people will not take the last few of any products in the grocery store, because of our instincts as humans for sharing, and grocers have to actively keep bins stocked full in order to prevent the better instincts of humans preventing us from buying products. But those of us which do not understand friendship become very fearful of stresses like financial insecurity, poverty, homelessness, etc., because we do not trust that anyone will help us when we encounter problems (notice I said when, not if) if we do not have money either to save ourselves or to control others and tempt or force their assistance. Indeed many people too earn lots of money and power simply to attract others, whom they mistake as friends, because the prospect of making real friends risks emotional vulnerability and requires surrendering of power and control in the bonds of intimacy, and is such a frightening exercise for those whom are very traumatized to their personal self worth, and money helps us avoid those realities since there is in turn also an unending supply of other humans whom are also frightened, insecure, and desirous of the security we think money can bring (it doesn't).

In reality, not a single person on this planet goes to work to earn money, build a career, or create wealth but to simply be engaged with other humans in common cause, through which we derive meaning, purpose, companionship, and (mostly) real security through association. To demonstrate this principle, consider if it were possible and ethical (it's not) to do an experiment offering a person one million dollars a day every day but only if they lived in isolation and are not allowed to leave isolation, see or talk to other people, or interact with or spend money on the products and work of other people, including no grocery store, no internet, no video games, books, media, etc. Not one person would take it because money is meaningless, and without other humans not even gold is precious, and the entire functions of economies, wealth, and markets serves the singular purpose of getting other humans simply to acknowledge us. As would also be expected, Neptune's transit through the eleventh house can be transmuted to a spiritual rather than delusional experience if, rather than selfishness, wealth is used in service of others, and friends are not seen as chattel but individual beings through whom we might experience God if we abandon selfishness, fear, control, defensiveness, and other manners of keeping people at length from us.

Friendship is thus the true wealth, but money and material things can bring comfort and security, and thus such fortunes are all also concerned by the eleventh house which follows the tenth house of society in which we lay the groundwork for earning friends and wealth, because without social structures there is also no context in which to make friends or create wealth. This can be rather frustrating for those who have a dearth not only of friendships but opportunity to even meet people which, in our car-centric cities and towns has become especially sparing, where once humans could not help but meet others now have to go far out of our way to do so. When I was a child the shopping mall was the only place besides school where kids our age met, and thankfully many young people today find this opportunity on the internet, somewhat, as even horrible malls and shopping centers decay and disappear. But the absence of third spaces such as coffee shops, parks, trails, bike paths, town squares, walkable communities, swimming pools, and other usual social infrastructure has prevented a normal human life which our ancestors enjoyed for millions of years, which was simply the opportunity to be around other people in a social setting which needs to be built back into our cities and towns so that ourselves and our children and later progeny are not continually robbed of a normal human life.

Similarly, the farming and agriculture communities here in the U.S. for a long time have been plagued by severe rates of depression and suicide, which is borne simply from dissociation of our lives with those of others, living out in the middle of nowhere as a direct consequence of the fucking psychotic "manifest destiny," phase of colonization by white Europeans which, as they drove and slaughtered the indigenous peoples, thought they were in turn getting a great opportunity for prosperity, owning all this open land, only to later kill themselves or their children and grandchildren from the sheer stress of continuous, undying loneliness, isolation, and exposure to toxic agrochemicals and industrial poisons. Although they arose as a product of aristocratic control and exploitation of the working class, small English villages like the Cotswolds which are so famous for their aesthetic and quaint nature are an ideal human habitation infrastructure in which people live in close proximity both to each other and nature and social infrastructure as pubs, parks, schools, etc., and travel out to their farms and agriculture jobs instead of living at the worksite, returning home every day to the community they know and love which supports them and to which members also directly contribute simply through the function of their presence, and the plague of depression which oppresses the American farmer will never abate until they reorganize their communities to center life instead of work and profit, literally in the infrastructure as well as behavior.

Such realities of human existence may seem separate from what can be seen in astrology but in fact these themes are directly affected by any cycle with Uranus, such as Uranus-Pluto or Uranus-Neptune, which impart thus themes of control and power or spirituality or delusion with Uranian themes of wealth, friendship, and unpredictability. The last conjunction of Uranus with Pluto occurred in the sign of Virgo in 1966, the house of labors and concerns our daily obligations for self-care responsibilities including both health and work, after the advent of the automobile and merciless destruction of cities, towns, and communities by car-centric infrastructure ripped through neighborhoods and emptied downtowns of their residents and paved over wonderful buildings to then create the populations of our parents and ourselves now sick with isolation, trapped in homes and apartments with nowhere to go and nothing to do except work and shop and the need to drive everywhere. The highway acts which built up those corridors for vehicle travel and bulldozed beautiful architecture and replaced them with parking lots so suburban middle class drivers could park when they drove downtown truly embodied the control behavior of Pluto in the devastation of communities. While this might have promoted material wealth and prosperity it came at the cost of our own satisfaction, participation, and friendships, so our children now grow up in landscaped prisons with nowhere to go and no one to meet, as a typical consequence of control. The prior conjunction of Uranus and Pluto in 1850 occurred in the sign of Aries, as did three of the six conjunctions before that beginning around 1090, and since Aries is the sign of war and aggression these unfortunate conjunctions of Uranus with Pluto likely encompassed much of the war and violence which occurred over the last thousand years (although since Pluto governs power its cycles often correlate with conflict).

Friendships are also at the heart of such structures as politics and other social leadership, so there is also a long history of Uranus cycles with leaders and politicking as it relates to nations and sociopolitical structures, and transits of other planets through the eleventh house likewise bring with them their influence upon all subjects of wealth as money and friendships. During Pluto transits in the eleventh house people often become controlling concerning friends or money, becoming suspicious, demanding, unreasonable, or even obsessive, which can and does then risk the destruction of friendships and loss of fortunes. Jupiter brings friendships and wealth but does not also determine the quality of those friendships, and during Jupiter transits here many of us pick up shallow, fair-weather associations we mistake as friendships whom later let us down when we really need them or experience misfortune. Saturn transiting this house demands structure and discipline in the maintenance of friendships and fortunes alike (not just taking them for granted), and failing that can result in their loss or diminishment. The North Node transiting this house likely means some fateful life event meaningful to our destiny will occur in terms of wealth or friendships, and Chiron similarly an experience of healing concerning friendship or wealth.

Having autism has always meant I fundamentally do not naturally make friends, and have been alone most of my life even when friends were nearby (though the instability of my youth also played a large role in that). When I was in my late thirties my attempts to make friends failed so often it became an amusing and ludicrous experiment in which I actually began writing down all the attempts or opportunities, every one of which failed, for instance one person in the recovery community I thought might be a good friend but whom refused me the few times I asked them to hang out. Almost a year later I received a text from him out of the blue with an enthusiastic invitation for coffee only for him to finally realize I was not the Nathan he had just met and was trying to sleep with, then embarrassingly tried to pass it off like he would still love to get coffee even though we both knew he would just not show up (so I flat out said 'no'). Before anyone feels pity for me I realized recently the problem with myself and my autism is not the stereotype of failing to connect emotionally, but the reverse, that my piercing gaze and deep connection I cannot help but feel with everyone oppositely makes *them* feel vulnerable. Many times in my adult life I even simply asked people to be my friend but was ignored or rebuffed, I guess never having realized that adults grow out of the childlike phase of being friends with everyone. But I kept my head down, took care of myself, and kept at my work and as Saturn began to rise to my Descendent ten years later I was eventually blessed with some truly wonderful friendships with people of common interest who were genuinely interested in me as much as I was in them (and not only in what I could do for them), and like all cycles those fallow seasons absent of friends and camaraderie do not last forever, even if it does seem long, as *all* cycles eventually end.

Later when meditating on this dilemma and trying to understand how friends are made, even doing research into it, it became apparent that the most important factor in the making of friends is simply *familiarity*. As children we are inherently around other children all the time and then have opportunity to observe and then approach if we desire. But as an adult this happens more infrequent the older we get, and social taboos in turn make it difficult for adults to simply be around strangers, so many of us get dogs so we can go to the dog park and meet people, or join cycling clubs or start social media channels just so that we can have an excuse to interact with others in a neutral environment where we can both observe and give others the chance to observe. When I was in my early twenties and searching for meaning and belonging I often went to a corner coffee shop in Salt Lake City which had lots of couches and benches where people congregated regularly, and unintentionally started becoming familiar with several people I saw there frequently but did not realize that friendship was forming, and I moved to Los Angeles thinking my happiness and success required a safer, better place to live than the hateful environment in Utah, and ironically found it more difficult to make friends because nobody moves to Los Angeles to make friends.

Familiarity begets friendship, and we can always take it upon ourselves to try to make friends and meet people but the fate of those decisions and the outcome of our intentions is not up to us, but instead to fate and life, and thus it is not our job to make people like us or event to make friends but only to show up for opportunity and do our best, and over time the transits of the planets will align with our desire to finally help us create the families and communities we need for our wellbeing, which are most likely to come with transits to the eleventh house or as what concerns Uranus.

Thusly, Uranus transits through the first house finds our ego, sense of identity, and life purpose unmoored by unexpected and surprising events, especially through friends, which may even rattle us to our core. For those whom are especially serious and insecure this may be quite upsetting, but this transit can be quite exciting and fun as we are awoken to realities of ourselves we may never have considered. Taurus was the most recent sign to have this transit and will have insight for Gemini following, and likely found their personalities evolving and growing in ways they never expected but likely now appreciate.

In the second house Uranus is quite unsettling because money and instability are inherently contradictory, and this transit benefits especially by an enlightened and unserious attitude about money. If this is lacking, do inventory on it.

The third house finds Uranus highly stimulating of communication and exchange of ideas and knowledge, especially through friends and close relationships through whom we may learn knowledge required to progress in life or solve problems we have been challenged with. Pay attention to your life at this time as it is easy to miss those opportunities as necessary, and afterward may be left wanting due to our intransigence and ego.

The fourth house transit of Uranus will unsettle our home and family life, and can take the form of unexpected behaviors from parents and children and even instability in our place of domicile as occurred to me when I was a child. Though we cannot control life we can behave responsibly and attempt to live as stably during this time as is possible. Because family is the place from whence we come there is also likely some personal evolution that will also occur during this transit.

The fifth house of creation may realize a Uranus transit as much exciting acts of creation such as in art or even of surprise children, but don't expect any of it to go the way you want and just go with the flow.

In the sixth house Uranus brings health and work surprises related to our daily labors and, like my friend, might find solutions to such problems through friends, so keep an open mind and try to meet your obligations as best you can.

Uranus in the seventh house is not conducive to stable relationships, and either yourself or your partner may become especially restless and unpredictable. While this can be understandably upsetting try to have empathy and compassion for your partner, and if things should end and are very sad remember that other surprising developments may come after like new romantic partners. Work and business partners can also be concerned with this transit.

The eighth house transit of Uranus will bring a few surprises with taxes and other shared resources (such as with a business or romantic partner) but may also have novel sexual encounters and experiences (since sex is a power dynamic), or even of regeneration and the letting go of control in certain areas. Most of all this house benefits from absolute relenting of control, as the eighth house is never lenient in its lessons, but taking these events less seriously or even with levity can take its power away.

Then Uranus ninth house transit will bring unexpected opportunities for enlightenment and empowerment, but we must be openminded to those experiences or they will pass and leave us none the better.

In the tenth house Uranus brings instability and unpredictability to our public life and public image, which can be especially upsetting for anyone married to what others think of us. Of course, we should do right by everyone, and as long as you are doing your best there is nothing to worry about with this transit, simply persist through the unwanted surprises and good ones will eventually also come.

In the eleventh house don't take money and people for granted.

Finally, the twelfth house transit of Uranus can and will bring surprise endings, rest, and opportunities for spirituality, likely often through friendships. Many of these will help set the stage for the upcoming transit through the first house, but that can be perilous if we do not act upright in the twelfth house transit and properly employ spirituality rather than delusion. If we chose the latter the transit of Uranus in the first house will be especially upsetting as we lose relationships, positions, status, beliefs, etc., which are not in harmony with reality.

Uranus is indeed a fun, trickster planet which keeps us on our toes and always guessing what might come next. If we are naturally reserved and serious Uranus can be especially upsetting, so try to be more openminded, lighthearted, don't take yourself too seriously, and have a little levity about things, most of all life itself, and Uranus transits will mostly be no big deal.

27. Ethical Scientific Experimentation

In 2020 during the United States elections it was revealed to the public that a prominent candidate and public figure whom was also regarded as a doctor had conducted so-called experiments on beagle dogs in which their heads were immobilized in a restraint device over a bed of biting sandflies. If this was not horrific enough the dogs' vocal chords had been surgically removed so that the depraved 'scientists' committing this heinous act would not have to hear the horrible screams and wails of their victims. To this day I have no idea what the purpose of such a stupid, immoral 'experiment' was meant to achieve, because there cannot possibly be any excuse for such inhuman disregard for the wellbeing of other creatures as to commit such horrors other than the simple sociopathic desire for sadism which is unfortunately a recurring human control behavior, the purpose of which is to assert control and power over anything we can in order to prove to our feeble and frightened minds that, yes, see, we do wield power over nature, see how we make it suffer for nothing more than our gratification?

While such scientists may delude themselves that there is no price for such acts the cost is our very soul, and if there are indeed laws of Karma as there appear to be we are all destined to have karmic justice meted upon us at some point in time, and I certainly would not want to be the recipient of whatever consequence arises from something so depraved.

Likewise, many of the studies I read over the decade of my research involved likewise shameful and horrific exploitation of the helplessness of other creatures, dosing rats and mice with infectious agents or toxic chemicals only to later rip them open to study the effects upon their little bodies as if the relatively shorter lifespan and smaller size of such creatures makes them deserving targets of such amoral and disgusting acts. It is true that many of my discoveries came from such studies, and every time I read about rats being poisoned, asphyxiated, or dissected it filled me with remorse, horror, and regret, especially because there was very often ways of obtaining results without literally killing or torturing the animal, which many scientists seemed determined to do regardless of the justification.

Because we are a predatory species with a history of hunting we also excuse such torment of animals as nothing more than an extension of our predatory nature. But killing for sustenance should be an act of humility, one in which we do only when necessary, and recognize the harm we cause in doing that in the first place, as a condition of our own mortality as another animal species. While it is an effective hunting strategy big cats even suffocate their prey first before eating them, a merciful method which helps spare animals worse suffering, a kindness that many supposed scientists do not even offer their own victims in their delusions of grandeur and quest for dominance over the laws of nature.

But, to be sure, there are laws of reality, even metaphysical ones which take as the price for this behavior our very humanity, and we are left soulless and unfulfilled as the screams and pain of animal test subjects haunt our dreams and waking nightmare. Many scientists I have seen which seek to justify the unjustifiable, including the contamination of our food, environment, and poisoning of other humans under the charade of progress are money-hunters desirous of fame and recognition as to supposedly transcend our mortal condition and cheat death, loss, and powerlessness. Yet the very consequences of such deeds is the inevitable loss of humanity, spent in denial of our culpability and responsibility which is the law of cause and consequence which no mortal can escape, and even if it is justified by powerful institutions and establishment traditions that will not save you from the immutable laws which govern the Universe.

Indeed such animal experiments have not even been the ones which most benefitted my research and achievements as the cure to cancer, depression, and alcoholism, the participation of silicon in mitochondrial respiration, the digestion of dietary fats, and uptake of vitamin C through aquaporin channels, that restoration of dopamine by limiting adrenaline production (through healthful means and not biohacking which instead raises cortisol) is the cure to depression, and that supporting the parasympathetic nervous system and acetogenic gut microbes is the cure to alcoholism and addiction. Instead, the most useful studies were those which use tissue cultures, conducted by scientists who actually know chemistry and biochemistry, molecules and math, or culture microbes in relevance to their role in our survival and wellbeing because the use of animal experimentation is nothing but a forced result of incurious and incompetent wannabes, poorly educated who cannot comprehend the chemical reactions of iron with vitamin C or the role of hydroxyl on immune cells. Most hypotheses are best studied in the more specific and accurate models as tissue cultures than entire, living organisms, and many scientists sacrifice life after life and come out the other side with nothing better than "maybe this very vague and unspecific thing might be true but it might also not be." Biology experiments need to change because the slaughter of innocent animals is not only immoral, but unproductive, and even counterproductive, and such errant detours from the progress of scientific discovery as fixation on serotonin which not only failed to cure the depressed but worsened it is a direct consequence of the murder of animals in the name of science, and it was in spite of this horror, not because of it, that I was able to find so many solutions to so many human maladies which long eluded researchers, picking through studies for morsels of information that was mostly found from the most competent of researchers which did not commit such atrocities. If scientists do not understand the mathematical nature of molecules and chemistry they should not be doing experiments in the first place. Simply loading an animal with toxic elements and seeing what happens is brutish, barbaric, and an insult to intelligence and scientific progress that indirectly costs human lives as well. Torturing and slaughtering living creatures because you're too stupid to infer how natural systems work does not make you a scientist, it makes you *vile*.

This goes without saying that dissection of animals as a regular practice in younger education is an amoral sin, and there is a big difference between dissecting a morally donated human body or animal which died from natural causes to mass slaughter of frogs, pigs, and sharks just to compel children to engage in the sociopathic exploitation of animals. While human experimentation is also often as equally morally depraved as those on animals, experiments on *ourselves,* personally, the researchers doing the research, are the most productive and moral, and while many self experiments can be stupid or ill informed (such as when I nearly went to the hospital from ingesting calcium and glycine together) it is not as immoral to treat our own diseases through experimentation as I have done, which allowed me, a man with no more than a High School education and no support from institutions, team members, or even equipment more than my kitchen stove and computer to make discoveries that not even vast institutions with unlimited wealth and tomes of knowledge were able to do, informed by the recognition that our body operates as a function of nature, not human ingenuity, with logic and reason, in which solutions are not found through exotic compounds but in the restoration of our own biology dependent on natural systems in which it evolved over millions and millions of years in a shared environment with other living beings (including plants) reliant in turn on nothing more than simple nutrients like sugars, protein, B vitamins, calcium, sodium, polyphenols, sunlight, etc.

One of the secrets as to why I could discover the cure to cancer and not the many institutions, billions of dollars, advanced technology, and thousands of other scientists invested in the problem is because although they think otherwise no cancer researcher has been searching for the cure to cancer—but for a *product* to sell to people with cancer, which is a very different thing. The latter is in fact a fantasy since cancer itself represents a fundamental breakdown of the holistic systems which make the human body which in fact cannot be restored with any product but an overhaul of the environment of those with cancer, including the diet and dietary behaviors and resolution of both opportunistic pathogens and the immune dysregulation which results from deficient diet and behavior. This holistic system for instance even includes microbes and the service they perform in supplying dietary silicon daily to our body from the food we eat which is otherwise locked behind indigestible cellulose, then acidified by the short chain fatty acids which makes it bioavailable to our biology, and the simple deficiency of mitochondrial silicon which results as a consequence of losing the ability to properly digest foods and assimilate dietary silicon is not a system which can be sold for profit, because even such products as silicic acid will be neutralized by ammonia and other disruptive compounds made by opportunistic pathogens, and requires that people understand how their own behavior and diet supports this system or ruins it.

This is the same reason why cancer plagued even ancient dinosaurs, as evidenced by the many tumor fossils which exist, long before there was ever industrial chemicals and manmade environmental toxins (though to be sure those can and do still contribute to cancer), yet researchers scour the globe for exotic compounds or novel processes which might achieve that which is already demonstrably occurring in every single fucking person and animal that does not have cancer, right under our own noses and in front of our face yet unable to recognize such simple realities because the motivation was always profit, fame, and control, *not* actually the cure of disease but as an opportunity to profit from it.

Because science is the fundamental law of the Universe, real science always bends toward nature and natural systems, not away from it, and the instinct to organize nature even to its extermination is instead a function of maniacal human survival and control instincts in response to fear in which we are driven not only to find but to establish order which underpins our instinct to farm as a means for our survival. In our overwhelming populations and dwindling resources fear rises further to drive us to insanity and reckless abandonment of fundamental truths such as that we will not survive upon this or any other planet without clean air, water, and intact ecosystems which sustain the microbes, pollinators, water cycle, halide cycle, phosphorus cycle, nitrogen cycle, etc., upon which we are wholly dependent.

Scientists must be stalwart bulwarks against the wanton destruction and exploitation of nature, not its doom-bringers, which is what we do when tormenting innocent, helpless creatures simply because we can, which not only takes the humanity of those which do such horrors but also results in errant, misguided, incorrect, distracting, unhelpful, useless data that risks sending research into entirely wrong directions away from truth and knowledge outright, condemning then humanity to continue in ignorance, suffering, and failure through our inability to recognize that we too are just a stupid animal on this planet like any other, and our motivations no less instinctual and potentially harmful not only to others, but to ourselves.

There is no excuse any longer to submit animals to scientific experimentation. Understand chemistry and biochemistry, math and natural systems. Abandon arrogance and ambition. Use cultures, computer models, microbes, and ethical human research to find the remaining answers to remaining questions. Not only will it return better results and actually achieve scientific progress and recognition, your soul will remain unburdened.

28. Death

Like many kids from my generation my entire childhood was spent watching *The Muppets*, created by the indomitable Jim Henson, and his masterworks *The Dark Crystal*, *Labyrinth*, *Fraggle Rock*, and of course, *Sesame Street*. But I was too young at nine-years of age to understand much what death was when Jim passed in 1990, but I distinctly remember feeling very sad at an older age when learning he had died a while ago, and that the man which had brought so much joy to my young life would never again. Like many humans Jim encountered a health crisis at the end of his life, succumbing to toxic shock syndrome from an infection of *Streptococcus* (strep throat). Jim initially refused to go to the hospital, but eventually did and although he received normally lifesaving medical attention it was too late and his organs shut down and he passed away.

Death has likewise always been one of the primary fixations of astrology as fearful humans try everything in our power to stay on this Earth beyond our appointed time. This is an especially ridiculous behavior, however, considering that all the trillions and trillions of every animal to ever live, including every human, have died and will always die. Yet fear of death is so overwhelming for some that we prefer to believe in fantastical mythology promising to save us from death rather than face it with humility and courage, or refuse to confront our fears such as in the reading of this chapter for the anxiety and trepidation we experience in doing so.

When Ray Peat, the biologist whose work on which mine is based whom also saved my life, died in 2022 at the old age of 86 many who exploited his work for their own control and power completely abandoned him in as little as a few weeks after his death, since their primary motivation for using his work was to wrest control of life and seek mastery over mortality, not scientific enlightenment, and since Ray did not cheat death as they demanded abandoned him entirely without even a second thought (to be fair Ray was not right about everything, as I will not be either, but the useful things about his work and life were still useful even after his death, because truth does not originate from any person and instead we are simply those who uncover it, if we do).

Similarly, many of us use astrology to anticipate such misfortune as death, even if we accept that death is inevitable and desire not to cheat it but simply to avoid being taken by surprise, that if we know the appointed time we can stop fearing the unknown and the unexpected, because knowledge is one of the most effective ways to eradicate fear. But even simply wanting to know the time of death for the sake of it and not to prevent it is still a fear-based control behavior, and being based in fear does not also actually bring the type salvation from fear which we expect, since we are acting on the very fear we are ostensibly trying to avoid and thus allowing fear to dictate how we live our life and the formation of our worldview, such as that we can know the time and nature of our passing. I will save you time and confirm that, yes, you will die. It's going to happen and you can rest assured that it will. No guessing required.

I talk about death this way as if this same problem has not been a personal experience, but in fact arrived at this realization after also fantasizing that knowing even the approximate time or nature of my own death might instead bring me relief from anxiety of it. But I found instead that a fear of death is in fact an inherent distrust of life, God, and our human condition, attempting to take into our hands the power to govern death as if to insist it should have our permission when it comes. In the end, we cannot do such things, and our unconscious mind knows this, so as we age and grow nearer to the end insanity most often takes us in our ridiculous desperation to avoid something that no one ever can, betraying our lack of faith and lonely worldview as we wrestle with the truth of our mortality and increasingly desperate fervor.

In truth the astrological appointment of our death is greatly misunderstood, even by those who do *not* search for it, as while yes fate is predestined it is still enacted *through* our free will, and death in astrology is not a specific alignment anymore than what occurs at the time of our birth, where there is no universal astrological aspect true of all births there is also not one which is associated with death. This does not mean, however, that there are not obvious patterns, as there are in all cycles of life, and in fact there are some very common associations with death which are very illuminating as to how the process works and what we can expect.

Yet fundamentally *denial* is our most common and powerful coping mechanism that is also associated with death, and more apparent and consequential than any astrological influence, which is employed by us in our journey through life in direct proportion to our ability and skillset to handle upsetting or uncomfortable experiences. Many of us live in constant denial about our eating habits, believing that starving our bodies or exercising excessively does not also come with significant consequences that increase the likelihood of disease and death. Others of us live in our completely destroyed relationships as if everything is fine and our lives are not in fact burned down all around us (I say this from experience). Death is no exception to this human survival coping mechanism, and like many people throughout history Jim's experience was one in which he desired not to be sick and in potential danger of dying, but Jim was no longer a young man and had worked a great deal in his life, and his immune system was clearly compromised such that his body was not able to fight off infection and it was not *Streptococcus* which killed him, but denial—and not just failing to go to the hospital in a timely manner, but also the need to eat regularly and rest more as Saturn was transiting his zodiac sixth house of labors in 1990 for Libras as it has again now in 2025.

Likewise was the experience of the young woman with breast cancer referred to me by her cousin whom, even though I gave a copy of my book and offered assistance, refused to even read the chapter on cancer let alone the entire thing and make requisite changes to her life, and she died several months after I last spoke with her in which she continued refusing to read my book or implement even one suggestion. But even people whom have done so when faced with the spectre of their mortality and the immensity of death will freeze with terror and refuse to do what is required, such as the other young woman whom was probably the first person reading my work to overcome cancer, around 2019 in which her late-stage cancer had moved to her bones in spite of extensive chemotherapy and radiation, and she and her family were simply making preparations for her to die, but after reading my work and incorporating simple steps like taking aspirin and vitamin C (the process is much more complex than that, though) suddenly received a few months later a no-evidence-of-disease diagnosis. In spite of this success I was surprised several years later when inquiring to her health that her cancer had returned, and immediately doubted my own work and asked what she had been doing, assuming that surviving cancer because of me would also result in a person continuing to follow my advice, instruction, and continued research and as such needed to identify and correct the errors in my work which had led to the return of her cancer. In fact, it turned out, she had not been doing *anything* thing as outlined in my book, not even taking vitamin C, and had also returned to her old diet dominated by meat and low in fruits and vegetables in complete defiance of every principle of health discussed in my writing.

At the time I did not admit (to myself) in fact how angry I was at her for having come so near a preventable death and recover due to my hard work and exhausting research only to then discard it as useless and irrelevant and thus also through association implicate me in her failure to defy nature and biology. She had gone through another round of chemo and the cancer was again in remission, and I admonished her to never again neglect her diet and therapy, as having cancer once means we can never, ever again take our health for granted and must always care for the body in specific ways to prevent relapse.

Of course, like me, you assume that story to be concluded but another several years after that in early 2025 she contacted me once again as her cancer had returned yet a *third* time and was worse than ever, with cancerous growths on her skull penetrating into her brain and causing seizures, and was too sick this time to do chemotherapy or radiation. Immediately I feared again all my work was wrong and my life's work a farce, and she asked for advice but it was too much to discuss in an email so I offered to meet virtually, assuming she was already following my advice and that we would have to spend more time exploring more deeply what might be wrong. But I was disturbed to again learn not only was she doing *nothing* to care for her health and body as required by my book but was actively eating *yogurt,* which in the book I describe as the act of slowly committing suicide due to the abundance of lactic acid and lactic acid producing microbes which downregulate our metabolic rate. Avoiding yogurt (and promoting short-chain-fatty-acid producing microbes instead) is a fundamental principle of my entire book and work of which she had been familiar for at least the last *six years*. She was not even using the solution to the Warburg Effect which I had shared publicly and for free a full year earlier, nor any of the advice about eating high silicon foods and suppressing ammonia producing pathogens at the root of cancer.

On our call it was immediately evident her mind was addled not only with the stress of having terminal cancer but the high ammonia burden which usually accompanies cancer, as ammonia disturbs neurological function, so I asked her to invite in a family member whom could take notes and help support her rehabilitation. But even after spending hours with them and without any compensation for my time and effort she continued to fight my admonitions for getting sunshine or reading my book, which is the entire source of all the information she requested of me in the first place, unwilling to put in the work and resolve required to recover from such conditions of mortality as cancer, and though she reported feeling a bit better after incorporating some of the recommended strategies I stopped hearing from her entirely no more than a month after and was later informed by a family member she had passed, and was extremely sorry and angry to be writing the conclusion to this story.

While the condition of cancer does indeed cause forgetfulness and impaired cognition the refusal even to continue reading my work and adhering to its guidelines in the care and maintenance required of our metabolic health even after having cancer *twice* before is a textbook example our denial coping mechanism as a response to fear and powerlessness in the face of nature and fate. Several months later I came to the realization that she had never read my book, but had only read my articles on my website, which is particularly delusional considering the severity of disease and requirement for thorough understanding to survive. In searching for her information I found that she and her family were also extremely religious, and her denial as an approach to handling such a traumatic life problem as having cancer became more clear as religionists are often taught that all we have to do in life is believe in God and things will be taken care of for us, no need to do the work ourselves or to learn about human biology and read books. Being so unempowered to care for our own needs can lead to feeling so powerless that people simply give up, which I think was sadly the problem in her case, because feeling so powerless in turn creates the condition of learned hopelessness discussed in *Fuck Portion Control*, so then someone battling issues like cancer that has returned several times tend to freeze and do nothing out of a sense of emotional and spiritual exhaustion and dearth of hope. Conditioning and indoctrination inherent to most religion also encourages and rewards passivity and compliance rather than action and personal accountability, which is not austere and disciplinary but instead empowerment and levity, to realize that life is not as serious as we often make it and that we can simply make progress by taking one step at a time, that our job is not to cheat death and overcome cancer but simply to eat high tannin foods regularly, avoid combining protein and tannin-free water in the gut, consume plenty of high silicon foods, get sunshine, etc., to take care of our physical health (and dieting and exercise is not that!), and so effectively incapacitates adherents by withholding life skills and behavior tools which would otherwise empower them to do otherwise, so they instead quiver in place waiting for God to do what they feel incapable.

Praising God for getting well or other auspicious developments in our lives is also an especially smarmy betrayal of faith and total failure to understand the true nature of God because it implies that God is not otherwise aware of our plight to begin with, even before things happen, and is especially puzzling of those who claim to believe in God otherwise they would understand his nature, which they do not. God does not let us suffer cancer and then save us from it only if we pray and believe sufficiently—that is a human anthropomorphisation of the powers which create the Universe, which instead allow cancer to happen as a natural law of causality so that we might experience the limitations of mortality and not only its joys. I cannot think of a more spiteful way to consider God than to accuse him of only helping us because we prayed to him, which instead describes a petty, egoist, maniacal control freak who does not actually care about His children but instead only those who worship him. God clearly has no ego, else humans would not be free to choose their conception of God, and there is a reason the saying *God helps those who help themselves* is not scripture, although it should be, because personal responsibility is the primary factor in the laws of cause and consequence. This is not a responsibility, but a gift, being able to exercise free will we can at any time do our best and are not dependent upon God to do that, and this empowers us to emancipation from religion and dependence on incorrect, useless, authoritarian conceptions of the power of the Universe. It's also a phrase which originates from ancient Greece, before the advent of monotheistic, authoritarian theology which has come to decimate the world of our culture, language, and traditions, but has survived thousands of years of chaos and changing sociopolitical dynamics *because it's true.*

However, discussing death, illness, and mortality in terms of helping ourselves and doing what is required also errantly still implies that we have *any* control over such things, and my ranting is not an admonition to those whom are unfortunate to experience disease to get well and defy cancer or to subvert their fate *because that is not possible,* and I did not get well from my own experiences because I have any more control over mortality than I did to begin with, but because I was fortunate enough to stumble across the work of Ray Peat and then conduct my own research and life choices within the apparent parameters of mortal limitations and dependence upon biology. Although it is entirely possible for anyone to take care of their health all we can ever do is show up for opportunity and do our best, and those whom would not do what was required to get well were never going to and were in fact destined for this life path, as we cannot save people from their fate anymore than ourselves, and blaming others for their illness and mortality *is also* the act of trying to control mortality by blaming others for not doing so, and it is none their fault being human with a mortal body beset by evolutionary pathogens like strep throat, parasites, *H. pylori*, or those which cause cancer. The existence of disease, disease causing microbes (yes, also those which are sexually transmitted), and the mortality of our biology is the domain of God and the powers which organize existence to which we are all beholden, but many of us blame others for the trappings of mortality like weight gain, aging, fatigue, herpes, and even cancer as we demand they defy mortality that our fears may be assuaged and thus make our fear everyone else's problem, or that God rescues us of the responsibility we have in caring for our health.

Resolving not to practice denial as a coping mechanism or to live presently are in fact ironic acts of denial because it is not possible to will ourselves to act against our human instincts and the coping and control behaviors which arise from experiences of trauma, which is instead only resolved by *unlearning* those control and coping behaviors and traumatic perspectives of life, reality, and ourselves which motivate the coping mechanisms of dissociation and denial, which reside largely in the unconscious mind and is a process only achieved through direct action as the practicing of inventory accomplishes. Of course inventory is inconvenient, but the desire that growth and change be convenient is also a control mechanism, which is also resolved by doing inventory, and ignoring the need to employ practices and behaviors to achieve goals, not simply intention or self-will, is itself willful denial of mortal limitations which then leads exactly to those ineffective choices and behaviors at the moment of action since we have not beforehand prepared ourselves adequately to meet the challenge, and in the moment of action it is too late.

As was the case with Jim Henson, Saturn in the sixth house is often an occasion for death, because Saturn is a strict disciplinarian which does not allow us to break the rules required of each house it enters, and during transits of Saturn in the sixth zodiac house which governs our daily labors as eating regularly, getting enough rest, and taking care of other obligations (like work) and not only those which are convenient is required. Steve Jobs was also born with Pluto retrograde in his sixth house, which portended a life obsessed with the control of work and health that was later reflected in his extreme behavior and fruitarian diet meant to cure his pancreatic cancer but which ironically fueled it due to the nutritional deficiencies such diets actually incur, *because we cannot control biology no matter how much we desire that power*. The beloved actors Matthew Perry, Luke Perry, and Michelle Trachtenberg also died during transits of Pluto or Neptune through their zodiac sixth house, because of how those influences challenge our experiences concerning our health obligations. Indeed being also a Libra I would similarly have died if I had not lucked upon Dr. Peat's work which helped me recover, but also admitted that I was powerless over biology and rather than fight my mortality submit to the work and humility required of learning, even to the point of creating knowledge which had not yet been created, through patience and perseverance, always asking nature what it is and never trying to impose upon it my desires, anxieties, or fears.

Deaths during sixth house transits also often occur because the obligations of both health and work and all other labors we must perform can become overwhelming, and because of our work-centric, success obsessed culture will, like Jim, prioritize work and make more time by neglecting our eating habits, by skipping meals or eating only as much as is required to treat hunger but which is insufficient to support our health, including rest which is required that the body can regenerate and sustain its systems to carry us through each day, and so we then succumb more easily to pathogenic injury or other trauma, when in fact it would be better for our long term prosperity to pare down work responsibilities to that which is instead manageable which also allows us to eat regularly, get enough daily sunshine and rest, and see to all our obligations, not only those demanded of our ego. I do not write about Jim to hold him as an example of errancy, but because of my love for him and gratitude of his work and life and all that he contributed to humanity, and the lesson not only in humanity he gave us through his art and good will but also that of his mortality, that it is okay to accept mortality and to pass when our time has come, and we all are appointed here for a brief time which, when it is over, is not up to us to decide.

Loss such as of those we hold dear can indeed be painful, made worse if we also hold trauma which characterizes death as somehow a punishment, insult, or failure, but in fact the limitation on the time that we have with each other is also the thing which makes it so special, and after loved ones have gone their memory all the more reinforces how important they were and the contribution their life gave to ours, which is truly one of the more awesome parts of reality that is an undeniable truth of existence, that we should be so important to others and they to us, a tangible and truthful aspect of reality that needs no support of philosophy or religion, just as the sunshine warms our skin and water quenches our thirst.

Indeed the solution to solving the astrological appointment of death is not any specific planetary aspect to our birth chart or alignment of planets but the recognition of the inherent value of life, what meaning comes from the mere fact of our existence one with another and our time here on this Earth which no force in heaven or Earth can invalidate, and our passing comes at fateful alignments of any of the planets which is greater than what we can overcome to instead describe the *manner* of our passing and not the time of it. Once when counseling someone over the passing of their grandma with which they were struggling they admitted to feeling responsibility for her death, because their grandma had refused to take her medication at the end of her life, and they did not otherwise encourage her to do so. Part of the stress of this experience was borne of the mistaken delusion that we have any control over the behavior of others or even that of fate, and it was fully in the grandmother's power whether to take her medications, or not, but not that of the grandchild. We also hold onto such presuppositions for our own benefit, however, even should such beliefs cause us much pain, because of the inherent control we feel it brings over mortality, to choose when we and others die, and the conflict between wanting to believe we have such control and not actually having it is what truly causes such stress. Of course, their grandmother was very important to them, and her loss also made it difficult to recognize such realities, and after such pain has lessened and we have space in our mind to do so the practice of inventory can sufficiently reveal the truth of existence and the error in our control mechanisms that cause such frustration, ironically this person desiring the control over life which also caused them to feel painful responsibility for the death of their grandmother, as to many of us control is more important even than life itself, but which we do not in fact control but which otherwise remains unknown to the unconscious mind since we cannot, by definition, consciously communicate with it and thus requires inventory to resolve.

The beautiful reality of losing someone is that we got to spend time with them at all, and the sweetness of their memory brings us comfort even in the emptiness of their loss *because* of the reality of their existence which occurred, which no force in existence can undo. Even if we are forgotten for some reason (such as old age and senility), that still does not negate that we did in fact exist, which is part of the fundamental laws of reality over which no man has control, which instead is the domain of the forces that control all reality which makes existence possible in the first place. Transits through the sixth house are thus only one portent of mortality, as any transit of any house can potentially lead to our demise—Many meet their end through violence when Pluto transits a house particularly concerned with control and power of others such as the seventh house of partnership or the eighth house of power, in which partners often wrestle for control of the other until one resorts even to violence. Or heads of state with Pluto in their fourth house of home murder and pillage to ethnically cleanse their home-land and often receive death in turn for such stupid and senseless behavior. Deaths during a third house transit such as what occurred to one of my uncles results from absence of knowledge required for healing, especially willful ignorance, or those in an eighth house transit can also occur from complications of diets or cosmetic surgeries, procedures, and pharmaceuticals in attempts to control mortality.

Because Pluto rules the eighth house and is concerned thusly with power and control, both Pluto, Scorpio, and the eighth house are often considered the domain of death, both literal and figurative in the metaphorical death such as of the ego or self and subsequent rebirth into a new life and purpose. Indeed with Scorpio rising in my birth chart my life path was destined, as are all with this influence, to experience some traumatic event which thereafter experience a rebirth of purpose and life goals. While this prospect can seem ominous and the experience which cata-lyzes this change may in fact be immensely stressful I can testify that it was indeed much preferable and completely auspicious, in the end, even in its loneliness and loss, because my mind and soul were for the first time unified and at peace, and peering into the depths of time and the expanse of the Universe such as arose from my experience better contextualizes my existence in the boughs of reality which were beforehand constantly shaking and breaking.

But this association of Pluto with death is not real, and is only associated with death because of Pluto's dominion over *power*, of which we have none, of which transgression can and does then sometimes result in death, where death is the Universe asserting ultimate power over us instead, and Pluto's influence such as its transits through our sixth house can result in death as a consequence but only because Pluto is the most unyielding planet. Indeed in about twenty years, start-ing in 2043, all Libras will have Pluto transiting our sixth house when I will be sixty-two, and will absolutely again encounter themes concerning health and work which has before taken many a soul when we are unable or unwilling to do those things required of our mortal obligations. Having recently survived Neptune's transit of this house, Libras should take special care to remember the lessons and not challenge Pluto, whom is even less yielding that Neptune (as will be the case for Scorpios after us, then Sagittarius, and so on). The key to surviving such transits, if Pluto wills it, is in the *everyday,* our behaviors, obligations, exposures, or other frequent variables, not in the exotic or unknown, since the sixth house simply rules our labors, making sure to get plenty of nutrition, rest, the daily eating of necessary foods, having a regular schedule, care, medical attention when needed, and all without attempting to wrest control of such things.

In fact, the sign of Leo recently finished this transit and working with one young Leo who had contracted lyme disease as a teenager, and evangelizing to them of our very inability to control biology and mortality they actually lost their temper and snapped at me, so angry were they that I should remove from them the power to control fate, believing their problems should be solved in that one thing they haven't yet discovered or that one thing which will give them the control they desire. Rejecting my advice and resenting my help they went on to continue tempting fate and trying to force their health back to submission, even doing such things as eating yogurt, and later admitted to experiencing the worst health effects from it as they had yet known in their desperation to recover and finally recognized the futility of wrestling with reality. The sign of Virgo at the time of this writing has currently stepped into the Pluto transit of their sixth zodiac house, and one such friend of mine from high school was diagnosed with stage-3 HER2 positive breast cancer in just the first months! Having lost several people to denial and ignorance I lured her to my home with promises of home-made pasta and proceeded to lecture her for hours about my personal experience with cancer, the deaths of those which came from denial, and the immutable consequences of Pluto transiting the sixth house which is the catalyst for such experiences as recently had by Leos of which there were endless examples. Unlike many people which had occasion to cross my path she fully listened to everything I said and incorporated most of it into her recovery and cancer treatment, and was declared cancer free about seven months later after undergoing only half a course of chemotherapy.

As they did to Dr. Peat, many people also reject my work when I also fail to liberate them from the laws of cause and consequence, as if such behavior somehow affects me in the least instead of their own lives which are bound every bit by the laws of reality as my own, in which we are subject not only to fallible biology but also an unreliable environment and assault by countless microorganisms seeking to colonize our bodies. Even when we know every last thing about human biology we still will have no more control over biology and death than we do now, which is absolute and unchanging because of the immutable physical and metaphysical laws which construct reality, and the desire to live forever and cheat mortality will never be anything more than a simple delusion borne of evolutionary animal biology whose purpose is to promote the survival of our species as a whole through instincts of fear, at the cost of the individual who then spends even an entire lifetime cowering at our humble station. Failure to recognize that decisions made in fear ironically hasten death as we attempt to wrest control from the Universe and God prevents us from being effective in the management and care of our health and wellbeing, which also includes psychological care which many are not willing to do since it does not endow us with a six-pack, and are ironically more *effective* when we accept our position in the Universe because that is in truth informed to the nature of realty which then empowers us to make more effective and informed decisions to achieve more desired outcomes because the laws of reality are not arbitrary and must operate on logical cause and consequence.

Of course, Pluto transits are long, lasting anywhere from about one to two decades, which is a long time indeed to maintain daily self-care consistency. But in truth all the transit through the sixth house requires is eating a healthy diet, going to sleep on time, and showing up for work, which is what we often already do anyway, although in some cases Pluto may throw us a curveball such as coming down with an uncontrollable or difficult illness, and it may be that I finally meet my end during this especially long, upcoming transit especially since I will at that time be much older. Or Pluto may cause a disturbing work environment in which we are subjected to the control behaviors of others, and may not cause health issues at all, and death is never an inevitability because of any particular transit but only because we are alive in the first place, and instead of any specific alignment which is always associated with death that can be predicted it instead comes from any transit which concerns mortality by which we may meet our inevitable end, which should not be handled by constant vigilance in anticipation of it but instead showing up for opportunity and doing our best, to live life while we are here, and leave such worries as death and the afterlife up to those powers in charge of such things, which under no circumstances is ever ourselves.

After years of wrestling with a fear of death and much inventory I realized we fail to recognize that the energy of which we are made and alive on this Earth is neverending, as energy can never be made or destroyed but instead only changes form, which is a direct refute of atheist dogma because we have literal scientific proof that what we are made of persists even after our bodies do not. The comfort which comes from this knowledge is likewise also not a confirmation of such delusional coping mechanisms as heaven or hell, which are nothing more than mythological constructs meant to help us cope with the spectre of death and obsession with retribution, but instead that such things are not within our power or responsibility and are instead those of the Universe and what we experience to be God, the reality that whatever happens after death is what happens, which is entirely outside of the ability of human mortals to even comprehend let alone control and manipulate, and since it is not our responsibility we have no need to try to control and manipulate it anyway. Part of the reason religion emphasizes the afterlife and insists we have control over it is that the disparity between such misguided ideology and reality causes people to become insane in our attempts to reconcile something which is inherently irreconcilable, thus making people easier to control and manipulate. Heaven and Hell exist here and now, with those given to peace and our eyes open to reality living in Heaven even as mortals before we have died, while those who scheme, hurt, and offend in the here and now living in a Hell of their own making in which they will never know anything more than constant discontent, regret, fear, and resentment. Insisting we control death brings false comfort and so no comfort, and the conquest of death continues even after claims to have done so are proclaimed. But acceptance of reality instead brings not only real peace and comfort but also the ability to exist for the first time in our lives in the present, not in fantasies of the afterlife or rumination of the past, because we are no longer concerned with assuming the power of God and can instead humbly let control of such things over to those forces and powers which actually do.

In terms of afterlife philosophy, a conclusion I also arrived after much inventory practice and meditation is that the concept of *reincarnation* makes the most logical sense. Considering the nature of energy and matter it seems that that we are perhaps borne from a great and unified consciousness, not necessarily homogenous, although it may be, but certainly connected to all else which also exists, and are born and reborn over and over as we experience new lifetimes to have new experiences which range the entire gamut of potential life and knowledge, and that the purpose of life may be the assimilation of knowledge from every possible combination of life and experiences from the infinitesimally small to the greatest all living creatures. Several religions such as Buddhism have of course long ago established the concept of reincarnation, and Shintoism or the beliefs of indigenous Americans and others recognizing the interconnectedness of all things, but even peaceable Buddhism is still a control mechanism which seeks to avert pain and suffering through control, by eschewing attachments to mortality, materialism, and even sometimes relationships, and ironically attempts to subvert one of the purposes of mortality which is to in fact experience its bounds in our dependency on such things as other people, food that we may survive, and even sometimes material possessions, and we cannot subvert the limits of mortality even when our intentions in doing so are honorable (such as achieving enlightenment), and instead the most effective and harmonious path is simply not trying to control things we cannot in opposition to our frequent instinct to do so.

Similarly, the act of fasting for spiritual reasons is a control behavior which seeks to discipline the mortal vessel and bend it to suit our will in the desire to transcend mortal limitations. But all religious practices which utilize fasting as a means to spirituality misunderstand its effects, because it is a control behavior which is not to uplift and empower the spirit but destroy and weaken the body, which is the temple of the spirit, which is why the bodies of chronic fasters and dieters alike become emaciated and gaunt as the body eats itself in its attempts to help us survive, often taking the mind with it or subjecting the body to increased risk of parasitism which further deranges the mental state through hyperammonemia or other effect. Excessive body worship is also not the same as care of the body, which itself also instead seeks to control mortality by avoiding disease and death or seeking validation from others, and is no less a product of fear than any religious practice. Even vegans desire liberation from fear and fear of death by refusing to kill other animals, but often our own survival in fact depends on predating other living creatures as a condition of our mortal biology. Most people understand that it would be ridiculous to demand that a jaguar or a shark refrain from killing other animals for food, but do this very thing to human animals which clearly use predation as one source of sustenance, and that the proper way to handle our fear of death is not trying to conquer it through absolute passivity but to instead have reverence for our requisite position as an animal which is not always capable of surviving on plants alone, recognizing that the taking of a life is a horrible thing which is part of our lowly position in mortality and should be treated with reverence and humility. It is also a delusion that eating plants is not the death and consumption of a living organism, and all plants have spirit and life and so do not desire to die any more than a cow or a chicken or another human, and reverence for life such as is the result of these experiences as a condition of our mortality that we in turn depend on other life to sustain our own.

Yet many people hunt for sport, and the slaughter of animals for our entertainment or the thrill of hunting is a vile practice that spites God and life on this Earth, taking of animals which must live a harsh and wild existence when we ourselves do not, or subjecting likewise farm animals raised for food to inhumane and disgusting existence in cages and concrete hellscapes, denied sunlight and the freedom to be what they are in the short time they have will surely condemn any human which causes this to come back as one of those very animals in their next life to experience the same. All animals require sunlight to be healthy, and plagues of bird flu which are often controlled by the mass, inhumane slaughter of millions of birds to be discarded in a heap of decay and waste is a result of being denied the immune requisite function of sun exposure and vitamin D synthesis such as occurs in the abhorrent, mass-factory farming of living creatures without regard to their wellbeing other than what directly benefits us, even the breeding and genetic manipulation of animals to be fatter and bigger that they cannot even walk, such heinous behavior by humans does not come without moral consequences as those who commit such amoral acts live in constant dissatisfaction and discontent and those whom consume them doomed to disease, disfigurement, fatigue, and early death as laws of causality are absolute and we must have certain conditions in order to thrive which are not met by such exploitative systems and selfish, inconsiderate treatment of nature.

While our species can consume meat, we are more naturally a vegetarian (or mostly vegetarian), as I have discovered from my work and research as discussed in *Fuck Portion Control*, because predation of other animals is not a primary biological function of human biology, which is instead an opportunistic behavior that occurred in response to nutrient deficiency which arose from the insufficient availability of our primary forage roots, fruit, nuts, and leaves as we spread across the globe, taking large megafauna as the means to sustain large populations of nomadic humans (to the extinction of many). Having no intention to eat vegetarian I was beyond surprised when in early 2025 after discovering that dietary silicon promotes the formation of micelles and liposomes in the gut for improved digestion of dietary fats and fat soluble nutrients to suddenly feel revulsion at the thought of eating meat and instead an overpowering craving for *minestrone* and even discarded some leftover chicken soup and instead made a giant pot of minestrone and proceeded to have four bowls of it to great satisfaction. While I did not eat meat extremely often, about several meals a week with chicken, pork, or beef and being as large as I am and requiring so many calories daily simply to stay alive I had previously craved plants, but not to the exclusion of meat, after discovering how commensal archaea produce cobalamin (vitamin B12) to supply our other commensal microbes which then empowers them to break down plant foods, for which we and all other plant-eating animals are dependent because we do not possess the enzyme *cellulase* and therefore cannot digest plants without the assistance of microbes. Then after discovering the role of dietary silicon in promoting better fat digestion and eating more high-silicon foods like green beans, dates, or spinach with healthy dietary fats as butter, coconut oil, or fatty nuts, or using a little silicon in boiling water for making tea taken with a low dose of those fats caused such extreme cravings for vegetables as I had never known and which has not become tired even after eight months now, never ideologically intending to become vegetarian but instead as a natural result of satisfying my human environmental niche (which includes the microbiome) and the ability to break down plant foods to restore my most primal function as a human animal, which is not a slaughterer of animals but of peaches, green beans, watermelon, and yam.

This is also why many vegans and vegetarians are often not actually healthy, when lacking sun exposure or other requisites for vitamin D production and being deficient in local microbially produced B12 and thus being unable to break down plant food, trying to force ourselves to be vegan or vegetarian is in fact just another control behavior, trying to insist that our bodies follow our desires when in reality it is achieved only by providing for our needs and to live within the parameters of biology, to then achieve the ability rely on plants but which is entirely separate from our will to do so, and when supplied with enough vitamin D and B12 this is indeed the most healthy diet possible (consumption of shellfish for boosts in zinc and copper are probably required though) because our commensal microbes absolutely thrive on plant food and then produce an abundance of short chain fatty acids, B vitamins, vitamin K2, and amino acids in beneficial ratio which is not possible from any exogenous food source. This state is thus not achieved by self-will but by resolving all the dietary and behavioral problems as discussed in my other book which prevents a thriving microbiome and good digestion, and when these problems are resolved it is natural for the human body to crave plants in preference of meat, and to feel extremely satisfied by a plant-based diet, while those who are sick or not able to resolve those problems will instead continue craving more easily digested foods like meat and refined grains until such time as they are able to resolve those problems. Although a plant based diet requires more frequent intake of calories, such as having two bowls of soup instead of one (or in my case four), there is no way to achieve optimal human health without adhering to the requisite structure and function of our evolutionary biology, which is not in fact based on meat consumption but primarily on roots, fruit, leaves, and nuts to naturally achieve a harmonious vegetarian diet with occasional meat consumption rather than the reverse, which can then relieve us of the need to consume so many other animals as a food source.

Reverence for the condition of mortality thusly includes accepting death as a constant reality and thus nothing to fear, since when our time also comes the energy and matter of which we are made will continue on as intended by the Universe, and whatever lies in store for us beyond the veil of this world is God's domain, not ours, and it is only our responsibility not to wrest control of such things as mortality from the God whose dominion this is, but to live life now, in the present, to show up for opportunity and do our best. Conceptualizing our return to the consciousness from which we arise, however, we often imagine a spiritual doppelgänger or clone of how we appear in this life. But if in fact we spring from some Universal consciousness this means that our current human form is simply the animal species in which we now live, and that our spirit is no more human than when we might have lived as a squirrel or a bug, if indeed that did occur, and is something else entirely that we cannot conceive. This does not mean we have no relationships or individuality, however, as the other people in our lives are very much the spiritual being that we know and are familiar, while the body is simply the vessel through which we then experience reality. Whether or not this is the case, however, is not really relevant for our experience here because the entire point of living is to have the experiences we are intended to have in the here and now, and have no consciousness of the before or after for a reason, and for the here and now it does not matter what we are or may be in the beyond, which similarly to the religionist takes us away from living in the present, which is the most important part of why we are here in the first place.

Not only can we not actually control mortality, trying to do so is a spiteful behavior which refuses the existence of God and the Universe as master of fate and mortality, and a fear of death is nothing more than hatred of the human form and a fear of being alone in this vast existence. We have each other in our shared struggle, in whom we can rely for comfort and companionship, but as God and the Universe have dominion of our fate this also means they are concerned too with our existence, and our death is as important to the Universe as our birth, and will see us out of this life just as it brought us into it, neither of which was our doing or design, so we are not at all as alone as we are often taught to fear, which arises simply from misunderstanding and ignorance that can be corrected by practicing inventory on such fears and resentment as death, illness, morality, loss, etc., as the entire Universe is constructed as such to be concerned with the existence of all life, and since we are life thus is it concerned with us, as we are as much a part of the Universe as any rock or drop of rain and not separate from it.

29. *In Awe of The Cosmos*

In the middle of 2022 NASA finally launched the much anticipated James Webb Telescope which, at the time, was the most powerful ever made and capable of peering into space back to the very beginning of time. Impressively, nothing went wrong with the complex telescope and its deployment a million miles out from the Earth, when absolutely so many things could have gone wrong, because of the many thousands of scientists, researchers, and officials who made it happen. When the first image started coming down from the telescope I was one of the many millions of people who excitedly viewed the most ancient galaxies we had ever seen in astonishingly crisp detail, complete with gravitational lensing and yet even more breathtaking number of galaxies that simply doesn't seem possible when looking so deep into the vastness of space.

James Webb was, like myself, an annoying Libra who oversaw much of the early years of NASA and its first crewed missions to space. In the naming of the telescope in honor of his legacy there was a brief controversy in which it was claimed that Webb oversaw some of the persecution of homosexuals in government which occurred in those years, but in fact some of that behavior was wrongly attributed to Webb and later proved to be from others while Webb oppositely appears to have never played any role in the persecutions of gay men. While Libras are just as capable of depraved and heinous behavior as any other sign, and in his diplomatic role Webb participated in the escalation of violence in the Korean War,

Webb also actively worked to undermine segregation and integrate black people into NASA through their active hiring into the agency, including direct recruitment from black schools and colleges at a time when racist sentiment was as high as ever, and openly challenged racists in front of the press, a behavior of seeking social justice which is common (but not universal) to Libras whom very often instinctually seek social justice, being the sign of justice and balance. Webb also took personal blame for the Apollo 1 accident instead of directing blame at others, helping to keep the integrity and structure of NASA intact which then subsequently led to its future resilience and later achievements that otherwise would not have occurred if a more disreputable leader had instead blamed subordinates and allowed it to fall apart.

Webb got his start in Washington immediately after the great depression wherein as a young man and newly barred lawyer he served as secretary to an influential congressman and assisted with the efforts to push through President Roosevelt's New Deal legislation which established new laws and regulations that helped rescue and stabilize the American economy, integrate socialist policies like social security and labor rights, and set the stage for the heretofore unseen wealth of the post-war generations unlike anything seen before in the entirety of human history, later serving in World War II before later continuing his career as a lawyer and bureaucrat which culminated in his leading of NASA.

One of the reasons I was so in awe of the images coming from the James Webb Telescope is that I fully appreciate just how magical and improbable it is that we not only have the ability to peer into the depths of space and see things like galaxies in their full and marvelous splendor but also the devices like computers and smart phones on which to view them, where most people have begun to lose the awe and gratitude for things that are in themselves so amazing as to have never even been imaged by nearly every one of our ancestors which came before us. As short as the time since my own birth almost a half-century ago typing was limited to a manual typewriter which could only punch letters onto paper and required constant refilling of the ink strip when it ran out, and yet now we don't hardly print anything anymore because other people can simply receive the information on their own phone or computer.

There was a time when even print had not yet been invented, and people wrote in fine handwriting entire books without automated assistance, and yet when we humans should also be in awe of the printing press and cotton gin the absolute marvels of innovations as smart phones—an entire personal computer that fit in a pocket—becomes as rote and mundane as taking a shit. The reason we are this way is because part of human evolutionary survival biology is to *always* be dissatisfied with life as a coping mechanism in response to trauma which motivates the continuous seeking of resources, in anticipation of their eventual absence, and because dissociation is another coping mechanism we literally are not even present in our lives when we are holding our phones browsing social media and texting with people on the other side of the world because our evolutionary human survival instincts have taken over and we ourselves operate as if by automation. But our biology could never have anticipated that humans would one day drive cars, fly planes, or call each other from miles and miles away, and driven by our basic instincts when burdened by unresolved trauma we become fully deprived of the pleasure and joy of true appreciation and gratitude, which is to fully understand just how amazing reality actually is and which far exceeds our wildest fantasies and mythologies that seek merely to assuage fear and excuse fallibility. All my life growing up my parents and others like them felt the need to invent magical and mythical lies to find some sense of wonder and purpose, which not only ultimately failed to do that but also destroyed relationships and cheapened experience, to whittle down love and joy to nothing more than mundane exercises in conformity and compliance, but in the mortality of reality and its fearsome instability I have discovered awe in a cup of coffee and heaven in the hand of my love.

To religionists like my parents the images of the deep of space are nothing but a smattering of twinkling lights on digital paper. Because they do not recognize the beauty in life nothing in life then holds beauty, not even the jewels of the very Cosmos itself and the incomprehensible wonders they contain. But upon their surfaces I see innumerous populations of extraterrestrial animals and other life-forms, vast civilizations as likely as our own, since life is a proven law of reality by nothing more than our own footsteps upon this Earth. The images of space from high atop mountains in the cold of night away from the light-fog of civilization spins around us so rapidly that laying on your back looking into the sky can cause a feeling of disorientation rather than the illusion of a steady ground beneath our feet, and the mind becomes suddenly aware of our Earth spinning on its axis in the middle of nowhere at the fringes of the Milky Way galaxy, glued to its surface by faithful gravity and all the amazing laws of nature that make it possible to be here. When I was younger the realization that our existence depended on this Earth filled me with a sense of dread, that if for some reason the Earth should suddenly disappear I would plunge down rapidly to the center. But in reality the presence of Earth itself is what creates gravity, so if it were to disappear (and we somehow not), we would simply float in space, and that like many aspects of reality our conception of existence is often an illusion created by the sheer vastness of existence and the incomprehensible laws responsible for our being here at all that our simple, animal minds find difficult to comprehend.

Unlike religious parents whose third baby crying in the crib fills them with frustration and angst, a thousand children could not make me numb to the wonder that is their life and newness, and I should love to adopt and care for all the unloved children of the world if such a thing were possible, yet while much of our desires for control of reality and the extinguishing of suffering is an impossibility it is through my works that I can care for the unloved and traumatized children of the world, even you reading this if that be the case, by doing myself what I desired of my own parents—to be a guide and a guardian who can effectively do my best and show up for opportunity, to make available and pass on the tools and skills I have learned which have made my life an amazement compared to the despair and pain I lived as a child and a young man. The knowledge also that others can have what I do simply by reading books and learning new ways of perceiving ourselves, the world, and life and existence is the heaven so dreamed by woeful religionists and others whom cannot see the truth they seek even as it stares them in the face like the night sky that circles ever above our heads, and the endless cycles of all the people and animals and things which have, do, and will exist upon the Earth.

Although I am nothing more than a single human animal I am allowed by my creator to see and smell the beautiful flowers in a meadow, or which grow along the roadside, hear birdsong as finches and sparrows flit about the trees and bushes, watch cormorants dive for fish and sunbathe on an old log, the geese honk and ducks nibble floating morsels, or watch the blue sky overhead give way to swirling, tempestuous clouds borne of water vapor rising from vast oceans thousands of miles away. In my lifetime the touch of a loving man has comforted me, the joy of conversation with a good person filled my heart, and held a new baby in my arms which looked into my eyes with unreserved adoration.

Daily I get to hold a cup of coffee, and feel the thrill of its goodness spread throughout my body while eating eggs and mushrooms, and cheese, pastry, asparagus, almonds and peaches, watermelon and pie, and all sorts of incredible things other foods made by nature and assembled by incredible cooks and chefs. I've been embarrassed, drunk, high, and sober, and had my heart torn from my chest and made to live without it, and nearly died several times, and got well and experienced all of life a second time. I've helped others find healing, and felt helpless watching those I could not.

Plants I've grown sprout on their own and produce fruit and other marvels I can eat which give me not just strength but joy. I've made art and written books, sung songs and listened to beautiful music played on incredible instruments invented by humans and artists whose talent rends the soul and uplifts the heart. I've celebrated holidays and given and received presents, and people have cooked for me and shared their hard earned effort, seen incredible movies with fire-breathing dragons and terrifying space aliens that made me contemplate my own mortality and dream even greater dreams.

Whales have come to my rescue, I've talked to sea turtles and pet a shark while breathing underwater. There are great buildings constructed by man, such beauty as which seems to defy all that is possible, and books written by amazing authors which showed me worlds and possibilities I never would otherwise known. I've traveled to other places and met and loved new people and saw how they live and love in turn. I've been in the freaking air, inside a giant metal tube that flies through the sky at breathtaking speed to emerge on the other side of the world in less than a day, and played video games that drew me into fantastical dreams never before seen by man, and had dogs that danced for joy around a piece of carrot or laid across my lap as if was the safest place in the world.

Trains are a thing of wonder, and I've ridden them across town and country, and had unexpected conversations with complete strangers which made me love humans and life. Children have learned facts in my presence, their eyes wide with the fascination that is life, history, and knowledge. People have made me laugh so hard it made my stomach hurt, and so lonely that suicide for a time seemed preferable to life. There are places in the world where rain freezes to intricate crystal snowflakes and accumulates on the ground and caps the mountains and tress, and plants which appear dead in the impossible cold of winter come back to life in the spring as if by magic.

I've lost my temper and hurt others, and made amends when I could to win their forgiveness and bury the pain of the past. There are eight billion other humans on the planet, every one of which could be my friend, and tens of millions a lover, and in my lifetime the deepest expanse of the cosmos has opened before us and become available through no more effort than clicking buttons on my keyboard. Though not all humans are likewise granted a dialogue with the Universe my life has that too, and nothing greater could be promised me than what I have already lived and seen and done and been yet I am supposed to believe life is hell and we are sorry creatures upon this planet? Get out of here. Life is a heaven and I know because I have lived it.

Like what is seen in the experience of Mr. Webb, myself, and any other person in the history of the world, one consistent law of reality which is as immutable as it is frightening and comforting is that we will all reach the end of our lives, but the duration from beginning to end will have within it an abundance of experiences both rich and heartbreaking, humbling and awesome, evidence of our place in the great Universe as part of something which is more than any one of us. And yet we are also a part of that greatness, which needs no justification or excuse or design of man but is evidenced by the mere fact of our existence, and of dogs and cats, birds and snakes, trees and flowers, whales and giraffes, the fossils of the Earth or the volcanoes which melt it.

There is no shortage of awe in the cosmos, and if you cannot see that it is not any fault of reality but simply the shrouding veil of human evolutionary biological psychology and trauma, which upon the mind of a human animal motivates base instincts for survival of our species through obsession with mundane, uninspiring foci as conflict, sex appeal, power, and materialism, which are boring indeed compared to the wonder of existence. Like any species on this Earth ours will ultimately at some point not survive, but in the time between then and now we get to experience the wonder that is life and living, which is not wondrous for being human but simply that we *are*, living on a planet out on the rim of the Milky Way galaxy so vast itself as to be incomprehensible, and there are incomprehensible number of galaxies just like it. When awe is needed and missing from life, just open a picture taken by the James Webb, or the Hubble, or learn some science or go out at night and look up at the stars on a moonless night away from the city as they swirl around us. We do not fall through the Earth because of the relative nature of time which solidifies matter, but grow and age because of the hurling of our planet through the stars in a system of organization which allows us to exist at all.

Our ancestors found truth in the heavens, measuring the cosmos and observing its behavior to therein find purpose, meaning, and truth, because the cosmos itself is the blackboard of God, and science is his tongue, and the great math and physics which comprise reality in which we live are his works, the incomprehensible vastness of space and innumerable worlds and lifeforms, within each of us here, now, reading together about our shared humanity and the nature of being alive, for the Universe was constructed such that we live here, at this moment, and have what life we will.

If you can't find comfort in all this then, well, *what*?

Index

Important Astrological Elements, Symbols, and their Meaning

Signs and their Associated House, Rulers, Element, and Modality

Major Aspects

The Zodiac

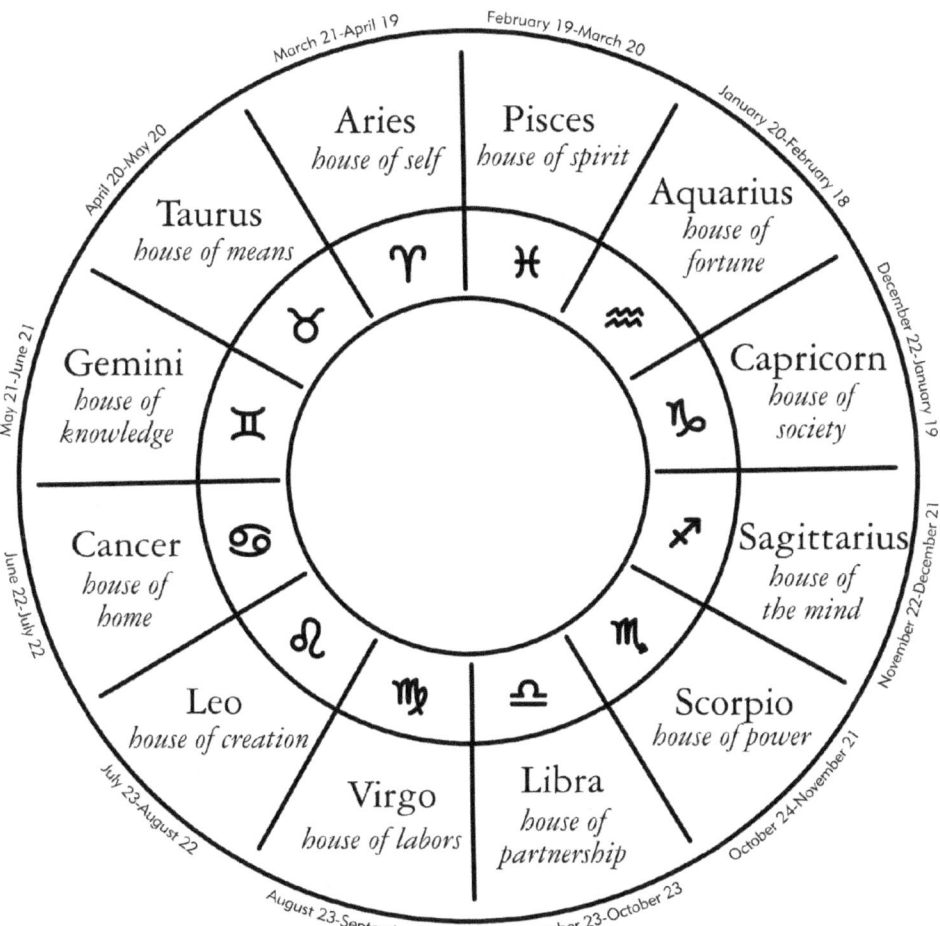

www.ingramcontent.com/pod-product-compliance
Lightning Source LLC
Chambersburg PA
CBHW061548120626
46550CB00004B/1407